John Ewald Siebel

Compend of mechanical Refrigeration

John Ewald Siebel

Compend of mechanical Refrigeration

ISBN/EAN: 9783337176730

Printed in Europe, USA, Canada, Australia, Japan

Cover: Foto ©ninafisch / pixelio.de

More available books at **www.hansebooks.com**

COMPEND OF

MECHANICAL REFRIGERATION

A COMPREHENSIVE DIGEST OF APPLIED ENERGETICS
AND THERMODYNAMICS FOR THE
PRACTICAL USE OF

Ice Manufacturers, Cold Storage Men, Contractors,
Engineers, Brewers, Packers and Others
Interested in the Application of
Refrigeration.

SIXTH EDITION

By J. E. SIEBEL

IRECTOR ZYMOTECHNIC INSTITUTE, CHICAGO

)

LLINS CO.

PRESS OF
ICE AND REFRIGERATION
CHICAGO

PREFACE

While in the third, fourth, fifth and sixth editions of the Compend the general arrangements of matter and the manner of treatment remain the same as in the first and second editions, it is nevertheless an entirely new book. Not only that the contents of the sixth edition cover nearly one hundred and fifty pages more than they did in the first edition, but also much of the former matter has been entirely rewritten and nearly every topic has received valuable additions.

This will be especially noticed in the practical chapters on the "Compressor and Its Attachments," "Ice and Distilled Water Making," "Cold Storage," "Piping of Rooms," "Insulation and Heat Leakage," "Brewery Refrigeration," "Absorption Machine," "Management and Testing of Machines," etc. On "Liquefied Air, Its Production and Uses," and on "The Carbonic Acid Machine" entirely new chapters have been added. The cold storage temperature tables and storage rates have again been thoroughly revised, and many important tables and many practical examples on various topics have been added to the book; and although it now covers over four hundred pages, it nevertheless retains its convenient shape, equally well adapted for pocket and table use.

Special attention has been given to the preparation of the table of contents, and more particularly to the topical index, which contains some fifteen hundred references, so that whatever has been said in the book on any subject can be readily found under any possible appellation.

Again, the hints and suggestions kindly offered by the engineering fraternity have been duly utilized in the present edition. Still many imperfections must necessarily remain, and for this reason the author solicits such further communications and criticism as may tend to render the work of the greatest possible utility to the profession.

PREFACE TO FIRST EDITION.

The object for which this book has been compiled is a two-fold one. In the first place it is intended to present in a convenient form those rules, tables and formulæ which are frequently needed by the refrigerating engineer. In the second place it is an attempt to present the subject in a simple yet systematic manner, so as to enable the beginner to acquire a more or less thorough insight into the matter and to understand the technical terms used in publications on the subject.

This course has been suggested or rather prompted by constant inquiries addressed to the publishers, and in order to best subserve this purpose the different paragraphs and chapters have been framed in such a manner, that while each paragraph may be consulted for the individual information which it contains, the whole forms a continuous chain of reading matter calculated to digest the entire subject of Energetics and Thermodynamics and their application to mechanical refrigeration.

Instead of making the futile attempt to describe the decorative details of the endless varieties of machines and appliances, the author has aimed to discuss the various methods of refrigeration and applications thereof for different purposes in such a manner as to enable every engineer, operator and owner of a plant to thoroughly understand all the vital points in the working of his machinery and in the handling of goods for cold storage, in the making of ice, in the refrigeration of breweries, packing houses, etc.

In this way it is thought that the familiar questions as to temperatures, say of brine and storage rooms, as to what a machine is able to do under given conditions, or what it might be made to do under others, as to the proper dimensions of different parts, and most other problems relating to the operation of refrigerating works, can be readily answered by turning to a paragraph or a table, and in cases of greater accuracy by doing some plain figuring.

The different amounts of space allotted to the different systems of refrigeration must not be construed into argument for or against the merits of one or the other system. The author is not interested in any one system in particular, and if his intention to be strictly impartial is not actually carried out in every respect, his judgment rather than his impartiality should be impeached.

As regards the mathematical treatment of the subject, it had to be strictly elementary and without the use of diagrams to subserve the desired purpose of a book for ready reference. In presenting the subject on this basis it has been the special object of the author to have the formulæ as plain and simple as they could be made without making an undue sacrifice in regard to accuracy. This is especially the case with all the formulæ relating to ammonia refrigeration, which subject, like some others, has been treated altogether on the basis of articles published by the author in *Ice and Refrigeration.*

In order to further enhance the usefulness of the book, and in forced recognition of the fact that many practical machinists have an aversion to even the simplest kind of a formula, a separate appendix has been devoted to the numerical solution of a number of varied examples, which it is thought will suffice to demonstrate that the formulæ in these chapters can be handled by any one versed in the simplest forms of common figuring.

Independent of the strictly practical issues, and in pursuance of the stated objects of the Compend, it has been sought to give so much of an elementary discussion of the terms and definitions of the science of energetics and of thermodynamics in particular, that its perusal will suffice to understandingly master the technical terms in treatises on refrigeration and kindred topics in *Ice and Refrigeration* and other publications.

In this attempt those definitions and concepts which are of more recent coinage and which have not as yet been generally accepted in text books, have for this reason received rather more attention in these pages than their direct relation to the main subject would seem to call for at first sight.

To those who possess the required practical and theoretical knowledge, the book will doubtless prove a welcome companion, as it contains in a very convenient form a prolific array of useful and indispensable tables, and a number of rules which are not usually committed to memory.

Aside from the works quoted in Appendix III. the author is indebted to many of the ice machine building fraternity for much of the information here presented, and he may also be allowed to mention in this direction the valuable contributions to *Ice and Refrigeration* by Wood, Denton, Jacobs, Linde, Sorge, Starr, Richmond, St. Clair, Post, Rossi, Kilbourn, Burns and others.

There naturally must be many imperfections and shortcomings connected with an attempt like this, and special pains have been taken to draw attention to them in the body of the book, and any further suggestions or hints in this direction by those using the same will be thankfully received by its author with a view to further improve and perfect the contents of this publication.

TABLE OF CONTENTS.

PART I.—GENERAL ENERGETICS.

CHAPTER I.—MATTER.

CHAPTER II.—MOTION, FORCE.

CHAPTER III.—ENERGY.

CHAPTER IV —HEAT.

CHAPTER VIII.—MODERN ENERGETICS.

PART II.—PRACTICAL APPLICATION.

CHAPTER I.—REFRIGERATION IN GENERAL.

CHAPTER IV.—THE AMMONIA COMPRESSION SYSTEM.

CHAPTER V.—ICE MAKING AND STORING.

CHAPTER VI.—COLD STORAGE.

CHAPTER VII.—BREWERY REFRIGERATION.

CHAPTER X.—THE CARBONIC ACID MACHINE.

CHAPTER XI.—OTHER COMPRESSION SYSTEMS.

CHAPTER XII.—AIR AND VACUUM MACHINES.

CHAPTER XIII.—LIQUEFACTION OF GASES.

APPENDIX II.—PRACTICAL EXAMPLES.

MECHANICAL REFRIGERATION.

PART I.

GENERAL ENERGETICS.

CHAPTER I.—MATTER.

MATTER.

Matter is everything which occupies space in three directions, and prevents other matter from occupying the same space at the same time. Matter is differentiated by its physical and chemical properties, color, hardness, weight, chemical changeability, etc.

GENERAL PROPERTIES OF MATTER.

The general properties of matter which are shared by all bodies are impenetrability, extension, divisibility, porosity, compressibility, elasticity, mobility and inertia.

CONSTITUTION OF MATTER.

To explain the different properties it is generally assumed that matter is ultimately composed of infinitely small particles called atoms, which aggregate or unite to form still infinitely small groups called molecules. Attractive and repulsive forces acting between the atoms and molecules, and their respective motions are made to account for the various physical and chemical phenomena.

SOLID MATTER.

Matter is solid when the molecules possess a sufficient degree of immobility to insure the permanence of shape.

LIQUID MATTER.

If the molecules of a body are sufficiently movable to allow of its being shaped by the surrounding vessel, and if the same can be easily poured, it is called a liquid.

GASEOUS MATTER.

The gaseous state of matter is characterized by almost perfect freedom of motion of the molecules, an unlimited tendency to expand and a great compressibility. The term fluid covers both the liquid and the gaseous states.

BODY.

A body is a limited amount of matter.

MASS.

Mass is the quantity of matter contained in a body.

UNIT OF MASS.

The unit of mass is the standard pound, which in the form of a piece of platinum is preserved by the government.

WEIGHT.

Weight, or absolute weight, is the pressure of a body exerted on its support. The unit of weight is the force necessary to support one pound *in vacuo*, and it differs with the latitude, as the gravity or the earth's attraction.

MASS AND WEIGHT.

The relations between mass and weight are expressed by the equation—

$$M = \frac{W}{g}$$

in which M stands for mass, W for weight and g for the acceleration caused by the attraction of the earth.

MEASUREMENT OF SPACE.

The unit of measurement of space is the cubic foot and its subdivisions (see tables of weight and measures in appendix, etc).

DENSITY.

Equal amounts of matter do not necessarily occupy the same space; in other words, the density of different bodies is not the same.

SPECIFIC WEIGHT.

The relative density of different bodies is expressed by their specific gravity, which is the figure obtained when the weight of a body is divided by the weight of an equal volume of water.

The specific weights used in the arts and industries are given in tables in Appendix 1.

FUNDAMENTAL UNITS.

The fundamental units of measurement are the units of distance, time and mass.

DERIVED UNITS.

From the fundamental units units for more complex quantities may be derived. As the fundamental units vary in different countries, the derived units vary also.

C. G. S. UNITS.

Besides our national units, the units derived from the French or metric system are also frequently employed. They are designated as the centimeter-gramme-second units; abbreviated C. G. S. units, and are also called absolute units.

CHAPTER II.—MOTION; FORCE.

MOTION.

The removal of matter from one place to another.

FORCE.

Any cause which changes or tends to change the condition of rest or motion of a body (in a straight line).

MEASUREMENT OF FORCE.

Force may be measured by the change of momentum it produces in a second. The unit of force is a dyne; it is based on the metric system, and represents that force which, after acting for a second, will give to a gram of matter a velocity of one centimeter per second.

GRAVITATION.

The tendency which is common to all matter, and according to which all bodies mutually attract each other with an intensity proportional to their masses and inversely as the square of their distances, is called *gravitation.*

The force of the earth attraction at its surface is equivalent to 981 dynes.

MOLECULAR FORCES.

The attraction and repulsion which exist between the minute and most minute parts or atoms of bodies are often referred to as the *molecular forces.*

COHESION.

Cohesion designates the attraction existing between the minute parts of the same body; and for solids it is measured by the force expressed in pounds to tear apart by a straight pull a rod of one square inch area of section. This measure is also called the tenacity of a body (tons).

The relative tenacities of the metals are given approximately in the table below, lead being taken as the standard.

Lead	1.0	Cast iron	7 to 12
Tin	1.3	Wrought iron	20 to 40
Zinc	2.0	Steel	40 to 142
Worked copper	12 to 20		

ADHESION.

Adhesion designates the attraction between the parts of dissimilar bodies.

CHEMICAL AFFINITY.

This expression generally stands for the relative attraction existing between the smallest particles (atoms and molecules) of different substances, which, if satisfied, brings about substantial or chemical changes.

WORK.

Work is the product of force by the distance through which it acts.

The unit of work is the product of the units of its factors, force and space. *Useful work* is that which brings about a specific useful effect, and *lost work* is that which is incidentally wasted while producing such effect.

UNIT OF WORK.

The unit of work is the foot-pound, *i. e.*, the work necessary to raise one pound vertically through a distance of one foot. One pound raised vertically through a distance of ten feet, or ten pounds raised through one foot, or five pounds raised through two feet, all represent the same amount of work, *i. e.*, ten foot-pounds.

TIME.

The interval between two phenomena or changes of condition. The unit of time is the hour and its subdivisions.

POWER—HORSE POWER.

Power is the rate at which work is done, and is therefore equivalent to the quantity of work done in the unit of time, expressed in foot-pounds, kilogrammeters, etc., per hour, minute or second. The unit commonly employed is the horse power, which is defined as work done at the rate of 550 foot-pounds per second, or 1,980,000 foot pounds per hour.

VELOCITY.

The length, l, of path traversed by a moving body in the unit of time, t; therefore—

$$V = \frac{l}{t}$$

V standing for velocity.

MOMENTUM.

Momentum is the product of mass (in motion) multiplied by its velocity or force multiplied by the time during which it acts.

INERTIA.

Inertia expresses the inability of a body to change its condition of rest or motion, unless some force acts on it.

LAWS OF MOTION.

Newton propounded the following laws of motion:

1. A free body tends to continue in the state in which it exists at the time, either at rest or in uniform rectilinear motion.

2. All change of motion in a body free to move is proportional to the force applied, and it is in the direction of that force.

3. The reaction of a body acted upon by the impressed force is equal, and directly opposed to, that force.

STATICS.

Statics is that branch of science which treats of the relation of forces in any system where no motion results from such action.

DYNAMICS OR KINETICS.

Dynamics or kinetics treats of the motion produced in ponderable bodies by the action of forces.

CHAPTER III.—ENERGY.

ENERGY.

Energy is the power or quality for doing work. We distinguish between different forms of energy, viz.:

VISIBLE ENERGY.

This is the energy of visible motions and positions, and is subdivided as follows:

KINETIC ENERGY.

Kinetic or actual energy is energy which a body possesses by virtue of its motion, such as the energy of winds, ocean currents, etc.

POTENTIAL ENERGY.

Potential or latent energy is that kind of energy which a body possesses by virtue of its position, a head of water, a raised weight, a coiled spring, etc.

MOLECULAR ENERGY.

The molecular energy comprises the energy of radiation or radiated matter, *i. e.*, electricity, light, heat, etc.; molecular, potential energy or energy of chemical affinity, etc.

C. G. S. UNIT OF ENERGY.

The unit of energy is one-half of the energy possessed by a gramme of mass when moving with a velocity of one centimeter per second. This unit is called the erg. The erg may also be defined as the work accomplished when a body is moved through a distance of one centimeter with the force of one *dyne*, that is a "Dyne Centimeter."

One million ergs is called a megerg.

CONSERVATION OF ENERGY.

The total amount of energy in the universe, or in any limited system which neither receives nor loses any energy to outside matter is invariable and constant.

TRANSFORMATION OF ENERGY.

The different forms of energy are convertible or transformable into each other, so that when one form of energy disappears, an exact equivalent of another form or kind of energy always makes its appearance. (See "Dissipation of Energy.")

PHYSICS.

Is the science which treats of the transformations and transference of energy, broadly speaking.

SUBDIVISIONS OF PHYSICS.

Physics, therefore, is subdivided into a science of optics or radiation, a science of heat, of mechanics, of electricity and of chemistry. Other distinct branches of science treat on the specific relations between two kinds of energies; for this reason we speak of thermodynamics, electro-chemistry, photochemistry, thermochemistry, electro-dynamics, etc.

DISSIPATION OF ENERGY.

In our efforts to transform one form of energy into another, a certain portion of the first energy always assumes a lower degree of tension; it is dissipated and now represents an amount of energy of less availability for useful purposes.

ENERGY OF A MOVING BODY.

The amount of kinetic energy possessed by a body by virtue of its motion may be expressed by the formula—

$$E = \frac{Mv^2}{2}$$

in which E stands for energy, M for mass and v for velocity.

MECHANISMS.

A machine or a mechanism is a contrivance enabling us to transform mechanical energy, by changing the direction, power and velocity of available forces to make them serviceable for useful proposes. The energy supplied to a machine is partly employed to do the useful work required, and partly it is consumed in doing what is called internal work, by overcoming friction, etc. It is the lost work of the machine, and the less the latter the more perfect is the machine.

CHAPTER IV.—HEAT.

HEAT.

Heat is a form of energy, and represented by the kinetic energy of the molecules of a body.

SOURCES OF HEAT.

As sources of heat we may quote: Friction, percussion and pressure, solar radiation, terrestrial heat, molecular action, change of condition, electricity, chemical combination, more especially combustion.

RADIANT HEAT.

The foregoing definition, while it accounts for the phenomena of bodily and conducted heat, does not account for the conditions which obtain when heat passes from one body to a distant other body without a ponderable intervening medium, or without perceptibly heating the intervening medium, *i. e.*, the radiation of heat. To explain these conditions in harmony with the mechanical or molecular theory of physics, it is supposed that the radiant heat is in the nature of a wave motion propagated by means of a hypothetical substance, the *ether*.

ETHER.

The hypothetical ether which is the supposed vehicle for the transmission of the supposed wave motion constituting radiant energy (radiant heat as well as light), in order to accomplish such transmission in accordance with the present conceptions of these phenomena would have to possess the following properties: "Its density would have to be such that a volume of it equal to about twenty volumes of the earth would weigh one pound; its pressure per square mile would be about one pound, and the heat required to elevate the temperature of one pound for 1° F would have to be equal to the amount of heat required to raise the temperature of about 2,300,000,000 tons of water for one degree. Such a medium would satisfy the requirements of nature in being able to transmit a wave of light or heat 180,000 miles per second, and to transmit some 130 foot-pounds of heat energy from the sun to the earth, each second per square foot of heat normally exposed, and also be everywhere practically non-resisting and sensibly uniform in temperature, density and elasticity." (Wood.)

RADIANT HEAT AND LIGHT.

Radiant heat follows the same laws regarding refraction, reflection, polarization, etc., as does light.

TEMPERATURE.

The temperature of a body is proportional to the average kinetic energy of its molecules, and is measured by the thermometer.

THERMOMETER.

The most prevalent form of thermometer consists of a body of mercury, enclosed in a glass tube so that slight variations of expansion due to change of temperature can be read of on the scale attached. Other substances, like alcohol, air, etc., are also used as thermometric substances instead of mercury.

THERMOMETER SCALES.

Three different scales are in use for thermometers, the "Fahrenheit" in England and United States, the "Reaumur" in Germany and the "Celsius" or "Centigrade" in France, and for scientific and technical purposes, more or less, all over the world.

The scales of the different thermometers compare as follows:

	Freezing point of water.	Boiling point of water.
Fahrenheit	32°	212°
Centigrade	0°	100°
Reaumur	0°	80°

If we designate the scales by their initials the following rules apply for the conversion of the degrees in one another·

$$C. = \frac{5}{9}(F. - 32) = \frac{5}{4} R.$$

$$R. = \frac{4}{9}(F. - 32) = \frac{4}{5} C.$$

$$F. = \frac{9}{5} C. + 32 = \frac{9}{4} R. + 32$$

COMPARISON OF THERMOMETER SCALES.

R.	C.	F.	R.	C.	F.
+80	+100	+212	+23	+28.75	+83.75
79	98.75	209.75	22	27.50	81.50
78	97.50	207.50	21	26.25	79.25
77	96.25	205.25	20	25	77
76	95	203	19	23.75	74.75
75	93.75	200.75	18	22.50	72.50
74	92.50	198.50	17	21.25	70.25
73	91.25	196.25	16	20	68
72	90	194	15	18.75	65.75
71	88.75	191.75	14	17.50	63.50
70	87.50	189.50	13	16.25	61.25
69	86.25	187.25	12	15	59
68	85	185	11	13.75	56.75
67	83.75	182.75	10	12.50	54.50
66	82.50	180.50	9	11.25	52.25
65	81.25	178.25	8	10	50
64	80	176	7	8.75	47.75
63	78.75	173.75	6	7.50	45.50
62	77.50	171.50	5	6.25	43.25
61	76.25	169.25	4	5	41
60	75	167	3	3.75	38.75
59	73.75	164.75	2	2.50	36.50
58	72.50	162.50	1	1.25	34.25
57	71.25	160.25	0	0	32
56	70	158	−1	−1.25	29.75
55	68.75	155.75	2	2.50	27.50
54	67.50	153.50	3	3.75	25 25
53	66.25	151.25	4	5	23
52	65	149	5	6.25	20.75
51	63.75	146.75	6	7.50	18.50
50	62.50	144.50	7	8.75	16.25
49	61.25	142.25	8	10	14
48	60	140	9	11.25	11.75
47	58.75	137.75	10	12.50	9.50
46	57.50	135.50	11	13.75	7.25
45	56.25	133.25	12	15	5
44	55	131	13	16.25	2.75
43	53.75	128.75	14	17.50	0.50
42	52.50	126.50	15	18.75	−1.75
41	51.25	124.25	16	20	4
40	50	122	17	21.25	6.25
39	48.75	119.75	18	22.50	8.50
38	47.50	117.50	19	23.75	10.75
37	46.25	115.25	20	25	13
36	45	113	21	26.25	15.25
35	43.75	110.75	22	27.50	17.50
34	42.50	108.50	23	28.75	19.75
33	41.25	106.25	24	30	22
32	40	104	25	31.25	24.25
31	38.75	101.75	26	32.50	26.50
30	37.50	99.50	27	33.75	28.75
29	36.25	97.25	28	35	31
28	35	95	29	36.25	33.25
27	33.75	92.75	30	37.50	35.50
26	32.50	90.50	31	38.75	37.75
25	31.25	88.25	32	40	40
24	30	86			

MEASURING HIGH TEMPERATURES.

Temperatures which are beyond the reach of the mercurial thermometers (over 500°) are measured by pyrometers constructed to meet the wants of specific cases. High temperatures may be estimated approximately by heating a piece of iron of the weight w up to the unknown temperature T, and then immersing the same into a known weight, W, of water of the temperature t. Then if t_1 is the temperature of the water after immersion and s the specific heat of the iron or other metal, T is found after the formula:

$$T = t_1 + \frac{(t_1 - t)W}{w\,s}$$

ABSOLUTE ZERO.

The zero points on the scales of thermometers mentioned are arbitrarily fixed, since the expressions of warm and cold have only a relative significance. The real zero point of temperature, that is, that point at which the molecules have lost all motion, the energy of which represents itself as heat, is supposed to be, and in all probability is over 460° F. below the zero of the Fahrenheit thermometer. At that temperature there is an entire absence of heat and demonstrations of heat phenomena, and above that the differences in temperatures are only such of degree, but not in kind. Hence the impropriety of speaking of heat and cold as such.

If t is a given temperature in degrees Fahrenheit the corresponding degrees T expressed in absolute temperature are found after the formula—

$$T = 461 + t.$$

UNIT OF HEAT.

The quantity of heat contained in a body is the sum of the kinetic energy of its molecules. Heat is measured quantitatively by the heat unit, which also varies in different parts like other standards. The unit used in the United States and England is the British Thermal Unit (abbreviated B.T.U.) and represents the amount of heat required to raise the temperature of one pound of water 1° F. The French unit is the calorie, and is the quantity of heat required to raise the temperature of one kilogram of water from 0° to 1° Celsius.

Some writers define the B. T. unit as the heat required to raise the temperature of one pound of water from 32° to 33°. Others make this temperature from 60° to 61°, and still others define it as that amount of heat required to raise $\frac{1}{180}$ pound of water from the freezing to the boiling point. The two last definitions give nearly the same result, and may be considered practically identical.

C. G. S. UNIT OF HEAT.

We have no unit for heat corresponding to the C. G. S. or absolute system. The small French calorie, being the heat required to elevate the temperature of one gram of water for 1° Celsius (from 17° to 18°) is equivalent to 41,-830,000 ergs.

CAPACITY FOR HEAT.

The number of heat units required to raise the temperature of a body for one degree is called its heat capacity. It gradually increases with the temperature.

SPECIFIC HEAT.

The ratio of the capacity for heat of a body to that of an equal weight of water is specific heat. Hence the figure expressing the capacity for heat of one pound of a body in B. T. U. expresses also its specific heat, and *vice versa*.

SPECIFIC HEAT OF METALS.

Antimony	.0507	Manganese	.1441
Bismuth	.0308	Mercury, solid	.0319
Brass	.0939	" liquid	.0333
Copper	.0951	Nickel	.1086
Cymbal metal	.086	Platinum, sheet	.0324
Gold	.0324	" spongy	.0329
Iridium	.1887	Silver	.0570
Iron, cast	.1298	Steel	.1165
" wrought	.1138	Tin	.0569
Lead	.0314	Zinc	.0959

SPECIFIC HEAT OF OTHER SUBSTANCES.

STONES.		**CARBONACEOUS—*Cont.***	
Brickwork and masonry	.20	Graphite, natural	.2019
Marble	.2128	" of blast furnaces	.197
Chalk	.2148	**SUNDRY.**	
Quicklime	.2169	Glass	.1977
Magnesian limestone	.2174	Ice	.504
CARBONACEOUS.		Phosphorus	.2503
Coal	.2411	Soda	.2311
Charcoal	.2415	Sulphate of lead	.0872
Cannel coke	.2031	" of lime	.1966
Coke of pit coal	.2008	Sulphur	.2026
Anthracite	.2017		

SPECIFIC HEAT OF LIQUIDS.

Alcohol	.6588	Turpentine	.4160
Benzine	.3932	Vinegar	.9200
Mercury	.0333	Water, at 32° F	1.0000
Olive oil	.3096	" 212° F	1.0130
Sulphuric acid:		" 32° to 212° F	1.0050
Density, 1.87	.3346	Wood spirit	.6009
" 1.30	.6614	Proof spirit	.973

SPECIFIC HEAT OF WATER AT VARIOUS TEMPERATURES.

Temperature.	Specific Heat.	Heat to Raise 1 lb. of Water from 32° F. to Given Temperature.	Temperature.	Specific Heat.	Heat to Raise 1 lb. of Water from 32° F. to Given Temperature.
°Fahr.		Units.	°Fahr.		Units.
32	1.0000	0.000	248	1.0177	217.449
50	1.0005	18.004	266	1.0204	235.791
68	1.0012	36.018	284	1.0232	254.187
86	1.0020	54.047	302	1.0262	272.628
104	1.0030	72.090	320	1.0294	291.132
122	1.0042	90.157	338	1.0328	309.690
140	1.0056	108.247	356	1.0364	328.320
158	1.0072	126.378	374	1.0401	347.004
176	1.0089	144.508	392	1.0440	365.760
194	1.0109	162.686	410	1.0481	384.588
212	1.0130	180.900	428	1.0524	403.488
230	1.0153	199.152	446	1.0568	422.478

USE OF SPECIFIC HEAT.

The amount of heat or cold necessary to elevate or lower the temperature of w pounds of a body having the specific heat c for t degrees is found after the following equation:

$$S = c \times t \times w$$

DETERMINATION OF SPECIFIC HEAT.

The specific heat of various bodies can be found from the table, and it may also be determined experimentally as follows for solid substances (to find the specific heats of liquids the same principle is followed, care being taken that the liquids to be mixed have no chemical affinity for each other): Take a known weight, w, of the substance whose specific heat is to be determined, and let it have a known temperature, t (above that of the atmosphere), then immerse it in a known weight, v, of water having the temperature t' and now observe the temperature, z, acquired by the mixture. From these quantities the specific heat, x, of the substance can be calculated after the formula

$$x = \frac{v\,(z - t')}{w\,(t - z)}$$

If the substance is soluble in water any other liquid whose specific heat is known may be used instead. This method, while it might answer for rough determinations, would have to be surrounded by special safeguards in order to allow for loss by radiation of the vessel, etc., in order to be applicable for exact determinations.

TEMPERATURE OF MIXTURES.

If two substances having respectively the weight w and w_1, the temperatures t and t_1, and the specific heat s

and s_1, are mixed without loss or gain of heat, the temperature, T, of the mixture is:

$$T = \frac{w\,t\,s + w_1\,t_1\,s_1}{w\,s + w_1\,s_1}$$

EXPANSION BY HEAT.

When a body becomes warmer it expands, when it becomes cooler it contracts, a rule of which ice, however, is one of the exceptions.

EXPANSION OF SOLIDS.

Amount of linear expansion of solids may be computed by the following formula for the Fahrenheit scale:

$$L_1 = \frac{L\left(1 + \frac{at_1}{180}\right)}{1 + \frac{at}{180}}$$

in which L_1 is the length of a bar at any temperature, t_1, knowing its length, L, at any other temperature, t, and a is a coefficient to be obtained from the following table:

COEFFICIENT OF EXPANSION FROM 32° TO 210° F.

Glass	0.000,861,30	Pine wood (lengthwise)	0.000,8
Platinum	0.000,884,20	Oak wood	0.000,7
Steel, soft	0.001,078,80	Granite	0.000,8
Iron, cast	0.001,125,00	Limestone	0.000,8
Iron, wrought	0.001.220,40	Antimony	0.001,1
Steel, hardened	0.001,239,50	Gold	0.001,4
Copper	0.001,718,20	Ebonite	0.001,7
Bronze	0.001,816,70	Nickel	0.001,8
Brass	0.001,878,20	Silver	0.001,9
Tin	0.002,173,00	Aluminum	0.002,3
Lead	0.002,857,50	Pine wood (crosswise)	0.005,8
Zinc	0.002,941,70	Mercury (in glass tube)	0.016,2

EXPANSION OF LIQUIDS.

The expansion of liquids by heat is expressed by the volume of a given quantity of liquid at different temperatures, as is done in the following table for water, showing also that at the point of maximum density.

The maximum density of water, as appears from this table, is between 32° and 46° F.; above 46° the volume increases, but below 32° it increases also. Apparently this is an exception to the general rule that all bodies expand by heat and contract when the temperature is lowered. This exception, however, may be accounted for when we assume that at 32°, when the water passes from the liquid to the solid state, its molecular constitution is changed also, which is also indicated by the change in specific heat at this point.

EXPANSION AND WEIGHT OF WATER AT VARIOUS TEMPERATURES.

Temperature.	Relative Volume by Expansion.	Weight of One Cubic Foot.	Weight of One Imperial* Gallon.
°Fahr.		Pounds.	Pounds.
32	1.00000	62.418	10.0101
35	.99993	62.422	10.0103
39.1	.99989	62.425 maximum dens'y	10.0112
40	.99989	62.425	10.0112
45	.99993	62.422	10.0103
46	1.00000	62.418	10.0101
50	1.00015	62.409	10.0087
52.3	1.00029	62.400 ordinary calculations	10.0072
55	1.00038	62.394	10.0063
60	1.00074	62.372	10.0053
62 mean temperature	1.00101	62.355	10.0000
65	1.00119	62.344	9.9982
70	1.00160	62.313	9.9933
75	1.00239	62.275	9.9871
80	1.00299	62.232	9.980
85	1.00379	62.182	9.972
60	1.00459	62.133	9.964
95	1.00554	62.074	9.955

Temperature.	Relative Volume by Expansion.	Weight of One Cubic Foot.	Weight of One Imperial* Gallon.
°Fahr.		Pounds.	Pounds.
100	1.00639	62.022	9.947
105	1.00739	61.960	9.937
110	1.00889	61.868	9.922
115	1.00989	61.807	9.913
120	1.01139	61.715	9.897
125	1.01239	61.654	9.887
130	1.01390	61.563	9.873
135	1.01539	61.472	9.859
140	1.01690	61.381	9.844
145	1.01839	61.291	9.829
150	1.01989	61.201	9.815
155	1.02164	61.096	9.799
160	1.02340	60.991	9.781
165	1.02589	60.843	9.757
170	1.02690	60.783	9.748
175	1.02906	60.665	9.728
180	1.03100	60.548	9.711
185	1.03300	60.430	9.691
190	1.03500	60.314	9.672
195	1.03700	60.198	9.654
200	1.03889	60.081	9.635
205	1.0414	59.93	9.611
210	1.0434	59.82	9.594
212	1.0466	59.64	9.565
250	1.06243	58.75	9.422
300	1.09563	56.97	9.136
400	1.15056	54.25	8.703
500	1.22005	51.16	8.204

The cubical expansion, or expansion of volume, of water, from 32° F. to 212° F. and upward, is given in the above. The rate of expansion increases with the temperature. The expansion for the range of temperature from 32° to 212° is .0466, or fully $4\frac{1}{2}$ per cent of the volume at 32°; or an average of .000259 per degree, or $\frac{1}{3865}$ part of the volume at 32° F.

EXPANSION OF LIQUIDS, FROM 32° TO 212° F.—VOLUME AT 32° = 1.

Liquid.	Volume at 212°.	Expansion.	Liquid.	Volume at 212°.	Expansion.
Alcohol......	1.1100	1-9	Sea water....	1.0500	1-20
Nitric acid...	1.1100	1-9	Water........	1.0466	1-22
Olive oil.....	1.0800	1-12	Mercury.....	1.018	1-56
Turpentine..	1.0700	1-14			

TRANSFER OF HEAT.

Heat is transferred from one body to another by conduction, radiation and convection.

*One imperial gal. is equal to 1.203 wine gals. (U. S. standard).

INSULATORS.

Insulators or non-conductors of heat are of special value in the construction of ice houses, cold storage rooms, etc., and the following table shows the retentive power of various substances, together with the percentage of solid matter in a given space (in first column). The figures in second column are for a covering one inch thick, and a difference of 100° F. on each side of the covering, and at temperatures of 176° F. on hot side of covering, except in some cases, in which it was 310° F., as stated.

Non-conductors One Inch Thick.	Net Cubic Inch of Solid Matter in 100.	Heat Units Transmitted per sq. Foot per Hour.
Still air	....	43
Confined air	....	108
" " =310°	....	203
Wool=310°	4 3	36
Absorbent cotton	2.8	36
Raw cotton	2	44
" "	1	48
Live geese feathers=310°	5	41
" " " "	2	50
Cat-tail seeds and hairs	2.1	50
Scoured hair, not felted	9.6	52.
Hair felt	8.5	56
Lampblack=310°	5.6	41
Cork, ground	....	45
Cork, solid	....	49
Cork charcoal=310°	5.3	50
White pine charcoal = 310°	11.9	58
Rice chaff	14.6	78
Cypress (*Taxodium*) shavings	7	60
" " sawdust	20.1	84
" " board	31.3	83
" " cross-section	31.8	145
Yellow poplar (*Liriodendron*) sawdust	16.2	75
" " " board	36.4	76
" " " cross-section	30.4	141
"Tunera" wood, board	79.4	156
Slag wool, best	5.7	50
Carbonate of magnesium	6	50
Calcined magnesia =310°	2.3	52
"Magnesia covering," light	8.5	58
" " heavy	13.6	78
Fossil meal =310°	6	60
Zinc white =310°	8.8	72
Ground chalk = 310°	25.3	80
Asbestos in still air	3	56
" " movable air	3.6	99
" " " " =310°	8.1	210
Dry plaster of Paris = 310°	36.8	131
Plumbago in still air	30.6	134
" " movable air =310°	26.1	296
Coarse sand = 310°	52.9	264
Water, still	....	335
Starch jelly, very firm, "	....	345
Gum-Arabic mucilage, "	....	290
Solution sugar, 70 per cent, "	....	251
Glycerin, "	....	197
Castor oil, "	....	136
Cotton seed oil, "	....	129
Lard oil, "	....	125

CONDUCTION OF HEAT.

The flow of heat from a warmer to a colder part of a body is called conduction. Some bodies conduct heat much more rapidly than others, hence we speak of good and bad conductors of heat. Very poor conductors and non-conductors of heat are also called insulators.

RELATIVE CONDUCTIVITY OF MATERIALS.

Material	Conductivity	Material	Conductivity
Gold	1.000	Bismuth	0.011
Copper	0.918	German silver	0.109
Brass	0.150	Iron	0.183
Zinc	0.305	Sandstone (Neumann)	0.007
Silver	1.096	Soft coal (Neumann)	0.0003
Cadmium	0.221	Granite	0.005
Tin	0.145	Ice	0.006
Lead	0.072	Marble	0.008

INSULATION OF STEAM PIPE.

With reference to the insulation of steam pipe, Norton estimates the loss through radiation of an uncovered steam pipe carrying steam of 200 pounds at 13.84 B. T. U. per square foot per minute. Covering the pipe as indicated in the following table, the radiation is reduced to the figures given. Box A is a ⅞-inch square pine box surrounding the pipe, leaving one inch minimum space at its four sides. The saving is calculated on above basis.

SPECIMEN.	B. T. U. per sq. ft. per min. at 200 lbs.	Saving on one year per 100 sq. ft. pipe.
Box A—		
1, with sand	3.18	$34.60
2, with cork, powdered	1.75	39.40
3, with cork and infusorial earth	1.90	38.90
4, with sawdust	2.15	37.90
5, with charcoal	2.00	38.50
6, with ashes	2.46	36.90
Brick wall, 4 inches thick	5.18	28.80
Hair felt, 1 inch thick	2.51	36.80
Pine wood, 1 inch thick	3.56	33.80
Spruce, 1 inch thick	3.40	33.90
Spruce, 2 inches thick	2.31	37.50
Spruce, 3 inches thick	2.02	38.50
Oak, 1 inch thick	3.65	33.10
Hard pine, 1 inch thick	3.72	32.90

NON-CONDUCTING COATING FOR STEAM PIPES.

M. Burnat's experiments were made with cast iron steam pipes, 4.72 inches in diameter externally, ¼-inch thick, in a large unheated hall free from drafts. They were in five groups differently coated:

First Group.—Coated with straw laid lengthwise, .60 inch thick, wrapped with straw rope.

Second Group.—Bare.

Third Group.—Each pipe laid in a pottery pipe, inclosing an air space, coated with a mixture of loamy earth and chopped straw, covered with tresses of straw.

Fourth Group.—Coated with cotton waste, one inch thick, wrapped in cloth bound with cord.

Fifth Group.—Coated with a plaster of clay and cow's hair, 2.36 inches thick.

The results are given in the following table.

CONDENSATION OF STEAM IN COATED PIPES.

Absolute Pressure of Steam per Square Inch.	Temperatures.			Steam Condensed per Sq. Foot of External Surface of Pipes per Hr.				
	Steam.	Air.	Difference.	Straw coat, 1st.	Bare 2d.	Pottery coat, 3d.	Waste coat, 4th.	Plaster coat, 5th.
Lbs.	°Fahr.	°Fahr.	°Fahr.	Lb.	Lb.	Lb.	Lb.	Lb.
16.5	218.0	46.4	171.6	.139	.496	.170	.217	.254
16.5	218.0	33.8	184.2	.152	.485	.166	.205	.262
18.4	223.4	33.7	189.7	.164	.555	.186	.229	.287
18.4	223.4	27.1	196.4	.182	.571	.264	.287	.344
22.0	233.2	41.5	191.7	.246	.576	.258	.244	.320
22.0	233.2	36.5	196.7	.164	......	.158	.250	....
22.0	233.2	36.1	197.1	.162	.557	.178	.260	....
22.0	233.2	28.9	204.3	.201	.586	.264	.328	.346
25.7	241.6	43.3	198.4	.244	.645	.301	.375	.389
25.7	241.6	36.5	205.1	.274	......	.285	.369	....
29.4	249.1	43.3	205.8	.252	.721	.270	.342	.379
29.4	249.1	30.6	218.4	.225	.621	.250	.328	.336
			Averages					
22.0	233.1	36.5	196.6	.200	.581	.229	.286	.324

The bare pipe was afterward coated with old felt, which had been treated with caoutchouc; and it condensed an average of .313 pound of steam per square foot per hour.

The rates of condensation and of emission of heat are summarized as follows:

SUMMARY RESULTS.

Coating of Pipe.	Steam Condensed per Square Foot per Hour.		Heat Emitted per Square Foot per Hour.	
	Total.	Per 1° F. Difference	Total.	Per 1° F. Difference
	Pound.	Pound.	Units.	Units.
Bare pipe....................	.581	.00300	552.8	2.812
Straw........................	.200	.00102	190.3	0.968
Pottery pipes with air space.	.229	.00115	224.8	1.108
Cotton waste.................	.286	.00146	272.1	1.384
Felt.........................	.313	.00159	297.8	1.515
Plaster......................	.324	.00165	308.3	1.568
The same, painted white......	.307	.00156	292.1	1.486

RELATIVE CONDUCTIVITIES OF METALS.

Gold	1,000	Zinc	360
Silver	973	Tin	304
Copper	878	Lead	180
Iron	374	Marble	25

RADIATION OF HEAT.

Heat is also transmitted from one body to another by radiation. In this case the temperature of the intervening medium remains unaltered.

THEORY OF HEAT TRANSFERS.

This theory asserts that all bodies are constantly giving out heat by radiation, at a rate depending on their substance and temperature, but independent of the substance and temperature of the bodies surrounding them; and that whether a body remains at the same temperature or alters its temperature depends upon whether it receives as much heat from other bodies as it yields up to them.

ABSORPTION OF RADIANT HEAT.

When heat rays fall upon a body a portion of them is reflected, a portion of them transmitted, and the rest of them is absorbed and increases the temperature of the body. In bodies, therefore, not transparent for heat rays the reflected and absorbed heat complement each other; that is to say, a good reflector is a bad absorber, and *vice versa.* By the same token, bodies which have a great absorbing power have also a great emissive or radiating power, but are bad reflectors for heat, and *vice versa.* Some bodies, however, are good reflectors for light, and at the same time excellent absorbers for heat, like white lead, for instance.

EMISSIVE AND ABSORBING POWER FOR HEAT.

Lampblack	100	Polished silver	2.5
White lead	100	Gold leaf	4.3
White paper	98	Copper foil	4.9
Crown glass	90	Polished platinum	9.2

The following table has evidently been compiled on the basis that the radiating and reflecting power are strictly complementary to each other. This holds good as a rule. However, it is one that is not without exceptions. Thus white lead has a good reflective, but also a good absorbing power, which it is well to note.

COMPARATIVE ABSORBING OR RADIATING AND REFLECTING PROPERTIES OF SOLIDS.

Substance.	Absorbing or Radiating Power.	Reflecting Power.
	Proportion, Per Cent.	Proportion, Per Cent.
Brass, bright polished	7	93
Brass, dead polished	11	89
Copper	7	93
Glass	90	10
Gold	5	95
Ice	85	15
Iron, cast, polished	25	75
Iron, wrought, polished	23	77
Marble	93 to 98	7 to 2
Mercury	23	77
Platinum, polished	24	76
Platinum, sheet	17	83
Silver leaf on glass	27	73
Silver, polished	3	97
Steel, polished	17	83
Tin	15	85
Water	100	0
Writing paper	98	2
Zinc, polished	19	81

CONVECTION OF HEAT.

Convection of heat takes place when heat is transferred from one place to another by the bodily moving of the heated substance, as it takes place when water is heated in a vessel, the hot water and the cold constantly exchanging places.

COMPLICATED TRANSFERS OF HEAT.

The phenomena of conduction, radiation and convection are complicated and combined in the transmission of heat through metal plates, tubes, jackets, etc., and the quantitative relations may be derived from the following data, after D. K. Clark's tables, etc.:

The heat radiated from incandescent coal or coke is expressed by the formula:

$$R = 144\, a\, \Theta\, (a^t - 1)$$

R = quantity of heat radiated per square foot of surface per hour, in British units.

Θ = temperature of the inclosure, in Fahrenheit degrees.

t = excess temperature of surface of hot body above the temperature of the inclosure, Θ, in Fahrenheit degrees.

a = constant, 1.00425.

According to the formula, the rate of radiation increases in a much more rapid ratio than the excess temperature, when the temperature of the inclosure is constant.

The heat radiated from a coal or a coke fire is estimated to be about one-half of the whole heat generated. It increases almost as fast as the rate of combustion of the fuel per hour per square foot.

CONVECTION OF HEAT FROM AN EXTERNAL SURFACE.

Surrounding Medium.	
Air	$C = .2849t^{1.233}$
Hydrogen	$C = .9827t^{1.233}$
Carbonic acid	$C = .2759t^{1.233}$
Olefiant gas	$C = .3817t^{1.233}$

C = quantity of heat, in British units, conveyed away from a solid body by a gas external to it, per square foot of surface per hour, under one atmosphere of pressure.

t = excess temperature of surface in Fahrenheit degrees.

CONDENSATION OF STEAM IN BARE PIPES EXPOSED TO AIR.

Tredgold found that steam of 17½ pounds absolute pressure per foot was condensed in cast iron pipes in a room at 60° F., at the rate of .352 pound per square foot of exposed surface per hour; or .0022 pound per degree of difference of temperature.

The following results were found by M. Clement. It is here assumed that the steam was of 20 pounds absolute pressure per square inch. The pipes were exposed in a room at 77° F.

Bare Surface.	Steam Condensed per Square Foot per Hour.
Cast iron pipe, horizontal	.328 pound.
Blackened pipe, horizontal	.308 "
Copper pipe, horizontal	.267 "
Blackened pipe, horizontal	.308 "
Blackened pipe, upright	.359 "

M. Burnat found that for steam of 22 lbs. absolute pressure, with 196°.6 F. difference of temperature, .581 lb. was condensed per square foot of a cast iron pipe, nearly horizontal, per hour.

Dr. William Anderson experimented with a tubular steam heater, of 2-inch wrought iron tubes, in a temperature of 59° F., with steam of 51 lbs. total pressure per square inch; .785 lb. was condensed per square foot per hour.

The foregoing results are collected in the following tablet:

Observer	Temperature of Surrounding Air.	Difference of Temperature.	Steam Consumed per Square Foot per Hour.		Heat Emitted per 1° F. Difference of Temperature.
			Total.	Per 1° F.	
	°Fahr.	°Fahr.	Pound.	Pound.	Units.
Clement…………	77	151	.328	.00217	2.07
Tredgold…………	60	161	.352	.0022	2.10
Burnat…………	36.5	196.6	.581	.0030	2.81
Anderson…………	59	223	.785	.0035	3.22

From these data the following approximate formulæ are deduced:

Condensation of steam in cast iron pipes, in air, per square foot of surface per hour at ordinary temperatures:

$$s = \frac{t^2}{55000} - .12$$

Heat emitted from cast iron pipes, in air, per square foot of surface per hour, at ordinary temperatures:

$$h = \frac{t^2}{58} - 114$$

Heat emitted from cast iron pipes, in air, per square foot of surface per degree of difference of temperature of steam and air, per hour, at ordinary temperatures.

$$h' = \frac{t}{58} - \frac{114}{t}$$

s = quantity of steam condensed in pounds.

h = quantity of heat emitted in units.

h' = quantity of heat emitted, per degree of difference of temperature.

t = difference of temperature, in Fahrenheit degrees.

The latent heat of steam of 22 pounds total pressure per square inch, 950 units per pound, is employed as the heat factor, as an average value.

The following table has been calculated by means of these formulæ:

STEAM CONDENSED IN BARE CAST IRON PIPES IN AIR, AND HEAT EMITTED, AT ORDINARY TEMPERATURES.

Steam.		Difference or Excess of Temperature of Steam above 62° Fahr.	Steam Condensed per Square Foot per Hour.		Heat Emitted. per Square Foot per Hour.	
Total Pressure per Square Inch.	Temperature.		Total.	Per 1° F. of Difference.	Total.	Per 1° F. of Difference.
Pounds.	° Fahr.	° Fahr.	Pounds.	Pounds.	Units.	Units.
14.7	212	150	29	.00193	276	1.84
18	222	160	346	.00216	329	2.05
21.5	232	170	.405	.00238	384	2.26
26	242	180	.47	.00261	446	2.48
31	252	190	.54	.00284	513	2.70
36.5	262	200	.607	.00303	577	2.89
43	272	210	.682	.00325	648	3.08
51	282	220	.76	.00345	722	3.28

For the increased rate of condensation induced by a draft of air, compared with that caused in the still air of a room, a bare steam boiler, in open air, was tested. Steam of 50 lbs. absolute pressure per square inch was condensed at the rate of 1.25 pounds per square foot of external surface per hour; or, for a difference of 236° of temperature, .0053 pound per degree of difference; showing that 4.79 units of heat per degree was emitted, or a half more than from a pipe in still air.

EXPERIMENTS NEEDED.

The foregoing and following data relate nearly all to the emission of heat from pipes, etc., filled with water or steam. It would of course be also highly desirable to have similar data for ammonia, especially for anhydrous ammonia, at the temperatures of condenser, freezing tank, brine tank and cold storage rooms. But such experiments have not been made so far. Numerical data on this topic have been abstracted from practical experience, and such as were attainable in this way have been mentioned in their place in the second part of this book, but are necessarily somewhat arbitrary.

COOLING OF WATER IN PIPES EXPOSED TO AIR.

Mr. Wm. Anderson experimented with 2-inch wrought iron pipes, $\frac{3}{16}$ inch thick, galvanized, and 4-inch cast iron pipes, $\frac{7}{16}$ inch thick, through which hot water was passed.

Results are given in the following table. The ultimate results harmonize with those for the use of steam in pipes.

COOLING OF WATER IN PIPES EXPOSED TO AIR.

	Two-inch Wrought Iron Pipes.				Four-inch Cast Iron Pipes.			
Number of experiment...........	1	2	3	4	1	2	3	4
Temperature of the atmosphere Fahr.....	53°	53°	52°.5	52°	60°	60°	60°	59°
Average difference of temperatures of the water and the air Fahr	103°.7	49°.4	25°.4	14°.3	62°.3	45°.8	33°.9	27°.3
Total heat emitted per square foot per hour. Units...........	233.7	104.4	46.45	19.7	99.5	69.9	49.5	38.2
Heat emitted per 1° F. difference of temperature Units...........	2.25	2.11	1.83	1.39	1.59	1.53	1.46	1.40

Tredgold experimented with small vessels of different materials, in which water was cooled from a temperature of 180° to one of 159°, in a room at 55°. The heat emitted per square foot per hour per degree of mean difference of temperature was as follows:

Tin plate....................................	1.37 units.
Sheet iron....................................	2.24 "
Glass	2.18 "

Also, in a 2½-inch cast iron pipe, ¼ inch thick, water was cooled from 152° to 140° F., in a room at 67°. The heat emitted per square foot per hour per degree of difference of temperature was as follows:

Ordinary rusty surface....................	1.823 units.
Black, varnished..........................	1.900 "
White (two coats of lead paint)..	1.778 "

TRANSMISSION OF HEAT THROUGH METAL PLATES FROM WATER TO WATER.

In a metal tubular refrigerator, hot wort was cooled by water at such a rate that, taking averages, 80 units of heat passed from the wort, and was absorbed by the water per square foot of cooling surface per 1° F. difference of temperature per hour. The water and the wort were moved in opposite directions.

M. Péclet proved experimentally that the rate of transmission of heat was directly as the difference of temperature at the two faces of metal plates.

TRANSMISSION OF HEAT THROUGH METAL PLATES FROM STEAM TO WATER.

The rate of transmission of heat from steam through a metal plate to water at the other side is practically uniform per degree of difference of temperature. The following table gives average results of performance, from which it appears that the transmission is much more effective for evaporating than for heating water, twice as much for flat copper plate, three times as much for copper pipe, one-fourth more for cast iron plate. Also, that pipe surface is one-fifth more effective than flat plate surface for heating, and more than twice as much for evaporation—the result of better circulation, no doubt.

HEATING AND EVAPORATING WATER BY STEAM THROUGH METALS.

Metal Surface.	Per Square Foot per 1° F. Difference of Temperature.			
	Steam Condensed.		Heat Transmitted.	
	Heating.	Evaporating.	Heating.	Evaporating.
	Pounds.	Pounds.	Units.	Units.
Copper plate..........	.248	.483	276	534
Copper pipe...........	.291	1.070	312	1034
Cast-iron boiler.......	.077	.105	82	100

Mr. Isherwood experimented with cylindrical metal pots, 10 inches in diameter, 21¼ inches deep; ⅛ inch, ¼ inch and ⅜ inch thick; turned and bored. They were placed in a steam bath of from 220° to 320° F. Water at 212° was supplied to the pots, and evaporated. The rate of evaporation per degree of difference of temperature was the same for all temperatures; and the rate was the same for the different thicknesses. The respective weights of water, and heats consumed per square foot of inside surface per degree of difference were as follows:

	Water at 212°.	Heat.
Copper................	.665 lb.	642.5 units
Brass.....................	.577 "	556.8 "
Wrought iron...........	.387 "	373.6 "
Cast iron................	.327 "	315.7 "

The differences of results for the same metal evidently arise in part from the comparative activity of circulation, and in part from the condition and position of the heating surfaces.

CONDENSATION OF STEAM IN PIPES OR TUBES BY WATER EXTERNALLY.

From the results of experiments with surface condensers, in which the steam was passed through the tubes, it appears that 500 units of heat by condensation were transmitted per square foot of tube surface per hour per 1° F. difference of temperature. The condensers were arranged in three groups of tubes successively traversed by the condensing water. In another case, where the condenser was arranged in two groups, from 220 to 240 units were transmitted.

Mr. B. G. Nichol experimented with an ordinary surface condenser brass tube, ¾ inch in diameter outside; No. 18 wire gauge in thickness; encased in a 3¾-inch iron pipe. Steam of 32½ pounds total pressure per square inch occupied the interspace, while cold water at 58° F. initial temperature was run through the brass tube. Three experiments were made with the tubes in a vertical position, and three in a horizontal position.

	Vertical Position.			Horizontal Position.		
	1,	2,	3,	4,	5,	6,
Velocity of water through tube, in feet per minute,	81,	278,	390,	78,	307,	415 feet.
Steam condensed per square foot of surface per hour, for 1° F. difference of temperature,	.335,	.436,	.457,	.480,	.603,	609 pound.
Heat absorbed by the water, per square foot per hour, per 1° F. difference of temperature,	346,	449,	466,	479,	621,	699 units.

The rate of condensation was greater in the horizontal position than in the vertical position. Also, the efficiency of the condensing surface was increased by an increase of velocity of the water through the tube, nearly in the ratio of the fourth root of the velocity for vertical tubes; and nearly as the 4.5 root for horizontal tubes.

TRANSMISSION OF HEAT THROUGH METAL PLATES OR TUBES, FROM AIR OR OTHER DRY GAS TO WATER.

The rate of transmission of convected heat is probably from 2 to 5 units of heat per hour per square foot of surface per 1° F. of difference of temperature.

In a locomotive fire box, where radiant heat co-operated with convected heat, the following results have been obtained in generating steam of 80 pounds pressure per square inch. The temperature of the fire is taken at 2,000° F.

	Water Evaporated per Square Foot per Hour.	Heat Transmitted per Square Foot per Hour per 1° F. Difference of Temperature.
Burning coke, 75 pounds per square foot of grate..............	25½ pounds.	14½ units.
Burning briquettes, 74½ pounds per square foot of grate	35 "	20 "

There are in practice little or no differences between iron, copper and lead in evaporative activity, when the surfaces are dimmed or coated, as under ordinary conditions.

COMPARATIVE RATE OF EMISSION OF HEAT FROM STEAM PIPES IN AIR AND IN WATER.

It appears that for equal total difference of temperature, the rate of emission of heat from steam pipes in water amounts, in round numbers, to from 150 to 250 times the rate in air, according as the pipes are vertical or horizontal.

COMPARATIVE RATE OF EMISSION OF HEAT FROM WATER TUBES IN AIR AND IN WATER AT REST AND IN MOTION.

It appears that the rate of emission from water-tubes in water was about twenty times the rate in air. Mr. Craddock proved it experimentally to be twenty-five times. When the water tube was moved through the air at a speed of fifty-nine feet per second, it was cooled in one-twelfth of the time occupied in still air. In water, moved at a speed of three feet per second, the water in the tube was cooled in half the time.

PASSAGE OF HEAT THROUGH METAL PARTITIONS.

From some recent observations made in Germany the following table, giving the transmission of heat through metal partitions per hour, per square foot and per one degree F. difference between each side, viz.:

Smoke or air through metal to air	1.20 to	1.70	B. T. U.
Steam through metal to air	2.40 to	3.40	"
Water through metal to air or reverse	2.15 to	3.15	"
Steam through metal to water..............	200.00 to	240.00	"
Steam through metal to boiling water.....	1,000.00 to	1,200.00	"
Water through metal to water.............	72.00 to	96.00	"

LATENT HEAT.

When a body passes from the solid to the liquid state, or from the liquid to the gaseous or vapor state, a certain amount of heat is required to bring about the change. As this heat is absorbed during the process of fusion or vaporization it is called latent heat of fusion and latent heat of evaporation (latent heat contained in the vapor).

LATENT HEAT OF FUSION.

The heat which becomes latent during the fusion or melting of a body is used or absorbed while doing the work of disintegrating the molecular structure, doing internal work as it is called.

TABLE SHOWING LATENT HEAT OF FUSION.

	Thermal units.		Thermal units.
Ice	142.5	Tin	25.5
Nitrate of ammonia	113.2	Cadmium	24.5
Nitrate of soda	104.1	Bismuth	22.7
Phosphate of potash	85.1	Sulphur	16.8
Nitrate of potash	78.4	Lead	9.5
Chloride of calcium	64.3	Phosphorus	9.0
Zinc	50.6	D'Arcet's alloy	8.1
Platinum	48.8	Mercury	5.1
Silver	37.8		

MELTING POINTS, ETC.

	°Fahr.		°Fahr.
Aluminum	Full red heat	Iron, cast, white	1992 to 2012
Antimony	1150	" wrought	2912
Bismuth	507	Lead	617
Bronze	1690	Mercury	—39
Copper	1996	Silver	1873
Gold, standard	2156	Steel	2372 to 2552
" pure	2282	Tin	442
Iron, cast, gray	2012	Zinc	773
Carbonic acid	—108	Spermaceti	120
Ice	32	Sulphur	239
Nitro-glycerine	45	Tallow	92
Phosphorus	112	Turpentine	14
Stearine	109 to 120	Wax, rough	142
		" bleached	154

EFFECT OF PRESSURE ON MELTING POINT.

Substances which expand during solidification, like water, have their freezing points lowered by pressure, and those which contract in solidification have their freezing points raised by pressure.

LATENT HEAT OF SOLUTION.

When a body is dissolved in water or in any other liquid, or if two solid bodies (salt and snow, for an example) mix to form a liquid, a certain amount of heat becomes likewise latent; it is called the latent heat of fusion. Since the latent heat of fusion in the case of such mixtures is taken from the mixture itself, the temperature falls correspondingly, as shown by the table on frigorific mixtures.

For practical purposes the mixtures of snow and hydrochloric acid, or, where acid is objectionable, the mixture of snow and potash, is very serviceable to produce refrigeration on a small scale. · The mixture of sodium sulphate, ammonium nitrate and nitric acid is also recommendable.

LIST OF FRIGORIFIC MIXTURES.

Mixture	Parts	Thermometer Sinks Degrees F.
Ammonium nitrate	1 part	From + 40° to + 4°
Water	1 "	
Ammonium chloride	5 parts	From + 50° to + 10°
Potassium nitrate	5 "	
Water	16 "	
Ammonium chloride	5 parts	From + 50° to + 4°
Potassium nitrate	5 "	
Sodium sulphate	8 "	
Water	16 "	
Sodium nitrate	3 parts	From + 50° to — 3°
Nitric acid, diluted	2 "	
Ammonium nitrate	1 part	From + 50° to — 7°
Sodium carbonate	1 "	
Water	1 "	
Sodium phosphate	9 parts	From + 50° to — 12°
Nitric acid, diluted	4 "	
Sodium sulphate	5 parts	From + 50° to + 3°
Sulphuric acid, diluted	4 "	
Sodium sulphate	6 parts	From + 50° to — 10°
Ammonium chloride	4 "	
Potassium nitrate	2 "	
Nitric acid, diluted	4 "	
Sodium sulphate	6 parts	From + 50° to — 40°
Ammonium nitrate	5 "	
Nitric acid, diluted	4 "	
Snow or pounded ice	2 parts	to — 5°
Sodium chloride	1 "	
Snow or pounded ice	5 parts	to — 12°
Sodium chloride	2 "	
Ammonium chloride	1 "	
Snow or pounded ice	24 parts	to — 18°
Sodium chloride	10 "	
Ammonium chloride	5 "	
Potassium nitrate	5 "	
Snow or pounded ice	12 parts	to — 25°
Sodium chloride	5 "	
Ammonium nitrate	5 "	
Snow	3 parts	From + 32° to — 23°
Sulphuric acid, diluted	2 "	
Snow	8 parts	From + 32° to — 27°
Hydrochloric acid	5 "	
Snow	7 parts	From + 32° to — 30°
Nitric acid, diluted	4 "	
Snow	4 parts	From + 32° to — 40°
Calcium chloride	5 "	
Snow	2 parts	From + 32° to — 50°
Calcium chloride, crystallized	3 "	
Snow	3 parts	From + 32° to — 51°
Potash	4 "	

HEAT BY CHEMICAL COMBINATION.

As one of the chief sources of heat chemical combination has been mentioned, which may be defined as the process which takes place when the ultimate constituent parts (atoms) of one or more elementary bodies unite with those of another elementary body or bodies to form a substance essentially different in its properties from those of the original bodies.

ELEMENTARY BODIES.

Substances which cannot be resolved into two or more different substances are called elementary bodies, elements or simple bodies.

CHEMICAL ATOMS.

Chemically considered, an atom is the smallest particle of matter entering into or existing in combinations. The atomic weight is a number expressing the ratio of the weight of the atoms of an element to the weight of an atom of hydrogen, the latter being taken as unit.

MOLECULES.

The smallest quantity of an elementary body, as well as of a compound body, which is capable of having an independent existence is called a molecule. A molecule, therefore, is a combination of several atoms of one and the same or of different elements.

CHEMICAL SYMBOLS.

The chemical elements are expressed by symbols which are the initial letters of their Latin or English name. The symbols also represent the relative quantity of one atom of an element.

The composition of the molecule of a body is indicated by the symbols of its constituents. The number of atoms of each element present is denoted by a number placed at the lower right hand end of the symbol. Thus H_2 represents a molecule of hydrogen which is composed of two atoms, and H_2O represents a molecule of water, which is composed of two atoms of hydrogen and one of oxygen. The atomic weight of hydrogen being 1 and that of oxygen 16, it is readily seen how the formula H_2O yields the percentage composition by a simple calculation.

ATOMICITY.

Atomicity or valence is that property of an element by virtue of which it can hold in combination a definite number of other atoms, the atomicity of an elementary body is measured by the number of atoms of hydrogen which can be held in combination by an atom of the elementary body in question, the atomicity of hydrogen being taken as unit. Thus by referring to the following table it is readily seen how one atom of chlorine will hold in combination one atom of hydrogen, one atom of oxygen two atoms of hydrogen, one atom of nitrogen three atoms of hydrogen, and one atom of carbon four atoms of hydrogen and form saturated compounds.

For obvious reasons the rare and new elements, argon, helium, atherion, etc., are not mentioned.

TABLE OF PROPERTIES OF ELEMENTS.

Element.	Symbol.	Atomicity.	Atomic Weight.	Specific Gravity.
Aluminium	Al	IV	27.5	2.56
Antimony	Sb	V	122	6.7
Arsenic	As	V	75	5.75
Barium	Ba	II	137	4.0
Bismuth	Bi	V	208	8.75
Boron	B	III	11	2.68
Bromine	Br	I	80	2.96
Cadmium	Cd	II	112	1.58
Calcium	Ca	II	40	1.65
Carbon	C	IV	12	2.33
Chlorine	Cl	I	35.5	....
Chromium	Cr	VI	52.5	6.5
Cobalt	Co	VI	58.8	....
Copper	Cu	II	63.5	8.958
Fluorine	F	I	19	....
Gold	Au	III	196.7	19.26
Hydrogen	H	I	1	....
Iodine	I	III	127	4.948
Iridium	Ir	VI	198	21.15
Iron	Fe	VI	56	7.79
Lead	Pb	IV	207	11.36
Lithium	Li	I	7	.594
Magnesium	Mg	II	24	1.70
Manganese	Mn	VI	55	8.03
Mercury	Hg	II	200	13.60
Nickel	Ni	VI	58.8	....
Nitrogen	N	V	14	....
Oxygen	O	II	16	....
Palladium	Pd	IV	106.5	11.40
Phosphorus	P	V	31	1.840
Platinum	Pt	IV	197.4	21.15
Potassium	K	I	39	.865
Rhodium	Rh	VI	104	12.1
Selenium	Se	VI	79	4.28
Silicon	Si	IV	28.5	2.49
Silver	Ag	I	108	10.53
Sodium	Na	I	23	.9722
Strontium	Sr	II	87.5	2.542
Sulphur	S	VI	32	2.07
Tellurium	Te	VI	128	6.180
Tin	Sn	IV	118	....
Titanium	Ti	IV	50	....
Tungsten	W	VI	184	....
Uranium	Ur	VI	120	18.4
Vanadium	V	V	51.2	5.5
Zinc	Zn	II	65	7.13

GENERATION OF HEAT.

The generation of heat by chemical combination is explained by the fact that the resulting compounds possess less energy than the constituent elements before they unite or combine. The difference of energy before and after combination appears in the form of heat, electricity, etc. By the same token heat is absorbed during the decomposition of chemical compounds.

MEASURE OF AFFINITY.

The amount of heat or other form of energy developed during a chemical change is a measure for the chemical work done or the amount of affinity displayed during the change.

TOTAL HEAT DEVELOPED.

The total amount of heat or energy developed during a chemical change depends solely upon the initial and final condition of the participating bodies (the initial or final condition of the system), and not on any intermediate conditions. In other words, the heat developed during a chemical change is the same whether the change takes place in one operation or in two or more separate processes.

MAXIMUM PRINCIPLE.

Of all chemical change which may take place within a system of bodies, without the interference of outside energy, that change will take place which causes the greatest development of heat, as a general rule.

According to the more modern conceptions it is held that that change will take place which will cause the greatest dissipation of energy, or by which the entropy of the system will suffer the greatest increase, or by which the greatest amount of energy will be dissipated. (For definitions of entropy see Chapters VII and VIII.)

EXPRESSIONS FOR HEAT DEVELOPED.

The amount of heat, expressed in units, developed or absorbed during a chemical process may be conveniently used in connection with the chemical symbols. Thus the formula

$$Pb + 2\,I = Pb\,I_2 + 7.1400\ U$$

signifies that 207 parts of lead combine with 254 parts of iodine to form 461 parts of iodide of lead, and develop thereby 7.1400 units of heat.

HEAT OF COMBINATION OF SUBSTANCES WITH OXYGEN.

Substances.	Product.	Units of Heat Evolved.		
		By 1 lb. of Substance.	By 1 lb. of Oxygen.	By 1 Atom of Substance in Pounds.
Hydrogen	H_2O	60,986	7,623	60,986
Wood charcoal.....	CO_2	14,220	5,332	170,640
Sulphate, native...	SO_2	3,996	3,996	127,872
Phosphorus(yellow)	P_2O_5	10,345	8,017	320,683
Zinc................	Zn. O	2,394	9,702	155,610
Iron................	F_3O_4	2,848	7,475	159,466
Copper..............	Cu O	1,085	4,309	68,947
Carbonic oxide.....	CO_2	4,325	....	121,111
Cuprous oxide......	Cu O	561	....	32,947

COMBUSTION.

Combustion is the rapid combination of combustible material (fuel) with oxygen.

SPONTANEOUS COMBUSTION.

In order to start the combustion of a combustible body it is generally necessary to elevate its temperature or to bring it in contact with a burning body. In other words, it must be ignited. If a body undergoes combustion without ignition it is a case of spontaneous combustion; and if combustion takes place without the appearance of a flame or light it is called *slow combustion*.

INFLAMMABLE BODIES.

Bodies which are able to undergo combustion as with the appearance of a flame are called inflammable.

EXPLOSIVE BODIES.

If combustion of a body takes place at once or simultaneously throughout its whole mass, an explosion generally takes place, especially if the body is confined in a limited space and if the products of the combustion are of a gaseous nature. Therefore such bodies are called explosives.

AIR REQUIRED IN COMBUSTION.

The volume of air consumed chemically in the combustion of fuel is expressed by the formula:

$$A = 1.52\ (C + 3H - .4O)$$

A = volume of air as at 62° F., and under one atmosphere of pressure, in cubic feet per pound of fuel.

A' = weight of air as at 62° F. per pound of fuel.

C = percentage of constituent carbon.

H = percentage of constituent hydrogen.

O = percentage of constituent oxygen.

The weight of the air thus found by volume is equal to the volume divided by 13.14. Or it is found directly by the formula:

$$A' = .116\,(C + 3H - .4O)$$

In these formulæ the heat evolved by the combustion of the sulphur constituent is not noticed, as it is trifling in proportion.

GASEOUS PRODUCTS.

The volume of the volatile or gaseous products of the complete combustion of one pound of a fuel, as at 62° F., at atmospheric pressure, is, by formula:

$$V = 1.52\,C + 5.52\,H$$

The weight of the gaseous products is, by formula:

$$w = .126C + .358\,H$$

V = volume of gaseous products, in cubic feet.
w = weight of gaseous products, in pounds.
C = percentage of constituent carbon.
H = percentage of constituent hydrogen.

The volume at any other temperature is found by the formula for expansion of gases, given elsewhere.

The proportion of free or unconsumed air usually present in the gaseous products is determined by multiplying the percentage of oxygen, found by analysis, by 4.35. The product is the percentage of free air in parts of the whole mixture.

HEAT GENERATED.

The heat generated by combustion is as follows:

Carbon	14,500	heat units per pound
Hydrogen	62,000	" "
Sulphur	4,000	" "

The heating power of fuels containing carbon and hydrogen is approximately expressed by the formula:

$$h = 145\,(C + 4.28\,H)$$

in which h is the total heat of combustion.

The evaporative efficiency for one pound of fuel is—

$$e = .15\,(C + 4.29\,H)$$

$$\text{or, } e = \frac{h}{966}$$

e = weight of water evaporable from and at 212°, in pounds, per pound of fuel.

The maximum temperature of combustion of carbon is about 5,000° F.; and that of hydrogen is about 5,800° F.

HEAT OF COMBUSTION OF FUELS.

Fuel.	Air Chemically Consumed per Pound of Fuel.		Total Heat of Combustion of One Pound of Fuel.	Equivalent Evaporative Power, from and at 212° F., Water per Pound of Fuel.
	Pounds.	Cub. Ft. at 62° F.	Units.	Pounds.
Coal of average composition	10.7	140	14,700	15.22
Coke	10.81	142	13,548	14.02
Lignite	8.85	116	13,108	13.57
Asphalte	11.85	156	17,040	17.64
Wood, desiccated	6.09	80	10,974	11.36
Wood, 25 per cent moisture	4.57	60	7,951	8.20
Wood charcoal, desiccated	9.51	125	13,006	13.46
Peat, desiccated	7.52	99	12,279	12.71
Peat, 30 per cent moisture	5.24	69	8,260	9.53
Peat charcoal, desiccated	9.9	130	12,325	12.76
Straw	4.26	56	8,144	8.43
Petroleum	14.33	188	20,411	21.13
Petroleum oils	17.93	235	27,531	28.50
Coal gas, per cubic foot at 62° F	....	....	630	.70

COAL.

Coal consists mainly of carbon, which varies from 50 per cent to 80 per cent, by weight, of the fuel. Lignite or brown coal contains from 56 to 76 per cent of carbon. The average composition of coal is, say, 80 per cent of carbon, 5 per cent of hydrogen, 1¼ per cent of sulphur, 1⅛ per cent of nitrogen, 8 per cent of oxygen, and 4 per cent of ash. The fixed carbon or coke averages 61 per cent. The average specific gravity is 1.279; average weight of a solid cubic foot, 80 pounds; and of a cubic foot heaped, 50 pounds; average bulk of one ton heaped, 44½ cubic feet; equivalent evaporative efficiency, 15.40 pounds of water per pound of coal, from and at 212° F.

Bituminous coals hold from 6 per cent to 10 per cent of water hygroscopically; Welsh coals from ⅝ per cent to 2¾ per cent.

COKE.

Coke contains from 85 to 97½ per cent of carbon; from ¾ to 2 per cent of sulphur, and from 1½ to 14½ per cent of ash. The average composition may be taken as 93½ per cent of carbon, 1¼ per cent of sulphur, 5½ per cent of ash. It weighs from 40 pounds to 50 pounds per cubic foot solid, and about 30 pounds broken and heaped. The volume of one ton heaped is from 70 to 80 cubic feet; average, 75 cubic feet. Coke is capable of absorbing from 15 to 20 per cent of moisture. There is ordinarily from 5 per cent to 10 per cent of hygrometric moisture in coke.

LIGNITE.

Lignite or brown coal consists chiefly of carbon, oxygen and nitrogen; averaging in perfect lignite, 69 per cent of carbon, 5 per cent of hydrogen, 20 per cent of oxygen and nitrogen, and 6 per cent of ash. The weight is about 80 pounds per cubic foot. Imperfect lignite weighs about 72 pounds per cubic foot.

CHIMNEY AND GRATE.

The quantity of good coal, C, in pounds, that may be consumed per hour with a chimney having the height, H, above the grate bars, a sectional area, A, in square feet at the top, may be expressed by the formula—

$$C = 16\,A\,\sqrt{H}$$

and the total area of fire grate G in square feet—

$$G = \frac{H^2\,\sqrt{H}}{1071}$$

HEAT BY MECHANICAL MEANS.

Mechanical work is also a source of heat, and in nearly all cases where work is expended, the appearance of an equivalent amount of heat is observed. The heat due to friction, percussion, etc., is an example of this kind, as also is the heat generated by the compression of gases and vapors (see *Thermodynamics*).

The height of chimney for a given total grate area, the diameter **at the top being equal to one-thirtieth** of the height, is

$$H - 16.3^{2.5}\sqrt{G}$$

The side of a square chimney equal in sectional area to a given round chimney is equal to the product of the diameter by 0.886; the equivalent fraction of the height for the side of a square chimney is one-thirty-fourth.

Conversely, the diameter of a round chimney equal in sectional area to a given square chimney is equal to the product of the side of the square by 1.13.

When the top diameter of the chimney is one-thirtieth of the height—a good proportion—the quantity of coal that may be consumed per hour is expressed by the formula—

$$C - .014\,H^{2.5}$$

CHAPTER V.—FLUIDS; GASES; VAPORS.

FLUIDS IN GENERAL.

Fluids may be generally defined as bodies whose molecules are displaced by the slightest force, which property is also called *fluidity*, and it is possessed in a much larger degree by gases than by liquids.

Gases are eminently compressible and expansible, while liquids are so but in a slight degree.

VISCOSITY.

The property of liquid to drag adjacent particles along with it is called viscosity (*Internal Friction*).

PASCAL'S LAW.

Pressure exerted anywhere upon a liquid is transmitted undiminished in all directions and acts with the same force on all equal surfaces in a direction at right angles to those surfaces.

BUOYANCY OF LIQUIDS.

The pressure which the upper layer of a liquid exerts on the lower layers, is consequently also exerted in an upward direction, causing what is termed the buoyancy of liquids. It is on account of the buoyancy of liquids that a body weighed under liquid loses a part of its weight, equal to the weight of the displaced liquid (*Archimedian principle*).

SPECIFIC GRAVITY DETERMINATION.

By ascertaining the loss in weight of a body immersed under water its volume may be readily ascertained, it being equal to the volume of water corresponding to the lost weight. . This principle is used to determine the specific gravities of bodies in various ways; for instance, for solid bodies, by dividing their weight in air by the loss of weight which they sustain when weighed under water.

HYDROMETERS.

From among the instruments frequently used to ascertain the specific gravity of liquids, and by inference their strength, we mention those called hydrometers as based on the Archimedian principle. They are generally made of a weighted body (usually of glass), having a thinner stem at the upper end provided with a scale divided in degrees. The degrees may be arbitrary or show specific gravities or the strength of some particular liquid or solution in per cents; in the latter case the instrument is called Saccharometer, Salometer, Alcoholometer, Acidometer, Alkalimeter, etc., according to the liquid it is designed to test. Hydrometers for different liquids or purposes, provided they cover the same range of specific gravities, may be used for either liquid when the relation their degrees bear to each other is known. For some of the more current hydrometers, these relations are shown in the following table:

TABLE SHOWING SPECIFIC GRAVITY CORRESPONDING TO DEGREES, TWADDLE, BEAUME AND BECK, FOR LIQUIDS HEAVIER THAN WATER.

Number of Degrees.	Corresponding Sp. Gr.			Number of Degrees.	Corresponding Sp. Gr.		
	Twaddle	Beaume.	Beck.		Twaddle	Beaume.	Beck.
0	1.000	1.000	1.000	21	1.105	1.166	1.1409
1	1.005	1.007	1,0059	22	1.110	1.176	1.1486
2	1.010	1.014	1.0119	23	1.115	1.185	1.1565
3	1.015	1.020	1.0180	24	1.120	1.195	1.1644
4	1.020	1.028	1.0241	25	1.125	1.205	1.1724
5	1.025	1.034	1.0303	26	1.130	1.215	1.1806
6	1.030	1.041	1.0366	27	1.135	1.225	1.1888
7	1.035	1.049	1.0429	28	1.140	1.235	1.1972
8	1.040	1.057	1.0494	29	1.145	1 245	1.2057
9	1,045	1.064	1.0559	30	1.150	1.256	1.2143
10	1.050	1.072	1.0625	32	1.160	1.278	1.2319
11	1.055	1.080	1.0692	34	1.170	1.300	1.2500
12	1.060	1.088	1.0759	36	1.180	1.324	1.2680
13	1.065	1.096	1.0828	38	1.190	1.349	1.2879
14	1.070	1.104	1.0897	40	1.200	1.375	1.3077
15	1.075	1.113	1.0968	45	1.225	1.442	1.3600
16	1.080	1.121	1.1039	50	1.250	1.515	1.4167
17	1.085	1.130	1.1111	55	1.275	1.596	1.4783
18	1.090	1.138	1.1184	60	1.300	1.690	1 5454
19	1.095	1.147	1.1258	65	1.325	1.793	1.6190
20	1.100	1.157	1.1333	70	1.350	1.909	1.7000

There is a slight difference between the indications of the Beaume scale in different countries. The manufacturing chemists of the United States have adopted the following formula for converting the Beaume degrees into specific gravity:

$$\text{Specific gravity} = \frac{145}{145 - \text{degrees Beaume}},$$

which gives specific weight slightly higher than those in the foregoing table. (See also table in *Appendix.*)

PRESSURE OF LIQUIDS.

The pressure exerted by a column of liquid at its bottom or base is proportional to the vertical height of the column of liquid, its specific gravity and to the area of the bottom, and independent of the shape or thickness of the column of liquid.

WATER PRESSURE.

The pressure in pounds, P, of a column of water h feet high is—

$$P = .4335\ h \text{ per square inch,}$$
$$\text{and } P = 62.425\ h \text{ per square foot.}$$

SURFACE TENSION OF LIQUIDS.

The layer of a liquid which separates the same from a gas or vacuum has a greater cohesion than any other layer of the liquid, owing to the fact that the attraction exerted on this layer by the interior of the liquid is not counteracted by any attraction on the outside. The surface is, as it were, stretched over by an elastic skin which exerts a pressure on the interior, which pressure is termed surface tension. It increases with the cohesion of the liquid.

VELOCITY OF FLOW OF LIQUIDS.

The velocity with which a liquid flows through an opening depends only on the height of the liquid above the orifice and is independent of the density of the liquid. The velocity, v, in feet per second is expressed by the formula—

$$V = \sqrt{2gh} = 8\sqrt{h}$$

g being the acceleration per second due to gravity, and h the depth of the orifice below the surface, both expressed in feet.

QUANTITY OF FLOW.

The quantity of a liquid, say water, discharged through an opening depends on the area of the opening, A (in square feet), and also on the shap etc., of the orifice. If the orifice is a hole in the thin all of a vessel, the quantity, E (in cubic feet), discharged is expressed by

$$E - 5A\sqrt{h}$$

A short cylindrical appendix to the opening would increase the discharge to—

$$E - 6.56A\sqrt{2h.}$$

and an appendix having the best form of a conic frustrum will nearly discharge the theoretical amount.

$$E - 8A\sqrt{h}$$

FLOW OF WATER IN PIPES.

The mean velocity, v, of water in a cast iron pipe of the length, l, and the diameter, d, under the head, h, is-

$$v = 48\sqrt{\frac{d\,h}{l}}$$

The velocity is affected by the surface of pipe, and the viscosity or interior friction of the liquid (hydraulic friction).

QUANTITY OF FLOW THROUGH PIPES.

Dawning's formula for the quantity, E, in cubic feet of water discharged by channel or pipe under the head. h, in feet is as follows:

$$E = 100\,a\sqrt{\frac{h\,D.}{l}}$$

l being the length of pipe in feet; a, sectional area o current in square feet; c, wetted perimeter in feet.

$$D = \frac{a}{c} = \text{hydraulic mean depth.}$$

HEAD OF WATER.

The head, h, approximately required to move water with a velocity of 180 feet per minute through a clean cast iron pipe, having a diameter D inches and the length l in feet, is—

$$h = \frac{l}{25\,D}$$

WATER POWER.

The theoretical effect of water power expressed in foot-pounds per minute, is equal to the weight of the water falling per minute, multiplied by the height through which the water falls. Divided by 33,000, it expresses horse powers. The practical effect depends on the efficiency of the motor (water wheel, turbine, engine, etc.). The power required to lift water is calculated in the same manner.

HYDROSTATICS AND DYNAMICS.

The science which treats of the condition of liquids while at rest is called hydrostatics, and that which treats of the motion of liquids is called hydrodynamics.

CONSTITUTION OF GASES.

In a general way the term gas has been defined in the foregoing. Speaking more specifically, a gas is a body in which the distance between the constituent atoms or molecules is so great that the dimensions of the molecules themselves may be neglected in comparison therewith. The atoms or molecules in a gas are constantly vibrating to and fro, and the average momentum or energy of this motion represents the temperature of the gas. The vehemence or force with which the atoms or molecules impinge on the walls of a surrounding vessel in consequence of this motion represents the pressure of the gas.

PRESSURE AND TEMPERATURE.

In accordance with the foregoing definition the pressure, volume and temperature of a gas are in direct connection, which is expressed by the laws of Boyle and St. Charles.

BOYLE'S LAW.

The law of Boyle or of Mariotte asserts that the volume of a body of a perfect gas is inversely proportional to its pressure, density or elastic force, if its temperature remains the same.

ST. CHARLES LAW.

If a gaseous body is heated while the pressure remains constant, its volume increases proportionally with the temperature. The increase of volume for every degree F. is equal to $\frac{1}{493}$ of its volume at 32° F.

UNIT OF PRESSURE.

The general unit of pressure is the pressure of the atmosphere per square inch, which is equal to that of a column of water of about thirty feet, or that of a column of mercury of about thirty inches, and also equivalent to a pressure of 14.7 pounds—in round numbers fifteen pounds per square inch.

ABSOLUTE AND GAUGE PRESSURE.

The pressure gauges in general use indicate pressure in pounds above the atmospheric pressure; it is called gauge pressure. To convert gauge pressure into absolute pressure 14.7 has to be added to the former.

Smaller pressures are designated by the number of inches of mercury which they will sustain, or, after the

French system, by millimeters of mercury, which are compared in the following table for ordinary pressures of the surrounding atmosphere.

COMPARISON OF THE BRITISH AND METRICAL BAROMETERS.

Inches.	Millimeters.	Inches.	Millimeters.	Inches.	Millimeters.
27.00	685.788	28.40	721.347	29.80	756.906
27.10	688.328	28.50	723.887	29.90	759.446
27.20	690.867	28.60	726.427	30.00	761.986
27.30	693.407	28.70	728.967	30.10	764.526
27.40	695.947	28.80	731.507	30.20	767.066
27.50	698.487	28.90	734.047	30.30	769.606
27.60	701,027	29.00	736.587	30.40	772.146
27.70	703.567	29.10	739.127	30.50	774.686
27.80	706.107	29.20	741.667	30.60	777.226
27.90	708.647	29.30	744.206	30.70	779.766
28.00	711.187	29.40	746.746	30.80	782.306
28.10	713.727	29.50	749.286	30.90	784.846
28.20	716.267	29.60	751.826		
28.30	718.807	29.70	754.366		

ACTION OF VACUUM.

The pressure of the atmosphere is the cause of the raising of water by suction pumps, the air in the pumps being removed by the movement of the piston, and its space occupied by water forced up by the pressure of the outside atmosphere. For the same reason such a pump cannot lift water higher than thirty-two feet, a column of water of this height exerting nearly the same pressure as the atmosphere at the earth's surface. For the same reason the mercury in a barometer (or glass tube from which the air is withdrawn) stands about twenty-nine inches high, varying with the pressure of the atmosphere, between twenty-seven and thirty inches at the earth's surface, but decreases with the height above the earth at the rate of 0.1 inch for 84 feet.

MANOMETERS—GAUGES.

The instruments for measuring higher gaseous pressures are usually called manometers or gauges.

WEIGHT OF GASES.

The weight of gases is determined by weighing a glass balloon filled with the same, and by subtracting from this weight that of balloon after the same has been evacuated by means of an air pump. One hundred cubic inches of air weighs 31 grains at a pressure of the atmosphere of 30 inches, and at a temperature of 60° F.; therefore the density of air is 0.001293 or $\frac{1}{773}$ that of water.

One hundred cubic inches of hydrogen, the lightest of the common gases weighs 2.14 grains.

MIXTURE OF GASES.

Two or more gases present in vessels, communicating with each other, mix readily, and each portion of the mixture contains the different gases in the same proportion. Mixtures of gases follow the same laws as simple gases.

DALTON'S LAW.

The pressure exerted on the interior walls of a vessel containing a mixture of gases is equal to the sum of the pressures which would be exerted if each of the gases occupied the vessel itself alone.

BUOYANCY OF GASES.

The Archimedian principle applies also for gases hence a body lighter than air will ascend (air balloons smoke, etc.).

LIQUEFACTION OF GASES.

If sufficient pressure be applied to a gas and the tem perature is sufficiently lowered all gases can be com pressed so as to assume the liquid state.

HEAT OF COMPRESSION.

When gases or vapors are being compressed, the energy or work spent to accomplish the compression appears in the form of heat.

CRITICAL TEMPERATURE.

There appears to exist for each gas a temperature above which it cannot be liquefied, no matter what amount of pressure is used. It is called the critical temperature. Below this temperature all gases or vapors may be liquefied if sufficient pressure is used.

CRITICAL PRESSURE.

The pressure which causes liquefaction of a gas at or as near below the critical temperature as possible, is called the critical pressure. Between these two temperatures—that is, in the neighborhood of the critical point—the transition from one state to another is unrecognizable.

CRITICAL VOLUME.

The critical volume of a gas is its volume at the critical point, measured with its volume at the freezing point, under the pressure of an atmosphere as unit. The critical temperature, pressure and volume are frequently referred to as *critical data.*

TABLE OF CRITICAL DATA.

Substance.	Critical Pressure in Atmospheres.	Critical Temperature, Degrees C.	Critical Volume.
Ammonia	115	130	....
Aethylen	51	10	0.00569
Alcohol	67	235	0.00713
Acetic acid	76.4	231.5	0.0110
Aethylic ether	37.5	200	0.01344
Acetate of aethyl	42.2	240	0.01222
Benzol	60	292	0.00981
Bisulphide of carbon	77.8	275	0.0096
Butyrate of amyl	23.8	332	0.03809
Carbonic acid	77	31	0.0066
Cumol	31.8	347.2	0.0258
Hydrogen	20.8	—240	....
Nitrous oxide (N_2O)	75	35.4	0.00480
Oxygen	50	—118	....
Propylic alcohol	53.2	256	0.00968
Sulphurous acid	79	155.4	....
Toluol	40	320.8	0.02138
Water	195	358	0.00187

SPECIFIC HEAT OF GASES.

A gas may be heated while its volume is kept constant and also while its pressure remains constant. In the former case the pressure increases and in the latter the volume increases. Therefore we make a distinction between specific heat of gases at a constant volume or at a constant pressure. In the former case the heat added is only used to increase the momentum of the molecules, while in the latter case an additional amount of heat is required to do the work of expanding the gas against the pressure of the atmosphere. The specific heat of all permanent gases for equal volumes at constant pressure is nearly the same and about 0.2374 water taken as unity.

TABLE OF SPECIFIC HEAT OF GASES.

For Equal Weights. (Water = 1.)	At Constant Pressure.	At Constant Volume.
Air	.2377	.1688
Carbonic acid (CO_2)	.2164	.1714
" oxide (CO)	.2479	.1768
Hydrogen	3.4046	2.4096
Light carbureted hydrogen	.5929	.4683
Nitrogen	.2440	.1740
Oxygen	.2182	.1559
Steam, saturated	......	.3050
Steam gas	.4750	.3700
Sulphurous acid	.1553	.1246

ISOTHERMAL CHANGES.

A gas is said to be expanded or compressed isothermally when its temperature remains constant during expansion or compression, and an isothermal curve or line represents graphically the relations of pressure and volume under such conditions.

ADIABATIC CHANGES.

As gas is said to be expanded or compressed adiabatically when no heat is added or abstracted from the same during expansion or compression, an adiabatic line or curve represents graphically the relations of pressure and volume under such conditions.

FREE EXPANSION.

When gas expands against an external pressure much less than its own, the expansion is said to be free. The refrigeration due to the work done by such expansion may be used to liquefy air. (See Linde's method.)

LATENT HEAT OF EXPANSION.

When a gas expands while doing work, such as propelling a piston, an amount of heat equivalent to the work done becomes latent or disappears. It is called the latent heat of expansion.

VOLUME AND PRESSURE.

The relations of volume pressure and temperature of gases are embodied in the following formulæ in which V stands for the initial volume of a gas at the initial temperature t and the initial pressure p. V^1, t_1 and p^1 stand for the corresponding final volume, temperature and pressure. For different temperatures—

$$V^1 = V \frac{t_1 + 461}{t + 461}$$

For different pressures—

$$V^1 = V \frac{p}{p^1}$$

For different temperature and pressure—

$$V^1 = V \frac{p\,(t_1 + 461)}{p^1\,(t + 461)}$$

If the initial temperature is 60° F. and the initial pressure that of the atmosphere, the final pressure may be found after the formula—

$$p^1 = \frac{V(t_1 + 461)}{35.38\ V^1}$$

If the volume is constant—

$$p^1 = \frac{t_1 + 461}{35.58}$$

If the temperatures in above formula are expressed degrees Fahrenheit above absolute zero, the figure 461 is to be omitted.

PERFECT GAS.

The above rules and formulæ apply, strictly speaking, only to a perfect or ideal gas, that is a gas in which the dimensions of the molecules may be neglected as regards the distance between them. Therefore when a gas approaches the state of a vapor, these laws do no more hold good.

ABSOLUTE ZERO AGAIN.

The expansion of a perfect gas under constant pressure being $\frac{1}{493}$ of its volume at 32° F. (freezing point), it follows that if a perfect gas be cooled down to a temperature of 493° below freezing, or 461° below zero Fahrenheit, its volume will become zero. Hence this point is adopted as the absolute zero of temperature. (See also former paragraph on this subject.)

VELOCITY OF SOUND.

The velocity, v, of sound in gases is expressed by the formula—

$$v = \sqrt{\frac{g\,h\,\delta}{d}(1 + at)\frac{c}{c_1}}$$

In which formula g is the force of gravity, h the barometric height, δ the density of mercury, d the density of the gas, t its temperature, c its specific heat at constant pressure, and c_1 its specific heat at constant volume. Hence the quotient, $\frac{c}{c_1}$, for a certain gas can be determined by the velocity of sound in the same.

FRICTION OF GAS IN PIPES.

The loss of pressure in pounds, P, sustained by gas in traveling through a pipe having the diameter d in inches, for a distance of l feet, and having a velocity of n feet, is—

$$P = 0.00936\ \frac{n^2 l}{d}$$

ABSORPTION OF GASES.

Gases are absorbed by liquids; the quantities of gases so absorbed depend on the nature of the gas and liquid, and generally increase with the pressure and decrease with the temperature. During the absorption of gas by a liquid a definite amount of heat is generated, which heat is again absorbed when the gas is driven from the liquid by increase of temperature or decrease of pressure. Solids, especially porous substances, also absorb gases. Thus charcoal absorbs ninety times its own volume of ammonia gas.

VAPORS.

As long as a volatile substance is above its critical temperature it is called a gas, and if below that it is called a vapor.

This definition, although the most definite is not the most popular one. Frequently a vapor is defined as representing that gaseous condition at which a substance has the maximum density for that temperature or pressure. Generally gaseous bodies are called vapors when they are near the point of their maximum density, and a distinction is made between saturated vapor, superheated vapor and wet vapor.

SATURATED VAPOR.

A vapor is saturated when it is still in contact with some of its liquid; vapors in the saturated state are at their maximum density for that temperature. Compression of a saturated vapor, without change of temperature, produces a proportional amount of liquefaction.

DRY OR SUPERHEATED VAPOR.

Vapors which are not saturated are also called dry or superheated vapors, and behave like permanent gases.

WET VAPOR.

A saturated vapor which holds in suspension particles of its liquid is called wet or moist vapor.

TENSION OF VAPORS.

Like gases, vapors have a certain elastic force, by virtue of which they exert a certain pressure on surrounding surfaces. This elastic force varies with the nature of the liquid and the temperature, and is also called the tension of the vapor.

VAPORIZATION.

A liquid exposed to the atmosphere or to a vacuum forms vapors until the space above the liquid contains vapor of the maximum density for the temperature.

EBULLITION.

If the temperature is high enough the vaporization takes place throughout the liquid by the rapid production of bubbles of vapor. This is called ebullition, and the temperature at which it takes place is a constant one for one and the same liquid under a given pressure.

BOILING POINT.

The temperature at which ebullition of a liquid takes place is called its boiling point, for the pressure then obtaining. When no special pressure is mentioned we understand by boiling point that temperature at which liquids boil under the pressure of the atmosphere.

DIFFERENT BOILING POINTS.

The boiling point varies with the nature of the liquid, and always increases with the pressure. It is not affected by the temperature of the source of heat, the temperature of the liquid remaining constant as long as ebullition takes place. The heat which is imparted to a boiling liquid, but which does not show itself by an increase of temperature, is called the latent heat of vaporization.

ELEVATION OF BOILING POINT.

Substances held in solution by liquids raise their boiling point. Thus a saturated solution of common salt boils at 214° and one of chloride of calcium at 370°. The boiling point of pure water may also be raised above the boiling point; for water free from gases to over 260° without showing signs of boiling. This retardation of boiling sometimes takes place in boilers, and may cause explosions, if not guarded against by a timely motion produced in the water.

LATENT HEAT OF VAPORIZATION.

The heat which becomes latent during the process of volatilization is composed of two distinct parts. The one part is absorbed while doing the work of disintegrating the molecular structure while doing INTERNAL WORK, as it is termed. The other part of heat which becomes latent is absorbed while doing the work of expansion against the pressure of the atmosphere, and is called the EXTERNAL WORK. In a liquid evaporized in vacuum, in which case no pressure is to be overcome, the external work becomes zero, and only heat is absorbed to do the internal work of vaporization (*free expansion*).

REFRIGERATING EFFECTS.

If liquids possess a boiling point below the temperature of the atmosphere the latent heat of vaporization is drawn from its immediate surrounding object, causing a reduction of temperature, *i. e.*, refrigeration.

LIQUEFACTION OF VAPORS.

When vapors pass from the aeriform into the liquid state, that is, when they are liquefied, the heat which became latent during evaporation appears again, and must be removed by cooling. Vapors of liquids the boiling point of which is above the ordinary temperature can be liquefied at the ordinary temperature without additional pressure (*distilling condensation*). Permanent gases require additional pressure, and in some cases considerable refrigeration, to become liquefied (*compression of gases*).

DALTON'S LAW FOR VAPORS.

The tension and consequently the amount of vapor of a certain substance which saturates a given space is the same for the same temperature, whether this space contains a gas or is a vacuum. The tension of the mixture of a gas and a vapor is equal to the sum of the tensions which each would possess if it occupied the same space alone.

VAPORS FROM MIXED LIQUIDS.

The tension of vapor from mixed liquids (which have no chemical or solvent action on each other) is nearly equal to the sum of tension of the vapor of the two separate liquids.

SUBLIMATION.

The change of a solid to the vaporous state without first passing through the liquid state is called sublimation (camphor, ice).

DISSOCIATION.

The term dissociation is used to denote the separation of a chemical compound into its constituent parts, especially if the separation is brought about by subjecting the compound to a high temperature.

CHAPTER VI.—MOLECULAR DYNAMICS.

MOLECULAR KINETICS.

It has already been stated that the laws of Boyle and St. Charles are in accordance with the molecular theory, by the consequent development of which a number of other relations have been established which are of the utmost importance in all discussions of energy, especially those of thermodynamical nature. Applied to gases, this theory means that the rectilinear progressive motion of the molecules, which constitutes the body of a gas, represents by its kinetic energy the temperature of a gas, and by the number of impacts of its molecules against the wall of the vessel containing the gas, its pressure.

DENSITY OF GASES.

If m represents the mass of a molecule and u the average velocity of its rectilinear progressive motion, the kinetic energy, E (*i.e.*, the temperature), of the molecule is expressed by—

$$E = \frac{m}{2} u^2$$

If the unit of volume, say a cubic foot of a gas, contains N molecules of the mass, m, the density of the gas, ρ, is—

$$\rho = m N$$

PRESSURE OF GASES.

The number of molecules which collide with the interior surfaces of a cube of above size is equal to $N u$, and hence the number which collide with one of the interior surfaces of the cube (one foot square):

$$\frac{N u}{6}$$

The number of impacts multiplied by the momentum of the impact of each molecule, $2 m u$, yields the pressure:

$$p = \frac{1}{3} N m u^2 = \frac{1}{3} \rho u^2$$

AVOGADRO'S LAW.

At the same temperature and pressure equal volumes of different gases contain the same number of molecules. Hence the molecular weights of gases are proportional to their densities.

MOLECULAR VELOCITY.

The average velocity of the molecules, u, is accordingly—

$$u = \sqrt{\frac{3p}{\rho}}$$

For hydrogen we find $u = 1,842$ meters per second.

If M is the molecular weight of a gas referred to hydrogen as unit (ρ being proportional to M) the average velocity of the molecules is expressed by—

$$u = 1,842 \sqrt{\frac{2}{M}} \text{ meters per second.}$$

The average distance, L, which a molecule travels in rectilinear direction before it meets another molecule is expressed by the formula—

$$L = \frac{\lambda^3}{1.41 \pi s^2}$$

in which λ is the average distance of the molecules, and therefore λ^3 the size of the cube which contains one molecule on an average.

L accordingly has been found to be for hydrogen 0.000185 millimeter; for carbonic acid, 0.000068 mm.; for ammonia 0.000074 mm.

INTERNAL FRICTION OF GASES.

The internal friction, η, of a gas is expressed by the equation—

$$\eta = \frac{L \rho \frac{12}{13} u}{\pi}$$

The velocity of sound in different gases is inversely proportional to the square root of their molecular weights (see page 49).

TOTAL HEAT ENERGY OF MOLECULES.

The total heat energy of a body is composed of the energy due to the progressive motion of its molecules, and the interior energy which is represented by possible rotatory motions of the molecules, or by motions of the atoms composing the molecule. In gases, and probably also in liquids and solid bodies, the former portion of energy is proportional to the absolute temperature, so that at the absolute zero —461° F.—the progressive motion of the atoms would cease.

LAW OF GAY LUSSAC.

Since chemical combinations between different elements take place in the proportion of their molecular weights, and since equal volumes of gases contain equal numbers of molecules, the chemical combination between gaseous elements must take place by equal volumes or their rational multiples, and the volume of the combination if gaseous bears equally a simple numerical relation to that of the elements.

EXPANSION OF GASES.

Since the same number of molecules of different gases occupy the same volume at equal temperatures and pressure, the expansion by heat of all gases under constant pressure must be the same, and for perfect gases it is the same for all temperatures, being equal to the $\frac{1}{493}$ part of the volume of a gas at the freezing point and at the pressure of one atmosphere. This is tantamount to saying that the volume of gas under constant pressure is proportional to its absolute temperature, $T = 461 + t$.

EQUATION FOR PERFECT GASES.

The increase of pressure of a gas heated at constant volume being likewise proportional to the absolute temperature and equal to $\frac{1}{493}$ of its volume at the freezing point, the product of pressure and volume, $p\,v$, must be likewise, and hence it can be expressed by the equation—

$$p\,v = R\,T$$

in which R is a constant factor, depending only upon the units used. T standing for absolute temperature, it may be written—

$$p\,v = \frac{p_0\,v_0\,T}{493} = \frac{T}{493}$$

p_0 and v_0 standing for pressure and volume at the temperature of 32° F., both being unit.

GENERAL EQUATION FOR GASES AND LIQUIDS.

This formula answers for a perfect gas in which the dimension of the molecules and their mutual attraction disappear in comparison with their volume and the expansive force due to the temperature. If the dimensions and mutual attraction are taken into consideration, the formula according to *Van der Waals* reads:

$$\left\{ p + \frac{a}{v^2} \right\} (v - b) = R\,T$$

In this formula the signs have the same meaning as in the former equation, except the two constants a and b, which differ with the nature of the gas $\frac{a}{v^2}$; atoning for the influence of the molecular attraction which may be derived from the deviation of the gas from Boyle's law; b stands for the influence of the volume of the molecule; it is equal to four times the volume of the molecules. Its value may be ascertained by inserting the value found for a into the formula of Van der Waals. However, it is generally more convenient and of more practical application to derive the values a and b from the critical data, as will be shown later on.

The formula of Van der Waals answers not only for all gases, but for the liquid condition as well, as far as changes of volume, pressure and temperature are concerned, provided, however, that the changes take place homogeneously and that the molecular constitution of the substance is not altered during the change.

CRITICAL CONDITION.

If this formula is elaborated numerically as to volume for given temperatures and pressures, we always obtain one real positive expression for volume except for pressures near the point of liquefaction at temperatures below the critical point.

Here the formula does not apply on account of the so-called critical condition (partly gas and partly liquid) which the substance maintains at this stage.

These conditions become readily apparent by an elaboration of the equation of Van der Waals, for if the equation—

$$\left\{ p+\frac{a}{v^2} \right\}(v-b)=R\,T$$

which may also be written—

$$\left\{ p+\frac{a}{v^2} \right\} (v-b)=\frac{\left(p_0+\frac{a}{{v_0}^2}\right)(v_0-b)}{493}T$$

is developed after powers of v (p_0 and $v_0=$ unit), we obtain—

$$v^3-v^2 \left\{ b+\frac{(1+a)\,(1-b)\;T}{493\,p} \right\} +v\frac{a}{p}-\frac{ab}{p}=0$$

This equation being a cubical one, it may be satisfied by three values, which may all be real or one of which may be real and the other two imaginary. Accordingly, we find for all temperatures above the critical point for any given pressure only one value for volume; except for temperatures below the critical point for certain values of p, *i.e.*, for pressures near the point of liquefaction for that temperature, or nearing the boiling point for that pressure. At these stages the substance is under so-called critical conditions, and here we find three different values for v, one of which may stand for the volume of the substance in its gaseous form, another for its volume as a liquid, and the third for an intermediate volume

CRITICAL DATA.

When, on increasing temperature and pressure these three values for volume converge into one, that is, if the three real roots of the equation become equal, we have reached the critical volume, that is, that volume which corresponds to the critical pressure and to the critical temperature. At this, the critical point, the substance passes gradually and without showing a separation into liquid and gas, that is to say homogeneously, from the gaseous into the liquid state; there is no intermediate stage at this temperature between the volume of the liquid and the volume of the gas, as is the case at temperatures below the critical point and at pressures corresponding to the boiling point.

The values of temperature pressure and volume at which the three roots of the above equation become equal is found by the following considerations: If in a cubical equation of the form—

$$x^3 - a\,x^2 + b\,x - c = 0$$

the three roots become equal to each other $= x_1$ the following relations obtain:

$$x_1 = \frac{a}{3} \qquad x_1{}^2 = \frac{b}{3} \qquad x_1{}^3 = c$$

Applying this to the above equation, which may also be written—

$$\varphi^3 - \varphi^2 \left\{ b + \frac{(1+a)\,(1-b)\,\vartheta}{493\pi} \right\} + \varphi\frac{a}{\pi} - \frac{a\,b}{\pi} = 0$$

by inserting the signs, φ, π and ϑ to stand for volume, pressure and temperature at the critical point we find—

$$\varphi^3 = \frac{a\ b}{\pi}$$

$$3\ \varphi^2 = \frac{a}{\pi}$$

$$3\ \varphi = b + \frac{(1+a)\ (1-b)\ \vartheta}{\pi\ 493}$$

which may be simplified thus:

$$\varphi = 3\ b$$

$$\pi = \frac{a}{27\ b^2}$$

$$\vartheta = \frac{8}{27}\ \frac{a \quad 493}{(1+a)\ b\ (1-b)}$$

We see from these formulæ how the two constants, a and b, which may be deduced from the deviations from Boyle's law, determine the critical pressure, temperature and volume.

APPLICATION OF GENERAL EQUATION.

On the other hand (and which is practically of more importance), it is readily seen how the two constants, a and b, and therefore the behavior of a homogeneous gas or liquid as to volume, temperature and pressure, may be derived from the critical data, viz.:

$$b = \frac{\varphi}{3}$$

$$a = 3\ \pi\ \varphi^2$$

$$\frac{3\ \vartheta}{8} = \frac{\pi\ \varphi\ 493}{(1+3\ \pi\ \varphi^2)\left(1-\frac{\varphi}{3}\right)}$$

The last equation may be rendered approximately by

$$\vartheta = \frac{8\ \pi\ \varphi\ 493}{3} = \frac{8\ a}{27\ b\ R}$$

$$\varphi = \frac{0.00076\ \vartheta}{\pi}$$

The numerical values for a and b for any substance having been found by these formulæ from the critical temperature and pressure, they may be inserted in the general equation of Van der Waals, which will then yield the relations of pressure and volume at different temperatures, etc.

UNIVERSAL EQUATION.

If the volume, pressure and temperature of a gas are measured by fractions of the absolute data, in other words, if $v = n\,\varphi$ and $p = e\,\pi$ and $T = m\,\vartheta$, the general equation may be written—

$$\left\{ e\,\pi + \frac{a}{n^2\,\varphi^2} \right\} (n\,\varphi - b) = R\,m\,\vartheta$$

If the values for π, φ and ϑ as found in the above, are inserted in this equation the same may be brought to the form—

$$\left\{ e + \frac{3}{n^2} \right\} (3\,n - 1) = 8\,m$$

This formula contains no terms dependent upon the nature of the substance, hence the equation establishing the relations between pressure, volume and temperature is the same for all substances, if volume, pressure and temperature are expressed in fractions of the critical data (provided $v > 4\,b$).

If v is smaller than $4\,b$ the formula may possibly give correct results, but when it does not such a result does not vitiate the admissibility of the theory in other respects, as *Van der Waals* has shown.

OTHER MOLECULAR DIMENSIONS.

In accordance with the foregoing the average space, γ, occupied by each molecule of a gas is expressed by—

$$\gamma = \frac{b}{4} = \frac{1\,\vartheta}{32 \times 273\,\pi}$$

and the specific weight, w, of a gas (water at 39° F. = 1):

$$w = \frac{M}{22350\,\gamma}$$

M being the molecular weight in grams, and 22,350 c. c. the volume occupied by the same at 32° F., and at the pressure of one atmosphere.

If the molecules are supposed to be of spherical form their diameter, s, is expressed by the formula—

$$s = 6\,\sqrt{2}\,L\,\gamma = 8.5\,L\,\gamma$$

L being the average distance which a molecule travels, as stated above, viz.:

$$L = \frac{\lambda^3}{\sqrt{2}\;\pi\,s^2}$$

ABSOLUTE BOILING POINT.

The definition of the boiling point as given heretofore fits only for a certain pressure, but in accordance with the critical conditions we can define an absolute boiling point as the temperature at which a liquid will assume the aeriform state, no matter what the pressure is, viz., the critical temperature.

CAPILLARY ATTRACTION.

Since capillary attraction (in consequence of which liquids rise above their surface in narrow tubes) and also the surface tension of liquids are both functions of the cohesion of liquids, and since the cohesion diminishes with the temperature, the capillary attraction must do likewise; and it has been shown that it becomes zero at the critical temperature or at the absolute boiling point.

CRITICAL VOLUME.

At the critical temperature the change from the liquid to the gaseous condition requires no interior work, and therefore the latent heat of vaporization at this temperature must be equal to zero.

The volumes of a certain weight of liquid or vapor of a substance at the critical temperature must likewise be the same.

GAS AND VAPOR.

If, with Andrews, we confine the conception of vapor to a fluid below its critical point, and that of a gas to a fluid above its critical point, we can also define as vapor such aeriform fluids as may be compressed into a liquid by pressure alone without lowering temperature; and by the same token a gas is an aeriform fluid which cannot be compressed into a liquid by pressure alone without lowering the temperature. By liquefaction we designate the production of a liquid separated from the vapor by a visible surface.

LIQUEFACTION OF GASES.

After the significance of the critical temperature had been duly understood and appreciated it became also possible to liquefy the most refrangible gases by pressure when cooled down below their critical temperature. A novel way for the liquefaction of such gases, more especially air, has been devised by Linde, and the process employed by him is so simple and successful that it will doubtless become of practical value in many respects, more especially also practical refrigeration.

CHAPTER VII.—THERMODYNAMICS.

THERMODYNAMICS.

Thermodynamics is the science which treats of heat in relation to other forms of energy, and more especially of the relations between heat and mechanical energy.

FIRST LAW OF THERMODYNAMICS.

This law is a special case of the general law expressing the convertibility of different forms of energy into one another. The first law of thermodynamics asserts the equivalence of heat and work or mechanical energy, and states their numerical relation. Accordingly heat and work may be converted into each other at the rate of 778 foot-pounds for every unit of heat, and *vice versa.*

SECOND LAW OF THERMODYNAMICS.

The foregoing law holds good without any limitation as far as the conversion of work or mechanical energy into heat is concerned. It must be qualified, however, with respect to the conversion of heat into work. It amounts to this, that of a certain given amount of heat at a given temperature only a certain but well defined portion can be converted into work, while the remaining portion must remain unconverted as heat of a lower temperature. This outcome is a natural consequence of the condition that heat cannot be directly transferred from a colder to a warmer body.

EQUIVALENT UNITS.

In accordance with the first law, we can measure quantities of heat by the heat unit or by the unit of work (foot-pound) and we can also measure it by its equivalent in heat units as well as by the units of work. The figure designating the number of foot-pounds equivalent to the unit of heat (778), *i. e.*, the mechanical equivalent of heat, is frequently referred to by the letter *J.*

When quantities of work and heat are brought in juxtaposition in equations, etc., it is always understood that they are expressed by the same units, *i. e.*, in either heat or work units.

SECOND LAW QUALIFIED.

In a system in which the changes are only such of heat and such of mechanical energy (work), the appearance of a certain amount of work is always accompanied by the disappearance of an equivalent amount of heat, and the appearance of a certain amount of heat is always accompanied by the expenditure of an equivalent amount of mechanical energy. From this, however, it must not be concluded that by withdrawing a certain amount of heat from a warmer body we can convert it into its equivalent amount of mechanical energy. This is only the case under exceptional conditions; but when, as in the case of practical requirements, the conversion of heat into work must be done by a continuous process it cannot be accomplished under conditions practically available.

CONVERSION OF HEAT.

The conversion of heat into mechanical work, and work into heat, takes place in many ways. Generally the change of volume or pressure brought about by heat changes mediates the conversion. The substance which is used to mediate the conversion is called the working medium or the working substance.

MOLECULAR TRANSFER OF HEAT ENERGY.

The manner in which heat is converted into mechanical work is readily understood on the basis of the molecular theory, when the working fluid is a gas, the pressure of which, due to its molecular energy (heat) is employed to propel a piston. The molecules of the gas by colliding with the piston impart a portion of their molecular energy to the piston, moving the same forward; at the same time the energy of the molecules grows less, and indeed the temperature of the gas decreases as the piston moves ahead. If the work done by the piston and the heat lost by the gas were measured in the same units, it would be found that they were practically alike (presupposing we employ a perfect gas, consisting of simple molecules, undergoing no internal changes).

GAS EXPANDING INTO VACUUM.

If there had been no pressure on the piston (and the piston supposed to have no weight) in the foregoing experiment, the piston would have been moved by the expanding gas, without doing work during the expansion, and hence the temperature of the gas, while expanding under such conditions (against a vacuum), remains constant and unchanged, at least practically so.

HEAT ENERGY OF GAS MIXTURES.

The same would happen if two vessels, containing the same or different gases at different pressures, are brought in communication; no change of heat takes place, while the pressures equalize themselves. *Hence, the heat energy of a gas is independent of its volume, and the energy of a mixture of gases is equal to the sum of the energy of its constituents.*

DISSIPATION OF ENERGY.

Accordingly we may allow a gas under pressure to dilate in such a way as to do a certain amount of work at the expense of an equivalent amount of heat, and we may allow it to expand without doing work. In the latter case the availability of the gas to mediate a certain amount of work has not been utilized, has been dissipated, as it were, since the original condition of the gas cannot be re-established again without the expenditure of outside energy.

ADIABATIC CHANGES.

In the former case, when the gas was allowed to expand while doing work, the greatest possible amount of work obtainable is produced when the pressure of the piston is always kept infinitesimally less than that of the gas. If this is being done the original condition of the gas can be established by making the pressure on the piston only infinitesimally more than on the gas, when the gas will be compressed to its original volume and temperature (no heat having been added to or abstracted from the gas during the operation). Both the operations of expansion and compression of the gas as conducted (without addition of heat, etc.) are therefore adiabatic changes, they are both reversible changes, and neither of them involves any dissipation of heat or energy. In the one change we have converted heat energy into work, and in the other work into heat.

ISOTHERMAL CHANGES.

The expansion of the gas while propelling a piston may be allowed to proceed while the energy imparted to the piston is replaced by heat supplied to the expanding gas from without. In this case the expanding gas is kept at the same temperature, and therefore it is said that the expansion proceeds isothermically. This operation may also be reversed and work converted into heat by applying the power gained by raising the piston, to push the piston back, and withdrawing the heat liberated by the work of compression as fast as it appears, so that the gas is always at the same temperature. (Isothermic compression.) If, during expansion, the temperature of the gas is always only infinitesimally smaller, and during compression infinitesimally greater than the outside temperature, both operations are considered to be reversible, and no dissipation of energy takes place during the performance of either of them.

MAXIMUM CONVERSION.

In conducting the operations in the foregoing (reversible) manner we obtain the maximum yield of mutual conversion of work and heat obtainable by the expansion or compression of the gas in question.

CONTINUOUS CONVERSION.

While a body of gas may be used in the above way to convert a certain amount of heat into work, and *vice versa*, it would not answer for the continuous conversion of work into heat, for if the operation of work production is reversed we simply re-establish the original condition without having accomplished any outside change whatever.

PASSAGE OF HEAT.

The fact that heat cannot of itself pass from a colder to a warmer body is also in harmony with the molecular theory. The molecules of bodies having the same temperature possess also the same average energy, and therefore cannot impart energy to one another; much less can energy of heat pass from a colder to a warmer body. The ability of heat to do work is due to its natural tendency to pass from a warmer to a colder body, and therefore, other circumstances being equal, is directly proportional to the difference of temperature between the warmer and colder body.

REQUIREMENTS FOR CONTINUOUS CONVERSION.

As stated, for the practical conversion of heat into work, we need a working medium that is a substance of some kind which mediates the conversion. As the heat which is communicated to this medium for the purpose of doing work is never entirely available for this purpose, but a portion of the heat always remains as heat of a lower temperature (not available for mechanical work except when it can pass to a temperature still lower), it follows as a matter of course and also of necessity, that when we desire to convert heat into work by a continuous process we need not only a working substance but also a warm body, a source of heat (boiler, generator, etc.), and a body of lower temperature, to which the heat not available for work in the operation may be discharged. The latter device is generally called a refrigerator or condenser; in the case of many heat engines it is the atmosphere. The same requirements, only in a reversed order, obtain for the continuous conversion of work into heat, *i. e.*, when heat is to be transferred from a colder to a warmer body, the work expended compensating for the transfer (lifting heat).

COMPONENTS OF HEAT CHANGES.

The changes produced in a body by heat may be divided in several parts, viz., the elevation of temperature, *i. e.*, the increase of energy of the molecules, the change produced by overcoming the interior cohesion, and by rearranging the molecular constitution of the body, and the change required to do outside work, overcoming pressure.

MAXIMUM CONTINUOUS CONVERSION OF HEAT.

The question as to the maximum amount of work which can be obtained from a certain amount of heat by continuous conversion, and the maximum amount of heat which can be obtained by or lifted by a certain amount of work, is one of the most important in thermodynamics. It has been solved with the same result in various ways, the following giving the outlines of one of them.

CYCLE OF OPERATIONS.

The contrivances which are required to perform the operations, by which through the aid of the working medium, etc., heat is continuously transformed into work, or work into heat, come under the general head of machines. A series of operations of the kind mentioned which are so arranged that the working substance returns periodically to its original condition is also called a cycle of operations.

REVERSIBLE CYCLE.

If a cycle of operations is conducted in such a manner that all the changes or operations can be carried out in the opposite direction the cycle is what is called a reversible cycle. Operations can generally be made reversible, at least in theory, if the transfers of heat follow only infinitesimally small differences in temperature and the changes in volume take place under but infinitesimally small differences of pressure. Not all changes can be performed in a reversible manner, however.

IDEAL CYCLE.

For the continuous conversion of heat into work we require the performance of a cycle, so that the working substance, which is generally not unlimited, may return periodically to its original condition, and may be used continuously over and over again. If at the same time the operations of the cycle are carried on reversibly the conversion of heat into work takes place at the greatest possible rate. In other words, the maximum amount of work obtainable from a given amount of heat is realized if the working substance is passed through the operations of a reversible cycle. Practically we can only approach the conditions of a reversible cycle, for which reason it is also called an ideal cycle of operations.

IDEAL CYCLES HAVE THE SAME EFFICIENCY.

The proof that a cycle of reversible operations for the transformation of heat into work yields the greatest return of work for a given amount of heat, and *vice versa*, may be based on the axiom that no energy can be created, or on the fact that heat cannot pass from a colder to a warmer body. For if one cycle of reversible operations would yield a greater amount of work for a certain amount of heat than another reversible cycle, the latter would also by reversing it require a lesser amount of work to produce that given amount of heat. Hence we could operate the first cycle to convert a given amount, C, of heat to produce a certain amount of work, B, and the second cycle, being operated in the reverse manner, would only need a portion of the work B, say B_1, to reproduce the heat C, which could be employed in the first cycle to again produce the work B. Therefore both devices or cycles co-operating in the manner indicated would during each co-operative performance create the work $B—B_1$, or rather, transfer an equivalent amount of heat from a colder to a warmer body, which is impossible. Hence both devices must operate with the same efficiency, and all reversible cycles devised for the mutual conversion of heat into work must, theoretically speaking, have the same efficiency, and the maximum efficiency at that.

INFLUENCE OF WORKING FLUID.

In the same manner it may be demonstrated that the nature of working substance has no influence upon the amount of work which can be obtained from a given amount of heat in a reversible cycle. For if one substance could be employed to yield a greater amount of work from the same amount of heat than another substance, and *vice versa*, a combination between two cycles, each one employing one of the two substances, could be formed like the above, which would create the same impossible results.

It should be noted that this deduction holds good only when the two cycles work between the same limits of temperature, and when no molecular changes take place in the working fluid, the mass of the latter remaining constant.

RATE OF CONVERTIBILITY OF HEAT.

The maximum amount of work derivable from a given amount of heat in a continuous cycle of operations, being accordingly independent of the nature of the working substance, and obtainable by every ideal reversible cycle, the rate of maximum conversion may be deduced from the working of any such cycle of operations. To do this we select as the working substance in our ideal cycle a perfect gas, since the laws governing the relation of pressure, temperature and volume in this case are not only well known but also comparatively simple. The first ideal reversible cycle of operations to determine the maximum convertibility of heat has been devised by Carnot, to whom the original elaborations of this subject are due. Of course any reversible cycle answers also. For simplicity's sake, following the example of Nernst, we use a cycle which is to be considered reversible when working between very small differences of temperatures (between boiler and refrigerator).

SYNOPSIS OF NUMERICAL PROOF.

Consequently we assume that the absolute temperature, T_1, of the boiler or generator is only a little higher than the temperature, T_0, of the refrigerator, when the working of our ideal cycle and its numerical theoretical result may be delineated as follows: The mechanical device consists of an ideal cylinder provided with a movable piston containing a certain amount of a permanent gas of the volume v_1. The cylinder is immersed in the refrigerator of the temperature T_0, and by forcing down the piston (reversibly) is compressed to the smaller volume v_2. The work, A, required to perform this change is expressed by—

$$A = R\,T_0\,ln\,\frac{v_1}{v_2}$$

R being the constant of the gas formula as above defined, and ln standing for natural logarithm.

As the temperature is to remain constant, an amount of heat, Q, equivalent to the work done must be imparted to the condenser, *i. e.:*

$$Q = R\,T_0\,ln\,\frac{v_1}{v_2}$$

Q being expressed in the same units as A. Now the cylinder is immersed into the generator or boiler and allowed to assume the temperature T_1, while the volume remains constant, v_2. The heat which is hereby conveyed to the gas is—

$$c\,(T_1 - T_0)$$

c being the specific heat of the gas at constant volume. At this juncture the gas is allowed to expand from the volume v_2 to the volume v_1, and the work A_1, which is done on the piston, is expressed by—

$$A_1 = R\,T_1\,ln\,\frac{v_1}{v_2}$$

while at the same time an equivalent amount of heat passes from the generator to the gas in the cylinder, *i. e.:*

$$Q_1 = R\,T_1\,ln\,\frac{v_1}{v_2}$$

Now the cylinder is brought back to the refrigerator, where, while the volume remains constant, the temperature is again reduced to T_0, the amount of heat, $c\,(T_1 - T_0)$, being transferred from the gas in cylinder to the refrigerator or condenser. The gas is now again in its initial condition, and the operations for one period of the cycle are completed.

The useful work, W, gained by this operation is—

$$A_1 - A = R\,(T_1 - T_0)\,ln\,\frac{v_1}{v_2}$$

while the amount of heat, H, which has been withdrawn from the boiler or source is equal to—

$$c\,(T_1 - T_0) + R\,T\,ln\,\frac{v_1}{v_2}$$

If we call W the total amount of work gained, and H the total amount of heat expended by the heat source to obtain the heat source, we can write—

$$\frac{W}{H} = \frac{R\,(T_1 - T_0)\;ln\,\frac{v_1}{v_2}}{c\,(T_1 - T_0) + R\,T_1\,ln\,\frac{v_1}{v_2}}$$

If we take $T_1 - T_0$, infinitesimally small, we can neglect the term $c\,(T_1 - T_0)$, as against the infinitely greater quantity $R\,T_1\,ln\,\frac{v_1}{v_2}$, and we can write—

$$\frac{W}{H} = \frac{T_1 - T_0}{T_1}$$

EFFICIENCY OF IDEAL CYCLE.

The term $\frac{W}{H}$, *i. e.*, the work obtained divided by the amount of heat (expressed in the same units) expresses what is termed the efficiency of the cycle.

Generally speaking, therefore, the convertibility of a certain amount of heat into work is the greater, the greater the difference of temperature between boiler and condenser, *i. e.*, the greater $T_1 - T_0$, and the lower this difference is located on the absolute scale of temperature, that is, the smaller T_1 under otherwise equal conditions. The limit is reached when T_0 becomes zero (absolute)= —493° F., and $W = H$, a condition which cannot even be approached in practical working.

CARNOT'S IDEAL CYCLE.

The ideal cycle originally devised by Carnot embraces four such operations. First, the cylinder with piston containing a given volume of a permanent gas is brought in contact with the heat source or boiler, and after it has attained that temperature and the pressure corresponding thereto, the piston is allowed to move forward against a resistance which is continually infinitesimally less than the pressure within (*i. e.*, reversibly). An amount of heat equivalent to the work done by the piston passes from the source of heat to the cylinder, so that the gas always maintains the temperature of the source, hence the expansion is isothermal.

Now the cylinder is removed from the source of heat to conditions which are supposed to be so that it can neither take in nor give out heat, and while under such conditions the piston is allowed to move forward again with the same precaution as to pressure. The expansion in this case is *adiabatic*, and it is allowed to proceed until the gas in the cylinder has attained the temperature of the colder body—the refrigerator, to which the cylinder is then removed. The piston is now forced inward reversibly, the heat of compression being withdrawn by the refrigerator; the temperature remains the same, thus constituting an isothermal compression. After this isothermal compression the cylinder is again brought under conditions where it can neither absorb nor discharge heat, and under these conditions is further compressed reversibly, until the gas within has acquired the temperature of the source of heat or boiler. With this fourth adiabatic operation, the cycle is completed, the working substance having been returned to its original condition, and each and all operations may be performed in the reversed order.

HEAT ENGINES.

A heat engine is a contrivance for the conversion of heat into mechanical energy, and in accordance with the above laws the efficiency of such a machine does not depend on the nature of the working substance (steam, hot air, exploding gas mixtures, etc.), but only on the temperature which the working substance has when it enters and when it leaves the machine.

AVAILABLE EFFECT OF HEAT.

The relation between a given amount of heat (H) employed in a heat engine and the greatest amount of work (W) which can be derived from same (expressed in units of the same kind) finds its expression in the said equation:

$$\frac{W}{H} = \frac{T_1 - T_0}{T_1}$$

in which T_1 is the temperature at which the heat is furnished to the engine, and T_0 the temperature of the refrigerator or condenser at which the heat leaves the engine. The temperatures are expressed in degrees of absolute temperature.

CONSEQUENCE OF SECOND LAW.

The above equation is a concise mathematical expression of the second law of thermodynamics. If in the same, T_0 becomes zero H will become W; in other words, in a machine in which the refrigerator or condenser temperature is absolute zero, the whole amount of the heat employed can be converted into mechanical energy, and it furnishes an important additional proof for the reality of an absolute zero of temperature, which is frequently looked upon as a mere scientific fiction.

IDEAL REFRIGERATING MACHINE.

A similar deduction can be made when the operations of the above cycle are reversed, the gas being allowed to expand at the lower temperature, taking heat from the refrigerator and its compression being performed at the higher temperature, discharging heat into the boiler. Instead of heat engine we have now a refrigerating machine, and one representing conditions of maximum efficiency which must find its expression in the same equation reversed, viz.:

$$\frac{H}{W} = \frac{T_0}{T_1 - T_0}$$

EFFICIENCY OF REFRIGERATING MACHINE.

The above equation signifies that by expending the amount of work W, we can withdraw the amount of heat H from a body (refrigeration) of the temperature T_0, and transfer the same to a body (boiler called condenser in the refrigerating practice) of the temperature T_1. The equation also shows that the efficiency of a refrigerating engine depends on conditions quite opposite to those applying to the efficiency of a heat engine, the conditions being, that the refrigeration which can be obtained by expending a certain amount of work is the greater the smaller $T_1 - T_0$, and the larger T_1, that is the higher $T_1 - T_0$ is on the scale of temperature.

FALL OF HEAT.

In analogy with the conversion of the energy of falling water into mechanical energy and still following Carnot, it is sometimes stated that the amount of heat W while falling from the temperature T_1 to T_0 is capable of doing the work H.

We see now that this expression is not correct; the amount of heat W leaves the source or boiler having the temperature T_1, but only the amount $W - H$ enters the refrigerator or falls to the temperature T_0 in a reversible heat engine.

On the other hand, in a reversible refrigerating machine the amount of heat W leaves the refrigerator at the temperature T_0 and the amount $W + H$ is brought over to the warmer body having the temperature T_1.

COMPENSATED TRANSFER OF HEAT.

When a certain amount of heat passes from a warmer to a colder body a portion of the same can be intercepted, as it were, to be converted into mechanical energy or work. If the maximum amount of work obtainable in this manner in accordance with the above equation has been produced, the transfer of heat from the warmer to the colder body is said to be fully compensated. The availability of the energy of the whole system participating in the transfer has not been changed, since the process is reversible and the former condition can be fully re-established, theoretically speaking.

UNCOMPENSATED TRANSFER.

When, however, heat passes from a warmer to a colder body without doing any work (as is the case in radiation of heat) or without doing the maximum amount of work obtainable, a corresponding amount of the availability of energy is wasted or dissipated, the heat at the lower temperature being lower on the scale of availability than it was before the transfer. In this case the transfer of heat is said to be not compensated, or only partially compensated. In the same way mechanical energy may be dissipated when expended without transferring the maximum amount of heat from a colder to a warmer body, as it is expected to do in the refrigerating practice.

ENTROPY.

This term is used to convey different meanings by different writers. It was originated by Clausius to stand for a mathematical abstraction expressing the degree of non-availability of heat energy for the production of mechanical energy under certain conditions.

LATENT AND FREE ENERGY.

That portion of energy present in a system which may be converted into its equivalent of mechanical work is called free energy, and the remaining energy is called latent energy. Hence when a transfer of heat takes place in a system without due compensation, the free energy decreases, and the latent energy of the system increases correspondingly. In accordance with this conception the latent energy of a body divided by the temperature is the entropy of the body; the increase of the latent energy in a body, divided by the temperature at which it takes place, yields the amount of increase of entropy, and *vice versa.*

FUTURE CONDITION OF UNIVERSE.

Only the changes of the entropy can be determined, not its absolute amount. As most changes take place without full compensation, not reversibly, it has been concluded that the entropy of the universe is constantly increasing, tending toward a condition when all energy will be latent, *i.e.*, not available for further conversion or changes. In reversible changes the entropy remains unchanged.

CHANGES OF FREE AND LATENT ENERGY.

The equation expressing the efficiency of an ideal reversible cycle of operations, viz.:

$$\frac{W}{H}=\frac{T_1-T_0}{T_1}$$

may also be written—

$$W=\frac{H(T_1-T_0)}{T_1} \text{ or}$$

$$\frac{H(T_1-T_0)}{T_0}-W=0$$

This equation furnishes also an expression for the change of free and latent energy in a system in which transfer of heat without compensation, or with only partial compensation, takes place. If the compensation is complete the expression $\frac{H(T_1-T_0)}{T_1}-W$ is zero, and the amount of free and latent energy remains the same; but if $\frac{H(T_1-T_0)}{T_1}-W>0$ that is, if W is smaller than $\frac{H(T_1-T_0)}{T_1}$, the equation covers all cases in which the changes are not reversible, and the conversion is incomplete. The free energy of the system has been decreased correspondingly in accordance with this equation. As W can never become larger than $\frac{H(T_1-T_0)}{T_1}$, the above difference can never be negative, which means that the free energy of a system can never increase. If in the equation, $W=\frac{H(T_1-T_0)}{T_1}$, T_1-T_0 is equal to 1, the equation becomes—

$$W=\frac{H}{T_1}$$

which means that the convertible energy of the amount of heat, H, while passing from one temperature to another one degree lower, with full compensation, is equal to that amount of heat divided by its absolute temperature.

INCREASE OF ENTROPY.

If an amount of heat, H, in a system is transferred from a higher temperature, T_1, to a lower temperature, T_0, without compensation, the free energy decreases, and the latent energy increases by an amount—

$$\frac{H(T_1-T_0)}{T_1}$$

and the increase of entropy, in accordance with a former definition, is expressed by the term—

$$\frac{H(T_1-T_0)}{T_1\,T_0}$$

Reversing the above argument, we can also say: If an amount of heat, H, leaves a body of the temperature T_1, the entropy of that body decreases by the amount $\frac{H}{T_1}$, and when this same amount of heat enters another body of the temperature T_0 (transfer without compensation), the entropy of the second body is increased by the amount $\frac{H}{T_0}$. The increase of the entropy of the system comprising the two bodies is therefore, as above—

$$\frac{H}{T_0}-\frac{H}{T_1}=\frac{H(T_1-T_0)}{T_1\,T_0}$$

ORIGIN OF HEAT ENERGY.

The source of nearly all, if indeed not of all, forms of energy applicable for the production of heat and power, is traceable to the sun, the radiant energy of whose rays has been converted into potential or chemical energy in the plants, whence it found its way into the deposits of coals, etc. The heat of the sun's rays also produces the vapors which reappear as water falls, etc.; it also brings about the commotion in the atmosphere which appears in the force of waves and in the useful applications of the wind as well as in the devastations of the storm.

SPECIFIC HEAT OF GASES AT CONSTANT VOLUME.

In accordance with the molecular theory, the specific heat or the increase of heat energy for an increase of one degree in temperature for a molecule of a gas, or a proportional quantity of the same of the weight, M, is expressible—

$$J C_v = \left\{ \frac{1}{2} \frac{M u^2}{T} + E \right\}$$

in which C_v is specific molecular heat at constant volume, T the absolute temperature, J the mechanical equivalent of heat, and E the heat required to increase the motion within the molecule, u the velocity of the molecule as above defined.

SPECIFIC HEAT UNDER CONSTANT PRESSURE.

If a gas is heated under constant pressure the volume increases, and a certain amount of work is done, the equivalent of which in heat must also be furnished to the gas when its temperature is elevated. If we express the work done by—

$$\frac{p v}{T} = \frac{1}{3} \frac{M u^2}{T}$$

the specific heat of a molecule (expressed in units of weight) of gas under constant pressure, C_p, is—

$$J C_p = \frac{1}{2} \frac{M u^2}{T} + E + \frac{1}{3} \frac{M u^2}{T}$$

hence—

$$\frac{C_p}{C_v} = K = \frac{\frac{5}{6} \frac{M u^2}{T} + E}{\frac{1}{2} \frac{M u^2}{T} + E}.$$

K must always be smaller than $\frac{5}{3} = 1.6667$, since E must always be positive, and when it is very small, K approaches this value, as for vapor of mercury (1.666), in which the molecule is probably composed of only one atom, while in gases of presumably very complex molecules, the value for K approaches the other limit, viz., 1, as for ether, $K = 1.029$.

COMPONENTS OF SPECIFIC HEAT OF GASES.

From the foregoing we know that the heat required to do the work of expansion, when a gas is heated under constant pressure, is always equal to two-thirds of the heat necessary to increase the energy of the molecule. We find the specific heat, c_1, for equal volumes of gases under constant pressure, to be composed as follows:

Heat to increase molecular motion......	$= 3 \times 0.034$
Heat to do work of expansion.....	$= 2 \times 0.034$
Heat to do internal work (in molecule)...	$= n \times 0.034$
Specific heat.......	$= (n + 5)\ 0.034$

n being the number of atoms composing the molecule.

As for perfect gases, we can substitute equal volumes for equal number of molecules (since the same volumes of different gases contain an equal number of molecules), we can also say that for equal volumes of practically perfect gases, the specific heat is the same (see page 47).

NEGATIVE SPECIFIC HEAT.

When the heat equivalent of the work required to compress a saturated vapor from a lower to a higher pressure is greater than the heat required to increase the energy of the molecules of that vapor, from the temperature corresponding to the low pressure to the temper ature corresponding to the higher pressure of the saturated vapor, then the specific heat of such saturated vapor is said to be negative. For heat must be abstracted during compression to keep it in a saturated condition, and when allowed to expand a portion of the saturated vapor will condense for the same reason.

AIR THERMOMETER.

As the expansion of liquids and solids by heat is not uniform throughout the thermometric scale, this constitutes a serious defect in all thermometers constructed by their aid. This difficulty does not exist when air or another gaseous body is used as the thermometric substance. Hence the air thermometer is used for exact determinations.

THERMODYNAMIC SCALE OF TEMPERATURE.

If a thermometer be graduated in such a way that each degree increase in temperature of the thermometric substance adds equal amounts of free heat energy or equal amounts of heat available for mechanical conversion to the thermometric substance, we have a thermodynamic scale of temperature as devised by Thomsen. The degrees of such a scale agree very nearly with those of the air thermometer.

HEAT WEIGHT.

In accordance with the terminology adopted by Zeuner, the "weight" or "heat weight" of a certain amount of heat, H, transferable at the absolute temperature T, is that portion or fraction of said amount of heat which is convertible into mechanical energy, viz.: $\frac{H}{T}$. If the same amount of heat, H, enters a body at the constant absolute temperature T (without compensation), the entropy of that body is said to increase by an amount $\frac{H}{T}$. Hence entropy and heat weight are expressions which are numerically synonymous. The terms thermodynamic function (Rankine), and Carnot's function are used in the same connection. Thomson's thermodynamic scale of temperature shows equal heat weights from degree to degree.

Thermodynamics also teaches that the difference between the specific heat of a gas at constant pressure, c_p, and that at constant volume, c_v, is a constant quantity, and equal to the constant R of the gas equation, viz.:

$$c_p - c_v = \frac{pv}{T} = R$$

ISENTROPIC CHANGES.

Adiabatic changes which are at the same time reversible are also called isentropic changes, because such changes do not alter the entropy.

LATENT HEAT AND ENTROPY.

The heat which enters a body at the same or at constant temperature is called latent heat. Hence entropy may also be defined as latent heat divided by the corresponding temperature. Accordingly during vaporization or fusion of a body its entropy is increased. The amount of increase may be expressed by $\frac{l}{T}$ when l stands for the latent heat of vaporization or fusion, and T for the boiling or melting point expressed in absolute degrees F.

If a gas expands at constant temperature while doing work, it absorbs an amount of heat equivalent to the amount of work done, and its entropy increases correspondingly. Chemical changes taking place at constant temperature with transferences of heat cause corresponding changes of entropy.

CHAPTER VIII—MODERN ENERGETICS

INTRODUCTORY REMARKS.

In the foregoing paragraphs mass has been treated as one of the fundamental units, and as the vehicle not only of mechanical energy, but also of molecular energy according to the atomistic or mechanical theory of natural phenomena, which is still more or less generally accepted, and therefore followed in this compend.

SYSTEM OF ENERGETICS.

More recently following the example of Ostwald, Gibbs and others, it has been found expedient to consider energy not as a function of mass, but as something real, tangible and unchangeable in itself, thus creating a new series of scientific conceptions in accordance with which mass appears in the role of a factor in mechanical energy.

The terminology of this system places many definitions in a plainer and clearer light, and is frequently used in discussions on questions of energy, so that a synopsis of its tenets will be welcome to those who desire to study them.

NEW DEFINITION OF ENERGY.

Energy may also be defined as that immaterial quantity which, while it causes the greatest variety of changes or phenomena between different objects, always maintains its value. This definition involves the principle of conservation of energy.

CLASSIFICATION OF ENERGY.

The different forms of energy may also be classified in the following groups:

1. Mechanical energy.
2. Heat.
3. Electric and magnetic energy.
4. Chemical or internal energy.
5. Radiated energy.

MECHANICAL ENERGY.

The mechanical energy may be subdivided into two classes, viz.:

The energy of motion or kinetic energy, and the energy of space, with the following subdivisions:

1. Energy of distance (force).
2. Energy of surface (surface tension).
3. Energy of volume (pressure).

ENERGY FACTORS.

According to Helm, etc., the different kinds of energy are expressible by two factors—one of intensity and the other of capacity. Equal increases or decreases of energy in a given system or configuration of bodies correspond to equal increases or decreases of intensity, or, in other words, the energy of a system is proportional to its intensity. This may be expressed by the formula

$$E = c\,i$$

in which E represents energy, i the intensity and c the factor of capacity which is a measure for the amount of energy which is present in a system at a given intensity, i, the latter being counted from $E = 0$. In other words, the capacity factor for energy, c, may also be termed the capacity of the system for energy.

The capacity and intensity factors of some of the various forms of energy are given as follows:

ENERGY.	CAPACITY.	INTENSITY.
A. Kinetic energy.	Mass (m).	Square of velocity $\frac{v^2}{2}$
	Quantity of motion (mv).	Velocity $\frac{v}{2}$
B. Energies of space.		
a. Energy of distance.	Length.	Force.
b. Surface energy.	Surface.	Surface tension.
c. Energy of volume.	Volume.	Pressure.
C. Heat.	Capacity for heat.	Temperature.
D. Electricity.	Quantity of electricity.	Potential.
E. Magnetism.	Quantity of magnetism.	Magnetic potential.
F. Chemical energy.	Atomic weight.	Affinity.

DIMENSIONS OF ENERGY.

The definitions of the conceptions relating to energy, by means of algebraical expressions, or their dimensions are rendered in the following manner:

If e stands for the unit of energy, t for time, l for length or distance and m for mass the dimensions of the different mechanical conceptions may be expressed as follows:

		OLD UNITS.	NEW UNITS.
1.	Energy,	$m\,l^2\,t^{-2}$	e
2.	Mass,	m	$e\,l^{-2}\,t^2$
3.	Quantity of motion,	$m\,l\,t^{-1}$	$e\,l^{-1}\,t$
4.	Force,	$m\,l\,t^{-2}$	$e\,l^{-1}$
5.	Surface tension,	$m\,t^{-2}$	$e\,l^{-2}$
6.	Pressure,	$n\,l^{-1}\,t^{-2}$	$e\,l^{-3}$
7.	Effect,	$m\,l^2\,t^{-3}$	$e\,t^{-1}$

The first three definitions belong to the domain of kinetic energy, 4, 5 and 6 represent potential energies, and 7 the effect corresponding to the mechanical conception of a horse power.

The dimensions as given in the second column differ from those in the first column in that the third fundamental unit energy is substituted for mass, in accordance with the foregoing definition of energy factors.

THE INTENSITY PRINCIPLE.

Energy will pass from places of higher intensity to such of lower intensity; but energy of a certain intensity cannot pass to such of the same or of higher intensity. A system containing but one kind of energy is in equilibrium if the intensity of energy is the same throughout the system. If the intensity is not the same changes will occur until the differences in intensity have been equalized. If two intensities are equal to a third intensity, they are equal among themselves.

COMPENSATION OF INTENSITIES.

If more than one kind of energy is present in a system the differences in intensity of one kind of energy may be balanced or compensated by differences in the intensity of other kinds of energy; hence, in order that a change may take place in such a system, there must be differences of intensities not compensated.

If in a system containing several forms of energy, there are sudden leaps or differences in the intensity of one energy they must be compensated by equivalent sudden leaps or differences in the intensity of some other form of energy in order that equilibrium may exist in the system.

REGULATIVE PRINCIPLE OF ENERGY.

Everything that happens, every change or phenomenon is the sensible demonstration of a transfer or transformation of energy.

Of different changes possible to take place in a system containing one or more kinds of energy, that change will take place which causes the greatest amount of transformation or transference of energy in the shortest time. The term "possible changes" implies such changes as would be in harmony with the general laws of energy.

STATE OF EQUILIBRIUM.

A change (compatible with the conditions of existence) in a system containing different kinds of energy in equilibrium must add and abstract equal amounts of energy if equilibrium is to be maintained. The algebraical sum of energy lost and energy gained is equal to zero, a relation providing an important criterion for the state of equilibrium.

ARTIFICIAL AND NATURAL TRANSFER.

Energy may maintain equilibrium or become transferred or transformed artificially by means of certain appliances or devices (machines) or without such means. The latter transfers may be called natural transfers.

ARTIFICIAL EQUILIBRIUM.

If in a system containing two kinds of energy in equilibrium, the compensation of the intensities is effected by artificial means, *i. e.*, a machine, then such a contrivance directly determines the relation of one factor of one energy to one factor of the other energy, and therefore indirectly also the relation of the other factors.

DISSIPATION OF ENERGY.

The difference in intensities of energy not compensated determines the ability of such energy to do work or bring about changes. Hence the difference in intensities is a measure of the availability of the respective energy to do work. After a change has taken place the sum total of energy (capacities multiplied by intensities) must be the same as before the change, but the availability of the energy for the production of further changes is generally lowered. This is due to the fact that after the change in one or more forms of energies has taken place the capacities have generally been increased and intensities decreased correspondingly. In other words, the difference in the intensity of one energy which has disappeared has not been compensated by the appearance of an equivalent difference in the intensity of another energy. The tendency which prevails in all natural as well as artificial processes or changes, to increase the capacity at the expense of the intensity of existing energy, or, in other words, to obliterate existing differences in intensities, is the cause of what is termed the dissipation of energy.

RADIANT ENERGY.

The state or condition of energy while on its way from one body to another without a ponderable intervening medium is called "radiant energy." Energy in this condition and connection is supposed to possess some of the qualities referable to the hypothetical medium ether, notably elasticity.

TRANSFORMATION OF ENERGY.

The compensation of a change in one form of energy by an equivalent change of another form of energy constitutes what is also termed the transformation of one kind of energy into another.

REVERSIBLE CHANGES.

If the change produced by decreasing the difference in intensity of a given quantity of one form of energy has been fully compensated by an equivalent amount of difference of intensities of some other form or forms of energies having made its appearance, such a change may be considered reversible (in the abstract, at least). Two co-ordinated reversible changes, if fully performed, re-establish the original condition of things before the change.

IRREVERSIBLE CHANGES.

Changes in which energy is dissipated are not reversible, and hence may be termed irreversible changes.

PERPETUAL MOTION.

Irreversible changes are inseparably connected with all practical operations, and hence a perfectly reversible operation is a practical impossibility. Such an operation, if it were possible, could be repeated without end, and would constitute what is termed a "conservative system," which would be a kind of perpetual motion akin to that of the heavenly bodies. Such a perpetual motion, while beyond the possibilities of human skill, is not in contradiction with the laws of energy.

Besides the perpetual motion of a conservative system, we make a distinction between attempts at perpetual motion of the first order and of the second order.

The first kind contemplates the actual creation of energy, or of power to do work, and is in direct conflict with the first law of energy proclaiming its absolute conservation and indestructibility and its transformability in equivalent proportions.

Perpetual motion of the second order involves the elevation of the intensity of energy without compensation, which is in direct conflict with the intensity principle or the second law of energy.

CONTINUOUS CONVERSION OF ENERGY.

As a rule nothing could be gained in a practical way by carrying out the two co-ordinate systems of reversible changes; the useful object generally is to produce changes in one definite direction, and not undo them by reversion. This is notably the case in our efforts to convert energy of one kind into energy of another kind by a continuous process, as when heat energy is converted into motive power or mechanical energy, etc.

In all such efforts a certain percentage of energy is dissipated, that is the energy expended cannot all be compensated for in the desired direction.

MAXIMUM CONVERTIBILITY.

It follows from the above that when energy is transformed by processes or operations which are reversible (in the abstract, at least) the greatest possible amount of transformation (*i. e.*, incurring no dissipation of energy) is effected, as otherwise perpetual motion of the second order could be produced by reversing the operations.

For the same reason the maximum amount of energy obtainable by transforming a certain amount of another energy depends solely upon the uncompensated difference in the intensity of the latter energy and on the position which it holds on an absolute intensity scale, counting intensity from its proper zero. Hence the maximum of transformation obtainable in a certain direction is independent of the special object with which the energy is connected, or which is instrumental in the transformation.

INTENSITY PRINCIPLE AND ENTROPY.

The intensity principle is a general form of the second law of thermodynamics. It broadly asserts that while energy of any kind may pass from places of higher intensity to such of lower intensity without compensation, the reverse change, *i. e.*, the passage of energy from places of lower intensity to places of higher intensity, can never take place without compensation.

In all natural changes, in all manifestations of energy, the changes are either so as to fully compensate each other, or when this is not the case, the deficiency in compensation must correspond to so much increase of latent energy, and to a corresponding increase of entropy. In other words, natural changes proceed either without changing the entropy or by increasing the same.

Hence the conception of the entropy function enables us to determine as to the possibility of any supposed change in a system of bodies. If the change involves a decrease of entropy, it must be deemed impossible. If, however, the change involves no decrease of entropy, but if the same would remain unchanged or increase, then the said change is not in conflict with the laws of energetics.

JUSTIFICATION OF CONCEPTS.

The importance and significance of the above somewhat fragmentary and abstract definitions and concepts becomes more apparent in the treatment of the different individual branches of energetics, and especially in thermodynamics. It is in this branch that the above principles have their origin and confirmation, and it is in this branch that they prove their adaptability and usefulness for the further development of science, which usefulness must plead the justification of these concepts. Moreover their unrestricted adaptability in all other branches of science appears to be only a question of time.

UNIFORM UNITS OF ENERGY.

One kind of energy being transformable into an equivalent amount of another, it is indicated to so select the units for different forms of energies as to represent equivalent quantities. This is accomplished in a manner by some of the C. G. S. units.

CHANGE OF ABSOLUTE ZERO.

In the foregoing thermodynamic discussions the point of absolute zero has been taken at 461 degrees below zero Fahrenheit, as it is universally accepted so far. Recently, however, in his experiments to liquefy helium (the new gaseous element discovered in the atmosphere) Olszewski reached a temperature as low as 443° below zero, and helium remained a gas still. But judging from the pressure, etc., it will become a liquid at a temperature of about 570° F. below zero. This temperature must therefore still be above absolute zero, although it is impossible to say how much. At any rate, it is more than likely that a different absolute zero point will have to be accepted in the future, and that then our conceptions in thermodynamics will also receive important additions. But the experiments mentioned must be further confirmed before any definite changes are advisable.

MECHANICAL REFRIGERATION.

PART II.

PRACTICAL APPLICATION.

CHAPTER I.—REFRIGERATION IN GENERAL.

REFRIGERATION.

The act of reducing the temperature of any body or keeping the same below the temperature of the atmosphere is called refrigeration.

MEANS OF PRODUCING REFRIGERATION.

Refrigeration may be produced in many ways:

1. By transferring heat from a warmer body to a colder one. (Refrigeration by cooled brine, etc.)

2. By the consumption of heat brought about by doing work. (Working a piston against resistance with compressed gas; air machines.)

3. By melting or dissolving solid bodies. (Melting of ice; solution of salts in water, etc.)

4. By evaporating liquids which have a low boiling point. The latent heat of evaporation represents the amount of cold that can be produced in this way. (Evaporating liquid ammonia, liquid carbonic acid, liquid sulphurous acid, ether, etc.)

AIR MACHINES.

The mode of production of refrigeration by doing work is exemplified in the air machines, as that of Windhausen, which was formerly much used on steamers for refrigeration. In this machine the atmospheric air is compressed in a compressor, the heat generated by compression being carried off by the cooling water. The compressed air is then used to propel an engine, whereby its temperature is reduced corresponding to the work done by it in the engine. The air cooled in this way is then introduced into the rooms to be refrigerated, ventilating them at the same time. The machine operates continuously, but the refrigerating agent is rejected along with the heat which it has taken up.

FREEZING MIXTURES.

The refrigeration obtainable by dissolving solid bodies in water (freezing mixtures) has been referred to on pages 31 and 32. This method may also be employed in a continuous process, but is too expensive to be employed on a large scale, and when done so is chiefly used as an expedient when other means fail. In such case a mixture or solution of salt in ice or snow is generally used.

ICE MACHINES.

The machines which are now used for the production of refrigerating effects on a large scale are nearly all based on the principle of production of cold by the evaporating of liquids. Preference is given to either ammonia, sulphurous acid or carbonic acid as the evaporating liquid, or a mixture of the latter two.

CONSTRUCTION OF MACHINES.

The construction of the machines is the same in principle, no matter what evaporating liquid is employed, but the sizes and strength of different parts of the system vary greatly with the physical properties of the liquid, principally the latent heat of evaporation, the temperature and pressure of liquefaction, etc.

VAPORIZATION MACHINES.

The machines which are employed to practically utilize the heat of vaporization for refrigerating purposes may be classified as vacuum machines, absorption machines, compression machines, and mixed absorption and compression machines.

VACUUM MACHINES.

In the vacuum machines water is used as the refrigerating medium, its volatilization at a temperature sufficiently low being effected by means of vacuum pumps, the working of which is assisted by sulphuric acid, which absorbs the vapors as soon as formed, thus making the action of the vacuum very effective. The sulphuric acid may be concentrated for repeated use.

ABSORPTION MACHINES.

The absorption machines are similar to the vacuum machine in their action, the difference being that not water but a liquid (such as anhydrous ammonia), which evaporates at a low temperature without the aid of a vacuum, is used as a refrigerating medium. The vapors, instead of being absorbed by sulphuric acid, are absorbed by water, and from this they are separated again by distillation, and liquefied by the pressure in the still and the aid of condensing water.

In this manner all the larger absorption machines are operated continuously, the solution of ammonia in water being subjected to distillation in a still heated by a steam worm, the vapors of ammonia entering a condenser where they are cooled and become liquefied into anhydrous ammonia. The anhydrous ammonia is kept in a liquid receiver, whence it enters the refrigerator coils in which it evaporates, causing a refrigeration corresponding to its heat of vaporization. The vapors after having done this duty are allowed to enter the absorber, where they come in contact with the weak solution of ammonia drawn from the lower portion of the still, and are reabsorbed by the same with generation of heat, which is carried away by cooling water. The rich and cold solution of ammonia coming from the absorber and going to the still, and the poor and hot solution coming from the still and going to the absorber, are passed through a device called the exchanger to equalize their temperatures as much as possible. A pump is required to pump the rich ammonia solution from the absorber into the still.

THE COMPRESSION MACHINE.

The compression machines which use the latent heat of vaporization of substances having a low boiling point, such as ammonia, sulphurous acid, carbonic acid, etc., work practically all on the same principle. The vapors created by vaporization of the refrigerating medium in the refrigerating coils enter a compression pump, which is operated by a steam engine, which forces the vapor into condenser coils, where they are liquefied with the aid of cooling water. The liquid enters a liquid receiver, from which it is allowed to enter the refrigerating coils, as required. The process is continuous, and represents a cycle of operations as the working substance returns periodically to its original state, in a manner which approaches reversibility more or less according to the modes of operating the different machines.

AMMONIA MACHINES.

Owing to its high latent heat of evaporation, its comparatively low vapor tension, admitting liquefaction at a comparatively low pressure and high temperature, its neutral chemical properties, ammonia is highly valued for refrigerating purposes, and ammonia machines are now mostly in use for refrigerating purposes in the United States.

PERFECT COMPRESSION SYSTEM.

In case of a perfect reversible compression system the operations would have to consist of the following changes:

First.—Evaporation of the liquid ammonia at the (constant) temperature of the refrigerator, constituting an isothermal change.

Second.—Compression of the vapor so formed without addition of heat, which is an adiabatic change.

Third.—Condensation of the compressed vapor at the (constant) temperature of the condenser, constituting another isothermal change.

Fourth.—Reduction of the temperature of the liquid from the temperature of the condenser to that of the refrigerator by means of vaporizing a portion of the liquid and doing work by moving a piston. This is the second adiabatic change, and it returns the working fluid to its initial condition, thus completing the cycle.

These changes are conceived to be carried on in such a manner that the transfers of heat follow only infinitesimally small differences in temperature, and the changes in volume take place under but infinitesimally small differences of pressure.

REVERSIBLE CYCLE.

Under these circumstances the changes can also be performed in the opposite direction, and therefore the cycle is what is termed a reversible cycle. A heat engine as well as a refrigerating apparatus (a heat engine reversed), if worked on the plan of reversible cycle, is working on the most economical plan that can be conceived.

For this reason the heat, H, removed by a refrigerating apparatus operated strictly on this basis has a certain and well defined relation to the work or mechanical power, W, required to lift the same in the cycle of operation. If in a refrigerating machine so operated t_1 is the temperature of condenser and t_0 the temperature of the refrigerator (T_1 and T_0 designating the corresponding absolute temperatures) thermodynamics teaches us that the following relations exist:

$$\frac{H}{W} = \frac{t_0 + 460}{t_1 - t_0} = \frac{T_0}{T_1 - T_0}$$

DEFECT IN CYCLE.

Thermodynamically speaking, there should be no difference in economy on account of the nature of the circulating fluid if a perfect cycle of operation was carried out, but practically this is not done. In all compression machines (barring some trials in the case of carbonic acid machines), the fourth operation, the reduction of temperature of the liquid while doing work, is not carried out, but the liquid is cooled at the expense of the refrigeration of the system. No work is attempted, as the amount obtainable would not be in proportion to the expense involved in procuring the same. This defect and other conditions in the working of a reversible cycle have some bearing on the choice of the circulating medium.

CHOICE OF CIRCULATING MEDIUM.

In the choice of a circulating medium, therefore, we should consider that its refrigerating effect depends on the latent heat of vaporization per pound.

That the size of the compressor depends on the number of cubic feet of vapor that must be taken in to produce a certain amount of refrigeration, and the strength of its parts on the pressure of the circulating medium.

And also that the loss of refrigeration on account of cooling the liquid circulating medium depends on the specific heat of the liquid as compared with the heat of volatilization.

The qualities chiefly involved in this question are compiled, approximately, in the following table for the principal liquids employed in refrigeration.

	Pressure in Lbs. per Square Inch at 0° F.	Heat of Vaporization per Pound at 0° F.	Volume Cubic Ft. per Pound at 0° F.	Specific Heat of Liquid.	Heat of Vaporization per Cubic Foot.	Relative Volume of Compressor for Equal Refrigeration.	Loss Due to Cooling Liquid.
							Per Ct.
Sulphurous acid.......	10	171.2	7.35	0.41	23.3	61.70	0.24
Carbonic acid..........	310	123.2	0.277	1.00	447.	3.24	0.81
Ammonia...............	30	555.5	9.10	1.02	61.7	23.3	0.18

This table explains itself and readily accounts for the preference generally given to ammonia as the circulating fluid. The loss due to the cooling of the liquid as shown in percentage for every degree difference in temperature of condenser and refrigerator, is less than in case of the other liquids, and the total refrigerating effect per pound of liquid is largest. The only instance speaking more in favor of sulphurous acid is the lower pressure of its vapor, while the compressor is smallest in case of carbonic acid, but the pressure and the loss due to heating of liquid is very large in the latter case.

SIZE OF ICE MACHINES.

The heat unit, as already stated, is used for measuring both heating and refrigerating effects. As a matter of convenience, however, the capacity of large refrigerating plants is expressed in tons of ice. By a ton of refrigerating capacity used in the above connection is meant a refrigerating capacity equivalent to a ton of ice at the freezing point while melting into water at the same temperature. This refrigerating capacity is equal to 284,000 units.

ICE MAKING CAPACITY.

The refrigerating capacity of a machine is different from the actual ice making capacity of a plant; the latter is considerably less, fifty per cent and upward, of the refrigerating capacity, according to temperature of water, etc.

USES OF REFRIGERATION.

The practical uses of mechanical refrigeration are so manifold that it is impossible to enumerate them all in a small paragraph. Foremost among them is cold storage, that is, the preservation of all kinds of articles of food and drink by the application of low temperature. Slaughtering, packing and shipping of meat can hardly be carried on nowadays without the use of mechanical refrigeration, and the days of the few breweries still working without this artificial appliance may be said to be numbered. Since ice has become an article of daily necessity, there are few towns that have not or will not have their artificial ice factory or factories.

Artificial refrigeration is or will be used for a great many other purposes, some of which will be mentioned later on.

CHAPTER II.—PROPERTIES OF AMMONIA.

FORMS OF AMMONIA.

The ammonia occurs in practical refrigeration in three different forms, as the liquid anhydrous ammonia, the gaseous anhydrous ammonia and solutions of ammonia in water of various strengths.

ANHYDROUS AMMONIA.

Ammonia is a combination of nitrogen and hydrogen expressed by the formula NH_3 which means that an atom of nitrogen (representing 14 parts by weight) is combined with three atoms of hydrogen (representing three parts by weight). At ordinary temperatures the ammonia, or anhydrous ammonia, as it is called in its natural condition, is a gas or vapor. At a temperature of —30° F. it becomes liquid at the ordinary pressure of the atmosphere, and at higher temperatures also if higher pressures are employed. The anhydrous ammonia dissolves in water in different proportions, forming what is called ammonia water, ammonia liquor, aqua ammonia, etc. At a temperature of 900° F. ammonia dissociates, that is, it is decomposed into its constituents, nitrogen and hydrogen, the latter being a combustible gas.

It appears that partial decomposition takes place also at lower temperatures, but probably not to the extent frequently supposed.

The liquid ammonia turns into a solid at a temperature of about —115° F. In this condition it is heavier than the liquid, and is almost without smell. At a temperature of —95° F. the chemical affinity between sulphuric acid and ammonia is zero, no reaction taking place between the two substances when brought in contact at or below this temperature.

Ammonia is not combustible at the ordinary temperature, and a flame is extinguished if plunged into the gas. But if ammonia be mixed with oxygen, the mixed gas may be ignited and it burns with a pale yellow flame. Such mixtures may be termed explosive in a certain sense.

If a flame sufficiently hot is applied to a jet of ammonia gas, it (or rather, the hydrogen of the same) burns as long as the flame is applied, furnishing the heat required for the decomposition of the ammonia.

Ammonia is not explosive, but when in drums containing the liquid ammonia not sufficient space is left for the liquid to expand when subjected to a higher temperature, the drums will burst, as has happened frequently during the hot season.

The ammonia vapors are highly suffocating, and for that reason, persons engaged in rooms charged with ammonia gas must protect their respiration properly.

PRESSURE AND TEMPERATURE OF AMMONIA.

The relation between pressure and temperature of saturated ammonia vapor is expressed by the formula:

$$\log._{10} p = 6.2495 - \frac{2196}{T}$$

in which p is the pressure in pounds per square inch, and T the absolute temperature.

DENSITY OF AMMONIA.

The density d of liquid anhydrous ammonia at different temperatures, water being 1, is approximately expressed by the formula:

$$d = 0.6502 - 0.00077\,t,$$

t being temperature in degrees Fahrenheit.

The density of the gas is 0.597 at 32° F., and at 760 mm. pressure. The volume, v, of the saturated vapor per pound may be calculated by the formula:

$$v = \frac{h\,T}{6.4993\,P} + v_1 \text{ cubic feet,}$$

in which P is the pressure in pounds per square foot, T the absolute temperature, h the latent heat of vaporization.

SPECIFIC HEAT OF AMMONIA.

The specific heat of liquefied ammonia is variously stated from 1 to 1.228. The specific heat of ammonia gas is given at 0.508 at constant pressure, and 0.3913 at constant volume. The coefficient of expansion of liquid ammonia is 0.00204.

The specific heat, s, of saturated vapor of ammonia is expressed by the formula:

$$s = 1 - \frac{555.5}{T}$$

This value is negative for all values of T less than 555° absolute, which means that if saturated ammonia vapor is expanded adiabatically a portion of it will condense, giving up its heat to the remainder of the vapor, thus maintaining the temperature corresponding to the pressure of saturation, and when compressed heat must be abstracted, if the temperature and pressure are continually to correspond to those of the state of saturation, otherwise it will become superheated.

SPECIFIC VOLUME OF LIQUID.

The specific volume, v_1, of liquefied ammonia may be found after the following rule:

$$v_1 = \frac{.0160}{0.6502-0.00077\,t} \text{ cubic feet.}$$

LATENT HEAT.

The latent heat, h, of evaporation of ammonia is

$$h = 555.5 - 0.613\,t - 0.000219\,t^2,$$

in which formula t stands for degrees F.

EXTERNAL HEAT.

That portion of the latent heat required to overcome external pressure or the external latent heat, E, is expressed by—

$$E = \frac{P\,(v-v_1)}{J}$$

in which formula P stands for external pressure in pounds per square foot, v for the volume of the vapor, and v_1 for the volume of the liquid (which is neglected in the calculations given in the accompanying table), and J the mechanical equivalent of heat.

WEIGHT OF AMMONIA.

The weight, w, of a cubic foot of the saturated vapor is—

$$w = \frac{1}{v}$$

And the weight, w_1, of a cubic foot of the liquid is—

$$w_1 = \frac{1}{v_1}$$

The weight of one cubic foot of liquid ammonia at a temperature of 32° F. is 39.017 pounds.

TABULATED PROPERTIES OF SATURATED AMMONIA.

The physical properties of anhydrous ammonia, both in the vapor and liquid state, which are of special use in the refrigerating practice, are laid down in the following table prepared by De Volson Wood, calculated by the above formulæ which have been elaborated by him also.

PROPERTIES OF SATURATED AMMONIA.

TEMPERATURE.		PRESSURE, ABSOLUTE.		Heat of Vaporization, Thermal Units.	External Heat. Thermal Units.	Internal Heat, Thermal Units.	Volume of Vapor per Lb., Cu. Ft.	Volume of Liquid per Lb., Cu. Ft.	Weight of a Cu. Ft. of Vapor, Pounds.
Degree F.	Absolute.	Lbs. per Sq. Ft.	Lbs. per Sq. In.						
— 40	420.66	1540.9	10.69	579.67	48.23	531.44	24.37	.0234	.0410
— 35	425.66	1773.6	12.31	576.69	48.48	528 21	21.29	.0236	.0467
— 30	430.66	2035.8	14.13	573.69	48.77	524.92	18.66	.0237	.0535
— 25	435.66	2329.5	16.17	570.68	49.06	521.62	16.41	.0 38	.0609
— 20	440.66	2657.5	18.45	567.67	49.38	518.29	14.48	.0240	.0690
— 15	445.66	3022.5	20.99	564.64	49.67	514.97	12.81	.0242	.0779
— 10	450.66	3428.0	23.77	561.61	49.99	511.62	11.36	.0243	.0878
— 5	455.66	3877.2	26.93	558.56	50.31	508.25	10.12	.0241	.0988
0	460.66	4373.5	30.37	555.50	50.68	504.82	9.04	.0246	.1109
+ 5	465.66	4920.5	34.17	552.43	50.84	501.59	8.06	.0247	.1241
+ 10	470.66	5522.2	38.55	549.35	51.13	498.22	7.23	.0249	.1384
+ 15	475.66	6182.4	42.93	546.26	51.33	494.93	6.49	.0250	.1540
+ 20	480.66	6905.3	47.95	543.15	51.61	491.54	5.84	.0252	.1712
+ 25	485.66	7695.2	53.43	540.03	51.80	488.23	5.26	.0253	.1901
+ 30	490.66	8556.6	59.41	536.92	52.01	484.91	4.75	.0254	.2105
+ 35	495.66	9493.9	65.93	533.78	52.22	481.56	4.31	.0256	.2320
+ 40	500.66	10512	73.00	530.63	52.42	478.21	3.91	.0257	.25 3
+ 45	505.66	11616	80.66	527.47	52.62	474.85	3.56	.0260	.2809
+ 50	510.66	12811	88.96	524.30	52.82	471.48	3 25	.0260	.3109
+ 55	515.66	14102	97.93	521.12	53.01	468.11	2.96	.0260	.33 9
+ 60	520.66	15494	107.60	517.93	53.21	464.72	2.70	.0265	.3704
+ 65	525.66	16993	118.03	514.73	53.38	461.35	2.48	.0263	.4034
+ 70	530.66	18605	129.21	511.52	53.57	457.85	2.27	.0268	.4405
+ 75	535.66	20336	141.25	508.29	53.76	454.53	2.08	.0270	.4808
+ 80	540.66	22192	154.11	504.66	53.96	450.70	1.91	.0272	.5262
+ 85	545.66	24178	167.86	501.81	54.15	447.66	1.77	.0273	.5649
+ 90	550.66	26300	182.8	498.11	54.28	443.83	1.64	.0274	.6098
+ 90	555.66	28565	198.37	495.29	54.41	440.83	1.51	.0277	.6622
+ 100	560.66	30980	215.14	491.50	54.54	436.96	1.39	.0279	.7194
+ 105	565.66	33550	232.98	488.72	54.67	434.03	1.285	.0231	.7757
+ 110	570.66	36284	251.97	485.42	54.78	430.64	1.203	.0283	.8312
+ 115	575.66	39188	272.14	482.41	54.91	427.40	1.121	.0235	.8912
+ 120	580.66	42267	293.49	478.79	55.03	423.75	1.041	.0287	.9608
+ 125	585.66	45528	316.16	475.45	55.09	420.39	.9699	.0289	1.0310
+ 130	590.66	48978	340.42	472.11	55.16	416.94	.9051	.0291	1 1048
+ 135	595.66	52626	365.16	468.75	55.22	413.53	.8457	.0293	1.1824
+ 140	600.66	56483	392.22	465.30	55.20	410.09	.7910	.0295	1.2642
+ 145	605.66	60550	420.49	462.01	55.34	406.67	.7408	.0297	1.3497
+ 150	610.66	64833	450.20	458.62	55.39	402.23	.6946	.0299	1.4396
+ 155	615.66	69341	481.54	455.22	55.43	399.79	.6511	.0302	1.5358
+ 160	620.66	74086	514.40	451.81	55.46	396.35	.6128	.0304	1.6318
+ 165	625.66	79071	549.04	448.39	55.48	392.94	.5765	.0306	1.7344

The critical pressure of ammonia is 115 atmospheres, the critical temperature at 266° F. (Dewar), critical volume .00482 (calculated).

VAN DER WAALS' FORMULA FOR AMMONIA.

As has been shown (page 56), the constants a and b of Van der Waals' formula can be derived from the critical data, which gave me the following values for ammonia:

$$a = .0079;\ b = .0016.$$

If the values for a and b thus found for ammonia are introduced in the general equation (page 56), setting p_0 and v_0 equal unit, the equation will read:

$$\left\{ p + \frac{0.0079}{v^2} \right\} (v - 0.0016) = (1 + .0079)\,(1 - .0016) \frac{461 + t}{493}$$

$$\text{or } \left\{ p + \frac{0.0079}{v^2} \right\} (v - 0.0016) = \frac{1.00627 \times (461 + t)}{493}$$

This equation may be used to establish the relations between pressure, volume and temperature for anhydrous ammonia, and in order to test the same we may compare the results so obtained with those derived from actual experiments for saturated ammonia vapor, the volume of which ought to satisfy one of the three values for v which are possible below the critical temperature at the pressure of liquefaction.

On this basis the values, p_1, for the pressure of ammonia gas for given volumes at given temperatures have been calculated in the following table:

t	p	v	$v_1 = \frac{v}{19}$	p_1
− 40	0.71	24.37	1.282	0.66
− 15	1.38	12.81	0.674	1.33
+ 32	3.96	4.57	0.24	4.02
+ 60	7.17	2.7	0.142	7.24
+122	20.3	1.0	0.052	20.4
+165	36.6	0.57	0.030	38.4

In this table the values for p and v_1 for the temperature t are in accordance with Wood's interpretation of Regnault's experiments for saturated ammonia vapor, and the values, p_1, are derived from the above formula for ammonia by inserting the value, v_1, obtained in measuring the volume by the volume of an equal weight of ammonia gas at the pressure of one atmosphere at 32° F. It will be noticed that p_1 agrees pretty closely with p between −15° and 165°, thus proving the approximate correctness of Waals' formula for saturated ammonia within these temperatures, and therefore the formula may doubtless also be safely used for superheated vapor of this substance within these limits for approximate estimation. Indeed, the agreement between the two sets of pressures obtained by entirely different experiments, and by an entirely different course of reasoning, is sufficiently close to inspire the greatest confidence in the experiments of Regnault and Dewar, as well as in the mathematical deductions of Van der Waals.

SUPERHEATED AMMONIA VAPOR.

Below its critical temperature (266° F.) ammonia in its volatile condition is to be termed a vapor, strictly speaking; but when it is not in a saturated condition, but in the condition of a superheated vapor, as it were, it behaves practically like a permanent gas and is also termed ammonia gas. In this condition one pound of ammonia gas, under a pressure of an atmosphere, and at the temperature of 32° F. occupies a volume of 20.7 cubic feet (one cubic foot of air weighing 0.0806 pound, and the specific gravity of ammonia being 0.597 of air under these conditions).

FORMULÆ FOR SUPERHEATED VAPOR.

On this basis the relations of volume, weight, pressure and temperature of ammonia gas or superheated ammonia vapor can be calculated after the general equation of gases on pages 46 and 51.

The volume v in cubic feet of one pound of ammonia gas at any temperature, t, and for any pressure, p, expressed in pounds per square inch below that which corresponds to the pressure of saturated vapor at that temperature, or for any pressure and for any temperature above that which corresponds to the temperature of saturated vapor at that pressure, can be found approximately after the formula—

$$v = \frac{20.7\,(461 + t)\,14.7}{493 \times p} = \frac{20.7\,(461 + t)}{33.5\,p} = \frac{0.62\,(461 + t)}{p}$$

If the volume, v, in cubic feet of one pound of ammonia gas at a certain temperature, t, is known, the pressure can be found after the equation—

$$p = \frac{20.7\,(461 + t)}{33.5\,v} = \frac{0.62\,(461 + t)}{v}$$

And if the volume, v, and the pressure, p, are known the temperature may be determined approximately after the equation—

$$t = 1.62\,p\,v - 461$$

As stated above, the formula of Van der Waals may also be used in this connection, but it is rather too cumbersome for this purpose. However, if the value of 20.7 in the foregoing formulæ is substituted by 19, which is the figure found in accordance with Van der Waals' equation, the results agree closer with the figures obtaining for vapor just saturated. The table on "Properties of Ammonia Gas or Superheated Vapor of Ammonia" in the appendix agrees practically with the formula given for v, on page 96, and for this reason gives only approximate values, since said formula considers ammonia a perfect gas, which it is not, as indicated by Van der Waals.

AMMONIA LIQUOR.

The solutions of anhydrous ammonia in water are employed in the so called absorption machines, and the properties of such solutions vary with their strength or the percentage of ammonia which they contain. The strength of such solutions, "ammonia liquor," as they are commonly called, is approximately determined by specific gravity scales or hydrometers, those of Beaumé being usually employed for this purpose.

STRENGTH OF AMMONIA LIQUOR.

Percentage of Ammonia by Weight.	Specific Gravity.	Degrees Beaumé Water 10.	Degrees Beaumé Water 0.
0	1.000	10	0
1	0.993	11	1
2	0.986	12	2
4	0.979	13	3
6	0.972	14	4
8	0.966	15	5
10	0.960	16	6
12	0.953	17.1	7
14	0.945	18.8	8.2
16	0.938	19.5	9.2
18	0.931	20.7	10.3
20	0.925	21.7	11.2
22	0.919	22.8	12.3
24	0.913	23.9	13.2
26	0.907	24.8	14.3
28	0.902	25.7	15.2
30	0.897	26.6	16.2
32	0.892	27.5	17.3
34	0.888	28.4	18.2
36	0.884	29.3	19.1
38	0.880	30.2	20.0

PROPERTIES OF AMMONIA LIQUOR.

On the following pages we publish a table prepared by Starr, and based on experiments made by him, which shows the relations between pressure and temperature for solutions of ammonia in water of different strengths.

The figures in the top row, marked P. P., indicate the "gauge" pressures, while those in the columns beneath "gauge" pressures give the temperatures in degrees Fahrenheit of the gas at the "gauge" pressures indicated at the head of each column, thus: Under "gauge" pressure 150 (lbs.), the temperature of the gas of 26° ammonia is 245.2° F.

POUNDS PRESSURE & DEGREES FAHRENHEIT.

% of NH_3 by weight and Deg. Baumé	0 Atmos.	5	10	15	20	25	30	35	40	45	50	55	60	65	70	75	80	85	90	95	100	105	110	115	120	125	130	135	140	145	150	155	160	165	170	175	180	P.P.
1.	206.3	223.6	234.9	247.4	256.2	263.5	270.4	277.1	282.2	288.1	292.9	297.5	301.9	306.3	310.6	314.0	318.2	321.8	325.0	328.5	331.7	334.0	337.8	340.5	343.5	346.1	348.1	351.3	353.7	356.0	358.2	360.3	362.3	364.2	366.1	367.4	369.5	
1.84–11°B	201.9	219.3	231.6	243.0	251.7	259.0	266.4	272.7	278.0	283.7	288.5	293.1	297.5	301.8	306.0	310.0	313.8	317.4	320.8	324.1	327.3	330.4	333.4	336.3	339.1	341.8	344.4	346.9	349.3	351.6	353.8	355.4	357.9	359.5	361.7	363.5	365.1	
2	201.1	218.5	230.8	242.1	250.9	258.6	265.5	271.9	277.6	282.8	287.7	292.2	296.7	300.9	305.2	309.2	312.4	316.6	320.0	323.2	326.5	329.0	332.6	335.4	338.0	341.0	343.6	346.1	348.2	350.8	352.9	355.0	357.3	358.9	360.9	362.7	364.3	
3	193.8	213.2	225.5	236.6	245.6	253.3	260.2	266.8	272.3	277.5	282.4	286.9	291.4	295.6	300.0	303.9	307.6	311.3	314.7	317.9	321.2	324.3	327.3	330.1	332.9	335.7	338.3	340.8	342.9	345.5	347.6	349.8	351.8	353.6	355.6	357.6	359.0	
3.10–12°B	191.5	208.8	221.0	232.9	241.0	248.7	255.7	262.0	267.7	272.9	277.8	282.6	286.5	291.1	295.3	299.3	303.1	306.7	310.1	313.4	316.6	319.7	322.7	325.4	328.4	331.1	333.7	336.2	338.6	340.9	343.1	345.2	347.3	349.3	351.1	352.8	354.4	
4.	190.5	207.7	220.4	231.2	240.4	247.6	254.7	260.9	266.7	271.8	276.7	281.4	285.7	290.1	294.2	298.3	302.1	305.6	309.1	312.4	315.5	318.7	321.6	324.5	327.4	330.0	332.6	335.2	337.6	339.8	342.1	344.1	346.2	348.2	350.1	351.7	353.4	
5.	185.2	202.4	214.6	225.8	234.6	242.2	249.3	255.6	261.4	266.5	271.4	276.1	280.4	284.8	288.9	293.0	296.8	300.3	303.8	307.1	310.2	313.4	316.3	319.2	322.0	324.7	327.5	329.9	332.3	334.5	336.8	338.8	340.9	342.9	344.9	346.4	348.0	
5.50–13°B	183.5	200.7	212.8	224.1	232.8	240.5	247.5	253.8	259.6	264.8	270.2	274.3	279.1	283.5	287.7	291.7	295.5	299.1	302.5	305.8	309.0	312.1	315.1	318.0	320.8	323.5	325.9	328.4	330.8	333.1	335.3	337.4	339.0	341.5	343.2	345.0	346.0	
6.	180.0	197.1	209.2	220.5	229.3	237.0	243.9	250.2	256.1	261.2	266.7	271.2	275.6	280.0	284.1	288.2	291.9	295.5	299.0	302.2	305.5	308.5	311.6	314.4	317.3	319.7	322.4	324.8	327.3	329.6	331.8	333.8	335.9	337.7	339.6	341.6	343.7	
6.80–14°B	175.8	143.0	205.0	216.2	226.4	232.6	239.6	246.0	251.3	257.0	262.1	266.7	271.1	275.0	279.6	283.6	287.4	291.0	294.0	297.7	300.4	304.0	307.0	309.9	312.7	315.2	317.8	320.3	322.7	325.0	327.2	329.8	331.3	333.3	335.2	336.9	338.5	
7.	174.9	192.1	204.0	215.3	223.9	231.7	238.6	245.1	250.8	256.1	261.1	265.8	270.1	274.5	278.6	282.7	286.4	290.1	293.5	296.7	300.0	303.0	306.1	308.9	311.8	314.2	316.9	319.3	321.8	324.0	[illegible]	328.3	330.6	332.3	334.1	335.9	336.6	
8.	170.0	187.2	199.1	210.3	218.9	226.9	233.7	240.1	245.9	251.2	266.2	260.8	265.2	269.6	273.7	281.7	281.5	285.2	288.6	291.7	295.1	298.1	301.2	303.9	306.9	309.3	312.0	314.3	316.9	319.1	321.4	327.3	325.6	327.6	329.5	330.9	331.7	
8.22–15°B	168.8	185.8	197.8	209.0	217.7	225.4	232.4	238.0	244.2	249.3	254.1	258.7	263.1	267.4	271.6	275.6	279.0	283.0	286.4	289.7	292.4	296.0	299.0	301.9	304.7	307.2	309.8	312.3	314.7	317.0	319.2	321.3	323.3	325.3	327.2	328.9	330.5	
9.	165.4	182.5	194.5	205.2	214.3	222.0	229.0	235.2	240.8	245.9	250.7	255.3	259.7	264.0	268.2	272.2	276.0	279.6	283.0	286.3	289.6	292.6	295.4	298.5	301.3	303.8	306.4	308.9	311.3	313.6	315.8	317.9	319.9	321.8	323.7	325.5	327.2	
10–16°B	160.8	177.7	189.6	200.6	209.2	216.9	223.9	230.1	235.5	240.6	245.4	250.0	254.4	258.7	262.9	266.9	270.7	274.8	277.7	281.0	284.2	287.8	290.3	293.0	296.0	298.5	301.1	303.6	306.0	308.2	310.5	312.6	314.6	316.6	318.5	320.2	321.8	
11.	156.4	173.2	185.1	196.1	204.7	212.4	219.4	225.6	231.0	236.1	240.9	244.5	249.1	254.2	258.4	262.4	266.2	268.8	273.0	276.5	279.7	282.8	285.8	288.7	291.5	294.0	296.0	299.2	301.5	303.8	305.0	308.2	310.2	312.1	314.0	315.8	317.5	
12.	151.9	168.9	180.6	191.9	200.6	208.3	214.8	221.0	226.4	231.5	236.4	240.6	245.4	249.8	253.9	257.8	261.7	264.3	268.7	272.0	275.2	278.3	281.3	289.2	287.0	289.5	292.1	294.7	297.0	299.3	300.5	303.8	305.8	307.6	309.5	311.3	313.0	
12.17–17°B	151.0	168.0	179.4	191.0	199.4	207.1	213.6	219.6	225.0	230.8	234.6	239.4	243.4	247.7	251.9	255.4	259.7	263.3	266.7	270.0	273.2	276.3	279.3	282.2	285.0	287.5	290.1	292.6	295.0	297.3	299.5	301.6	303.6	305.6	307.5	309.2	310.8	
13.	147.5	164.9	176.4	187.4	198.1	203.7	210.1	216.1	221.4	226.8	230.8	235.5	239.9	244.2	248.4	251.8	256.2	259.8	263.1	266.5	269.0	272.8	275.7	278.4	281.5	284.0	286.5	289.1	291.5	293.7	296.2	298.1	300.1	302.0	304.0	305.7	307.3	
13.88–10°B	143.7	160.5	172.3	183.4	192.0	199.7	206.0	212.1	217.6	222.7	227.8	231.8	236.2	240.5	244.0	248.7	252.5	256.5	259.5	262.8	266.0	269.1	272.1	275.0	277.8	280.3	282.9	285.4	287.8	290.1	292.3	294.4	296.4	298.3	300.1	302.0	303.6	
14.	143.2	160.0	171.8	182.9	191.3	199.2	205.1	211.6	217.1	222.2	226.7	231.3	235.7	240.0	243.5	248.1	252.0	255.6	259.0	262.3	265.5	268.6	271.6	274.5	277.3	280.0	282.4	284.9	287.3	289.4	291.6	293.9	295.9	297.5	299.6	301.5	303.1	
15.	139.[illegible]	155.8	167.6	178.7	187.3	195.0	201.3	207.4	212.9	218.0	222.5	227.1	231.5	235.8	239.3	244.0	247.8	251.4	254.8	258.1	261.3	264.4	267.4	270.3	273.1	275.8	278.2	280.7	283.1	285.4	287.6	289.7	291.7	293.8	295.4	297.3	298.9	
16.	134.8	151.6	163.4	174.5	183.1	190.3	197.1	203.2	208.7	213.8	218.3	222.9	227.3	231.6	235.1	239.8	243.6	247.2	250.8	253.7	257.1	260.8	263.2	266.1	268.9	271.6	274.0	276.5	279.9	281.8	283.4	285.5	287.5	289.0	291.3	293.1	294.7	
16.22–19°B	131.8	150.4	162.3	173.3	181.4	189.3	196.0	201.5	207.1	212.3	217.1	221.7	226.1	230.4	234.6	238.6	242.4	246.0	249.4	252.7	255.4	259.0	262.0	264.4	267.1	270.0	272.8	275.3	277.7	280.0	282.2	284.3	286.3	288.3	290.3	292.0	293.6	

DEGREES FAHRENHEIT

POUNDS PRESSURE & DEGREES FAHRENHEIT.

% of NH_3 by weight and Deg. Beaumé	0	5	10	15	20	25	30	35	40	45	50	55	60
17.	130.6	147.4	159.1	170.1	178.2	186.3	192.5	198.6	203.9	209.1	213.9	216.5	222.4
18.93–20°B	116.2	142.9	154.6	165.6	174.2	181.9	188.4	195.1	200.7	205.7	209.5	214.1	218.5
19.	122.3	138.9	150.7	161.6	170.3	177.9	185.4	191.1	196.5	201.7	205.6	210.1	214.6
19.87–21°B	119.4	135.9	147.6	158.6	167.2	174.0	181.5	187.2	192.5	197.5	202.3	206.9	211.3
20.	118.9	135.5	147.1	152.2	166.7	174.4	181.1	186.7	192.1	197.0	201.9	206.4	210.8
21.	115.2	131.8	143.4	154.5	163.0	170.7	177.4	183.0	188.0	193.3	198.2	202.7	207.1
21.75–22°B	112.9	129.4	141.0	151.4	160.3	168.2	174.6	180.1	185.3	190.3	195.1	199.7	204.1
22.	112.0	128.3	140.1	151.0	159.6	167.3	173.7	179.2	184.4	189.4	194.0	198.8	203.2
23.03–23°B	108.0	124.5	136.1	147.0	155.6	163.3	170.0	175.4	180.2	185.2	190.0	196.6	199.0
24.	114.3	121.3	132.9	143.8	152.4	160.1	166.8	172.2	177.0	182.0	186.5	191.4	195.8
24.99–24°B	101.5	117.8	129.3	140.1	148.6	156.3	162.0	168.4	173.6	178.6	183.2	187.8	192.2
26.	98.3	114.6	126.2	136.9	145.5	153.1	159.8	165.3	170.4	175.5	179.9	184.7	189.1
27.	95.1	111.4	123.1	133.7	142.3	150.0	156.6	162.1	167.2	172.4	176.7	181.5	185.4
27.64–25°B	93.8	109.4	121.0	131.7	140.1	147.9	154.5	159.9	165.1	170.3	174.0	178.9	183.3
28	92.0	108.3	120.0	130.6	139.1	146.8	153.4	158.9	164.0	169.3	173.3	177.9	183.2
29.	88.9	105.2	117.0	127.5	136.0	143.8	150.3	155.8	161.0	166.2	170.2	174.5	180.2
29.60–26°B	87.0	103.3	114.7	125.4	133.9	141.6	148.2	153.8	159.2	164.3	168.1	172.7	178.1
30.	85.1	102.1	113.5	124.1	132.7	140.4	147.0	152.6	157.2	163.1	166.9	171.6	176.9
31.03–27°B	82.6	98.8	110.3	120.9	129.4	137.4	143.5	149.2	154.5	159.8	163.4	168.5	173.5
32.	80.1	96.2	107.6	118.3	126.3	134.6	140.9	146.6	151.9	157.2	161.0	165.7	170.9
33.	77.4	93.5	104.9	115.6	124.1	131.8	138.7	143.9	149.1	154.5	158.3	163.0	166.2
33.25–28°B	76.5	92.6	103.9	114.6	123.1	130.8	137.3	143.0	148.2	153.6	157.4	162.1	167.3
34	74.6	90.7	102.0	112.7	121.2	128.9	135.9	141.1	146.4	151.7	155.5	160.3	165.4
35	72.0	88.1	99.4	110.1	118.6	126.3	133.3	138.5	143.0	149.1	152.9	157.6	162.5
35.60–29°B	70.4	86.5	97.8	108.5	117.0	128.7	131.7	137.9	142.3	147.5	151.3	156.0	161.2
36.	60.5	85.6	96.9	107.5	116.1	123.8	130.8	137.0	141.7	147.2	151.0	155.7	160.8
37.	67.2	83.3	94.6	105.2	113.8	121.5	128.5	134.7	140.7	146.8	150.2	154.9	159.7
38	65.0	81.0	92.3	104.4	111.5	119.2	126.2	132.5	138.4	149.9	149.4	154.0	158.6
38.80–30°B	64.5	80.5	91.8	102.5	111.0	118.7	125.7	132.0	138.1	143.6	149.3	153.9	158.3

% of NH_3 by weight and Deg. Beaumé	65	70	75	80	85	90	95	100	105	110	115	120	125
17.	227.2	231.8	235.4	239.3	242.8	246.3	244.5	252.7	255.8	258.8	261.7	264.5	267.0
18.93–20°B	222.5	227.0	231.0	234.1	238.4	241.8	245.1	248.3	251.4	254.0	257.3	260.1	262.6
19.	218.8	223.1	227.0	230.9	234.0	237.9	241.1	244.4	247.4	250.5	253.4	256.1	258.7
19.87–21°B	215.6	219.8	223.8	227.6	231.2	234.6	237.9	241.1	244.2	247.2	250.1	252.9	255.6
20.	215.2	219.3	223.4	227.1	230.7	234.1	237.4	240.5	243.8	246.7	249.8	252.0	255.0
21.	211.5	215.6	219.7	223.3	227.0	230.4	233.7	237.0	240.1	243.0	245.9	248.7	251.3
21.75–22°B	208.4	212.6	216.6	220.4	224.0	227.4	230.7	233.9	237.0	240.0	242.9	245.7	248.2
22.	207.5	211.7	215.2	219.5	223.1	226.5	229.8	233.0	236.1	239.1	242.0	244.8	247.3
23.03–23°B	203.3	207.5	211.5	215.3	218.9	222.3	225.6	228.8	231.9	234.9	237.8	240.6	243.1
24.	200.1	204.1	208.3	212.1	215.7	219.1	222.4	225.6	228.7	231.7	234.6	237.4	240.0
24.99–24°B	196.5	200.7	204.7	208.5	212.1	215.5	218.8	222.0	225.1	228.1	231.0	233.8	236.3
26.	193.3	197.5	201.6	205.3	208.9	212.2	215.6	218.9	221.9	225.4	227.8	230.6	233.1
27.	190.2	194.3	198.4	202.2	205.7	209.0	212.5	215.2	218.7	221.8	234.1	227.4	230.0
27.64–25°B	187.6	191.8	195.8	199.6	203.2	206.6	209.9	213.1	216.2	219.2	222.1	224.9	227.4
28	186.6	190.7	194.8	198.5	202.2	205.6	208.8	212.1	215.1	218.2	221.0	223.9	226.3
29.	183.5	187.6	191.8	195.4	199.1	202.6	205.7	209.0	212.1	215.1	217.9	220.9	223.3
29.60–26°B	181.4	185.6	189.6	193.6	197.0	200.4	203.7	206.9	210.0	213.0	215.9	218.7	221.2
30.	180.2	184.4	188.4	192.2	195.6	199.2	202.5	205.7	208.8	211.8	214.7	217.8	220.0
31.03–27°B	177.0	181.7	185.2	189.0	192.6	196.0	199.3	202.3	206.6	209.6	212.3	215.3	217.8
32.	174.4	178.6	182.6	186.4	190.0	193.4	196.7	199.9	204.0	207.0	209.9	212.7	215.4
33.	171.7	175.4	179.9	183.7	187.3	190.7	194.0	197.2	201.3	204.3	207.2	210.0	213.5
33.25–28°B	170.8	175.0	179.0	182.8	186.4	189.8	193.1	196.3	200.4	203.4	206.3	209.1	211.6
34	168.9	173.1	177.1	180.9	184.5	187.4	191.2	195.4	198.5	201.5	204.4	207.2	209.7
35	166.3	170.5	174.5	178.3	181.9	185.5	188.6	192.8	195.9	198.9	201.8	204.6	207.1
35.60–29°B	164.7	168.9	172.9	176.7	180.3	183.7	187.0	191.2	194.3	197.3	200.2	203.0	205.5
36.	164.5	168.7	172.7	178.5	180.1	183.5	186.8	191.0	193.9	196.9	199.8	202.6	205.1
37.	163.7	167.9	171.9	175.8	179.3	182.7	186.0	190.2	192.8	195.8	198.7	201.5	204.0
38	162.9	167.1	171.1	175.0	178.5	181.9	185.2	189.4	191.7	194.7	197.6	200.4	202.9
38.80–30°B	162.6	167.0	171.0	174.3	178.4	181.8	185.0	188.3	191.4	194.0	197.3	200.1	202.6

% of NH_3 by weight and Deg. Beaumé	130	135	140	145	150	155	160	165	170	175	180	F.P.
17.	269.6	272.1	274.5	276.9	279.0	281.1	283.1	285.1	287.1	288.3	290.4	
18.93–20°B	265.2	267.7	270.1	272.4	274.6	276.7	278.7	280.7	282.6	284.3	285.9	
19.	261.2	263.8	266.0	268.5	270.6	272.8	274.7	276.3	278.6	280.4	282.0	
19.87–21°B	258.0	260.5	262.9	265.2	267.0	269.5	271.8	273.1	275.0	277.1	278.7	
20.	257.5	260.1	262.4	264.5	266.9	269.0	271.1	273.0	275.0	276.6	278.2	
21.	253.8	256.4	259.7	261.0	263.2	265.3	267.8	269.7	271.3	273.9	278.5	
21.75–22°B	250.8	253.3	255.7	258.0	260.2	262.3	264.9	266.3	268.2	269.8	271.5	
22.	249.4	252.0	254.8	257.1	259.3	261.8	263.4	265.4	267.3	269.0	270.6	
23.03–23°B	245.7	248.2	250.6	252.9	255.1	257.3	259.2	261.1	263.1	264.8	266.4	
24.	242.8	246.0	247.4	249.7	251.9	254.0	256.0	257.4	259.2	261.5	263.1	
24.99–24°B	238.4	241.4	243.8	246.1	248.3	250.6	252.4	254.4	256.3	258.0	259.6	
26.	235.8	238.3	240.6	242.4	245.2	247.3	249.3	251.2	253.1	254.9	256.5	
27.	232.6	235.2	237.4	239.8	242.1	244.2	246.1	248.0	249.9	251.7	253.4	
27.64–25°B	230.0	232.5	234.4	237.2	239.4	241.5	243.5	245.3	247.2	248.9	250.5	
28	229.0	231.4	233.4	236.1	238.4	240.4	242.6	244.3	246.1	247.8	246.5	
29.	225.9	228.4	230.8	233.0	235.4	237.3	239.4	241.3	243.1	240.8	243.5	
29.60–26°B	223.8	226.3	228.7	231.0	233.2	235.3	237.3	239.3	241.2	242.9	244.5	
30.	222.6	225.1	227.3	229.8	232.0	234.0	236.1	238.1	240.0	241.7	243.3	
31.03–27°B	220.6	222.9	225.3	227.6	229.8	231.4	233.4	235.8	237.1	239.2	240.0	
32.	217.8	220.3	222.7	225.0	227.2	229.8	231.3	233.2	234.9	236.4	238.2	
33.	215.1	217.6	220.0	222.7	224.5	226.6	228.0	230.5	232.1	233.8	236.5	
33.25–28°B	214.2	216.7	219.1	221.8	223.6	225.6	227.5	229.3	231.0	232.8	234.4	
34	212.3	214.3	217.2	219.4	221.7	223.7	225.6	227.4	229.1	230.9	232.5	
35	209.7	212.2	214.6	217.3	219.1	221.1	223.1	224.8	226.8	228.3	231.3	
35.60–29°B	208.1	210.6	213.0	215.7	217.5	210.5	221.4	223.3	224.0	226.6	230.3	
36.	207.7	210.2	212.6	215.2	217.1	219.2	220.9	221.9	224.6	226.4	229.0	
37.	206.6	209.4	211.5	214.0	216.0	218.8	219.7	221.1	224.1	225.8	227.9	
38	205.5	208.0	210.4	213.8	214.9	217.4	218.5	220.5	222.2	224.1	225.7	
38.80–30°B	205.3	207.7	210.1	212.8	214.6	216.7	219.1	220.1	222.0	223.7	225.3	

DEGREES FAHRENHEIT

BEAUMÉ SCALES.

It should be noted that there are three Beaumé specific gravity scales, or hydrometers; one of liquids which are heavier than water, and two for liquids lighter than water. Of the latter two the scale of the one designates pure water 10, and the other designates pure water zero. As ammonia liquor (comprising mixtures of water and ammonia in all proportions) is lighter than water, only the latter two Beaumé scales come into question in this respect, and generally the one which designates pure water 10 is referred to when mentioned in connection with ammonia liquor, and the degrees given in this connection correspond to a certain specific gravity, *i. e.*, to a certain percentage of water and ammonia contained in the ammonia liquor as shown in the table on page 97.

SATURATED SOLUTION OF AMMONIA.

The amount of ammonia which can be absorbed by water decreases with the temperature, as is shown in the following table.

SOLUBILITY OF AMMONIA IN WATER AT DIFFERENT TEMPERATURES (ROSCOE).

Degrees Celsius.	Degrees Fahrenheit.	Pounds of NH_3 to one pound water.	Degrees Celsius.	Degrees Fahrenheit.	Pounds of NH_3 to one pound water.
0	32.	0.875	28	82.4	0.426
2	35.6	0.833	30	86.	0.403
4	39.2	0.792	32	89.6	0.382
6	42.8	0.751	34	93.2	0.362
8	46.4	0.713	36	96.8	0.343
10	50.	0.679	38	100.4	0.324
12	53.6	0.645	40	104.0	0.307
14	57.2	0.612	42	107.6	0.290
16	60.8	0.582	44	111.2	0.275
18	64.4	0.554	46	114.8	0.259
20	68.	0.526	48	118.4	0.244
22	71.6	0.499	50	122.	0.229
24	75.2	0.474	52	125.6	0.214
26	78.8	0.449	54	129.2	0.200
			56	132.8	0.186

The heat H_n developed when one pound of ammonia is dissolved in as much poor liquor containing one pound of ammonia to n pound of water, in order to obtain a rich liquor which will contain $b+1$ pound of ammonia for each n pound of water (see pages 101 and 102) is—

$$H_n = 925 - \frac{284 + 142b}{n} \text{ units.}$$

The figures in the following table on the solubility of ammonia in water at different temperatures have been obtained by Sims:

Degrees Fahr.	Lb. of NH_3 to 1 lb. of Water.	Volume of NH_3 in 1 Volume of Water.	Degrees Fahr.	Lb. of NH_3 to 1 lb. of Water.	Volume of NH_3 in 1 Volume of Water.
32.0	0.899	1,180	125.6	0.274	359
35.6	0.853	1,120	129.2	0.265	348
39.2	0.809	1,062	132.8	0.256	336
42.8	0.765	1,005	136.4	0.247	324
46.4	0.724	951	140.0	0.238	312
50.0	0.684	898	143.6	0.229	301
53.6	0.646	848	147.2	0.220	289
57.2	0.611	802	150.8	0.211	277
60.8	0.578	759	154.4	0.202	265
64.4	0.546	717	158.0	0.194	254
68.0	0.518	683	161.6	0.186	244
71.6	0.490	643	165.2	0.178	234
75.2	0.467	613	168.8	0.170	223
78.8	0.446	585	172.4	0.162	212
82.4	0.426	559	176.0	0.154	202
86.0	0.408	536	179.6	0.146	192
89.2	0.393	516	183.2	0.138	181
93.2	0.378	496	186.8	0.130	170
96.8	0.303	478	190.4	0.122	160
100.4	0.350	459	194.0	0.114	149
104.0	0.338	444	197.6	0.106	139
107.6	0.326	428	201.2	0.098	128
111.2	0.315	414	204.8	0.090	118
114.8	0.303	399	208.4	0.082	107
118.4	0.294	386	212.0	0.074	97
122.0	0.284	373	....	....	....

HEAT GENERATED BY ABSORPTION OF AMMONIA.

The questions regarding the heat generated by the absorption of ammonia in water, as well as in water containing a certain percentage of ammonia, have been experimentally studied by Berthelot, whose results may be expressed by the following formula:

$$Q = \frac{142}{n} \text{ units.}$$

in which Q stands for the units of heat (pound Fahrenheit) developed when a solution containing one pound of ammonia in n pounds of water is diluted with a great amount of water. This equation fully suffices to solve the different problems arising in refrigerating practice. Assuming 925 units (the values of different experimenters differ) of heat to be developed when one pound of ammonia is absorbed by a great deal (say 200 pounds) of water, the amount of heat, Q, developed in making solutions of different strengths (one pound of ammonia to n pounds of water) may be expressed by the formula—

$$Q_1 = 925 - \frac{142}{n} \text{ units.}$$

The heat, Q_2, developed when b pounds of ammonia are added to a solution containing one pound of ammonia to n pounds of water, is expressible by the formula:

$$Q_2 = 925b - \frac{142\,(2b + b^2)}{n} \text{ units.}$$

Let the poor liquor enter the absorber with a strength of 10 per cent, which is equal to one pound of ammonia to nine (n) pounds of water. Let the rich liquor leave the absorber with a strength of 25 per cent, which is three ($1+b$) pounds of ammonia per nine (n) pounds of water. Inserting these values, $n=9$ and $b=2$, in the above equation, we have—

$$Q_2 = 925 \times 2 - \frac{142\,(4+4)}{9} = 1724 \text{ units.}$$

Hence by dissolving two pounds of ammonia gas or vapor in a solution of one pound of ammonia in nine pounds of water, we obtain twelve pounds of a 25 per cent solution, and the heat generated is 1,724 B. T. units.

SOLUBILITY OF AMMONIA IN WATER AT DIFFERENT TEMPERATURES AND PRESSURES. (SIMS.)

One Pound of Water (also Unit Volume), Absorbs the Following Quantities of Ammonia.

Absolute Pr's'ure in Lbs. per Sq. Inch.	32° F.		68° F.		104° F.		212° F.	
	Lbs.	Vols.	Lbs.	Vols.	Lbs.	Vols.	Gr'ms.	Vol
14.67	0.899	1.180	0.518	.683	0.338	.443	0.074	.97
15.44	0.937	1,231	0.535	.703	0.349	.458	0.078	.102
16.41	0.980	1.287	0.556	.730	0.363	.476	0.083	.109
17.37	1.029	1.351	0.574	.754	0.378	.496	0.088	.115
18.34	1.077	1.414	0.594	.781	0.391	.513	0.092	.120
19,30	1.126	1.478	0.613	.805	0.404	.531	0.096	.126
20,27	1.177	1.546	0.632	.830	0.414	.543	0.101	.132
21.23	1.236	1.615	0.651	.855	0.425	.558	0.106	.139
22.19	1.283	1.685	0.669	.878	0.434	.570	0.110	.140
23.16	1.336	1.754	0.685	.894	0.445	.584	0.115	.151
24.13	1.388	1.823	0.704	.924	0.454	.596	0.120	.157
25.09	1.442	1.894	0.722	.948	0.463	.609	0.125	.164
26.06	1.496	1.965	0.741	.973	0.472	.619	0.130	.170
27.02	1.549	2.034	0.761	.999	0.479	.629	0.135	.177
27.99	1.603	2.105	0.780	1.023	0.486	.638	....	....
28.95	1.656	2.175	0.801	1.052	0.493	.647	....	....
30.88	1.759	2.309	0.842	1.106	0.511	.671	....	....
32.81	1.861	2.444	0.881	1.157	0.530	.696	....	......
34.74	1.966	2.582	0.919	1.207	0.547	.718	....	......
36.67	2.070	2.718	0.955	1.254	0.565	.742	....	......
38.60	......	......	0.992	1.302	0.579	.764	....	....
40.53	......	......	....	....	0.594	.780	....	....

The ammonia does not follow the absorption laws of Dalton, inasmuch as the quantity of ammonia absorbed by water does not vary directly with the pressure.

DIFFERENT SYSTEMS OF REFRIGERATION.

Both the anhydrous liquor and the ammonia are used in refrigeration, the former in what is known as the Linde or compression system, and the latter in the Carré or absorption system.

TESTS FOR AMMONIA.

As the boiling point of pure anhydrous ammonia is at 29° below zero at a pressure of the atmosphere (30 inches of mercury), the purity of anhydrous ammonia may be tested by means of an accurate thermometer. The same is inserted into a flask containing the ammonia in a boiling condition, and provided with a tube to carry off the obnoxious vapor. If the boiling temperature differs materially from the above (allowance being made for the barometric pressure), it demonstrates that the ammonia is not pure. If after the ammonia is evaporated, an oily or watery residue is left in the flask, the same is also attributable to impurities. Ammonia leaks are generally easily detected by the smell or by the white fumes which form when a glass rod moistened with hydrochloric acid is passed by the leak.

If traces of ammonia are to be detected in water or in brine it is best to use "Nessler's Reagent," which is prepared as follows:

Dissolve 17 grams of mercuric chloride in about 300 cc. of distilled water; dissolve 35 grams of potassium iodide in 100 cc. of water; add the former solution to the latter, with constant stirring, until a slight permanent red precipitate is produced. Next dissolve 120 grams of potassium hydrate in about 200 cc. of water; allow the solution to cool; add it to the above solution, and make up with water to one liter, then add mercuric chloride solution until a permanent precipitate again forms; allow to stand till settled, and decant off the clear solution for use; keep it in glass stoppered blue bottles, and set away in a dark place to keep it from decomposing.

The application of this reagent is very simple, a few drops of the same being added to the water or brine in question, contained in a test tube or a small glass of any other kind. If the smallest trace of ammonia is present a yellow precipitation of the liquid will take place, which turns to a full brown when the quantity of ammonia present is larger.

TESTING AMMONIA.

The purity of anhydrous ammonia is practically tested by allowing the same to evaporate from a flask placed in water and provided with a cork and bent tube to carry off the obnoxious water. If after the evaporation a notable oily or watery residue is left it is attributable to impurities. The boiling point may be observed at the time (it is 29-30° F. below zero), and if any permanent gases are given off when the tube carrying off the ammonia vapor is discharged into water they may be tested for their inflammability. However, these latter two tests will hardly prove satisfactory except in the hands of an experienced chemist.

In order to test the liquid residue in anhydrous ammonia, Faurot used a glass tube about six and one-half inches deep and one and one-eighth inches in diameter, and drawn out to a narrow tube at the bottom, the latter being divided in fractions of a centimeter, while the whole tube contains about 100 cubic centimeters. The open top may be closed with a rubber cork having a vent tube of glass, the outer portion of which is bent down close to the large tube, so that the whole may be placed in a glass of water after the tube has been filled to about half with the anhydrous ammonia to be tested. The ammonia will now boil away and be absorbed by the water in which the vent tube dips, and the amount or percentage of any residue that may be left can be readily estimated by the readings on the graduated portion of the tube. Permanent gases in the ammonia will manifest themselves by bubbles passing through the water.

Ammonia liquor is tested for its strength by the hydrometer, as shown. For chemical tests it should be diluted with two times its volume of distilled water when, after acidification with hydrochloric acid, the addition of chloride of barium solution will show the presence of sulphates by a white precipitate. In the same diluted ammonia liquor clear lime water will show the presence of carbonates by a similar precipitation. Chlorides may be detected by acidifying the diluted ammonia solution with nitric acid and the addition of nitrate of silver solution by the formation of white precipitate. If on the addition of nitric acid to the ammonia a red color appears it indicates traces of organic bases.

CHAPTER III.—WATER, STEAM, ETC.

Water is a combination of one atom of oxygen with two atoms (one molecule) of hydrogen, consequently to be designated by H_2O, which means that two parts by weight of hydrogen are combined with sixteen parts by weight of oxygen to form eighteen parts (one molecule) of water.

FORMATION OF ICE.

Water solidifies at 32° F., but in very fine capillary tubes the freezing point may be depressed for 20° or more. If rigidly confined or placed under pressure, the freezing point is depressed likewise. For a pressure of n atmospheres the freezing point is depressed for $n \times 0.0135°$ F. Latent heat of ice, 142 B. T. units.

PROPERTIES OF ICE.

The ice which freezes out of solutions of salt or other substance, consists of pure water, the impurities remaining in the unfrozen portion. Ice melts at 32° F., but by a pressure sufficiently high it can be converted into liquid at a temperature of 4° F. One cubic foot of ice weighs 998.74 ounces, avoirdupois.

STEAM.

Water volatilizes like any other liquid in accordance with the tension of its vapor, which at a temperature of 212° is equal to the tension of the atmosphere when the water boils, and is converted into steam, which occupies about 1,700 times the volume of the water. The water dissociates completely at a temperature of about 4.500°, but a partial decomposition takes place at a lower temperature.

SATURATED STEAM.

When steam is still in connection with water, or if it is in such condition that a slight decrease of temperature will cause liquefaction of some of the steam, it is called saturated steam.

The pressure of saturated steam depends on its temperature in a manner approximately expressed by Rankine's formula:

$$log.\, p = A - \frac{B}{T} - \frac{C}{T_2}$$

In which p is the pressure in pounds per square inch at the absolute temperature T in degrees F., the value of constants being: $A = 6.1007$, $log.\, B = 3.43642$, $log.\, C = 5.69873$.

TOTAL HEAT.

By total heat of steam we understand that quantity of heat required to raise the temperature of unit weight of water from the freezing point to any given temperature, and to entirely evaporate it at that temperature. The total heat, l, for any temperature, t, may be expressed by the formula:

$$l = 1091.7 + 0.305\,(t-32)$$

LATENT HEAT OF VAPORIZATION.

If the heat of the liquid, g (*i. e.*, the amount of heat required to raise the temperature of unit weight of water from the freezing point to the temperature t) is subtracted from the total heat, l, at that temperature, we find the heat of volatilization, h, viz.:

$$h = l - g$$

EXTERNAL LATENT HEAT.

That portion of the latent heat required to overcome external pressure, or the external latent heat, E, is expressed by—

$$E = \frac{P(v-v_1)}{J}$$

In which formula P stands for external pressure, v for the volume of the saturated vapor, v_1 for the volume of the liquid, and J for the mechanical equivalent of heat.

INTERNAL LATENT HEAT.

The heat required to bring about the change from the liquid to the gaseous state, *i. e.*, to perform the work of disintegration, or the so-called internal latent heat, F, is expressed by the equation—

$$F = h - E$$

SPECIFIC HEAT OF WATER.

The specific heat, c, of water at any temperature, t (expressed in degrees Celsius), is—

$$c = 1 + 0.00004\,t + 0.000000\,t^2$$

See also table, page 16.

SPECIFIC HEAT OF STEAM.

The specific heat of superheated steam is 0.3643 at constant volume and 0.475 at constant pressure. The specific heat of saturated steam, s, is expressed by the equation—

$$s = 1 - \frac{1436.8}{T}$$

which is negative for all values of T less than 1436° F., above absolute zero.

SPECIFIC HEAT OF ICE.

The specific heat of ice is about half of that of water, or 0.504.

PROPERTIES OF SATURATED STEAM, AT PRESSURE FROM ONE POUND TO 200 POUNDS ON THE SQUARE INCH.

Pressure Absolute.		Heat, in Degrees, Fahr.					Volume, that of an Equal Weig't of Water at Its Greatest Density Being 1.	Weight of One Cubic Foot in Decimals of a Pound.	Specific Gravity, the Atmosphere at 32° Being 1.
[illegible] one Sq. Inch.	In Inches of Mercury at 32°.	Temperature.	Dif. pr lb	Latent Heat.	Total Heat.	Dif. pr lb			
1	2.0375	102.	....	1,043.05	1,145.05	....	20,890	.0029	.037
5	10.1875	162.37	9.26	1,001.9	1,163.46	2.82	4,627	.0135	.167
10	20.375	193.29	4.93	979.00	1,172.89	1.50	2,429	.0257	.318
15	30.5625	213.07	3.47	965.85	1,178.92	1.05	1,669	.0373	.463
20	40.75	228.	2.8	955.5	1,183.5	.8	1,280	.0487	.604
25	50.9375	240.2	2.3	947.	1,187.2	.7	1,042	.0598	.742
30	61.125	250.4	2.	939.9	1,190.3	.6	881	.0707	877
35	71.3125	259.3	1.7	933.7	1.193.	.5	764	.0815	1.012
40	81.5	267.3	1.5	928.1	1,195.4	.4	676	0921	1.142
45	91.6875	274.4	1.4	923.2	1,197.6	.4	608	.1025	1.272
50	101.875	281.	1.3	918.6	1,199.6	.4	552	.1129	1.402
55	112.0625	287.1	1.2	914.4	1,201.5	.4	506	.1232	1.529
60	122.25	292.7	1 1	910.5	1,203.2	.3	467	.1335	1.654
65	132.4375	298.	1.1	906.8	1,204.8	.3	434	.1436	1.779
70	142.625	302.9	1.	903.4	1,206.3	.3	406	.1536	1.904
75	152.8125	307.5	.9	900.3	1,207.8	.3	381	.1636	2.029
80	163.	312.	.9	897.1	1,209.1	.2	359	.1736	2.151
85	173.1875	316.1	.8	894.3	1,210.4	.3	340	.1833	2.271
90	183.375	320.2	.8	891.4	1,211.6	.2	323	.1930	2.391
95	193.5625	324.1	.8	888.7	1,212.8	.3	307	.2030	2.511
100	203.75	327.8	.7	886.1	1,213.9	.2	293	.2127	2.631
105	213.9375	331.3	.7	883.7	1,215.0	.2	281	.2224	2.751
110	224.125	334.6	.6	881.4	1,216.0	.2	269	.2319	2.871
115	234.3125	338.	.6	879.	1,217.0	.2	259	.2410	2.990
120	244.5	341.1	.6	876.9	1,218.0	.2	249	.2503	3.105
125	254.6875	344.2	.6	874.7	1,218.9	.2	239	.2598	3.227
130	234.875	347.2	.6	872.6	1,219.8	.2	231	.2693	3.347
135	275.0625	350.	.5	870.7	1,220.7	.1	223	.2788	3.467
140	285.25	352.9	.6	868.6	1,221.5	.1	216	.2883	3.582
145	295.4375	355.6	.6	866.8	1,222.4	.2	209	.2978	3.697
150	305.625	358.3	.5	864.9	1,223.2	.2	203	.3073	3.809
155	315.8125	360.9	.5	863.1	1,224.	.2	196	.3168	3.927
160	326.	363.4	.5	861.4	1,224.8	.2	191	.3263	4.042
165	336.1875	365.9	.5	859.7	1,225.6	.2	186	.3353	4.157
170	346.375	368.2	.4	858.1	1,226.3	.2	181	.3443	4.270
175	356.5625	370.6	.5	856.4	1,227.	.1	176	.3533	4.383
180	366.75	372.9	.4	854.8	1,227.7	.1	172	.3623	4.495
185	376.9375	375.3	.5	853.1	1,228.4	.1	168	.3713	4.607
190	387.125	377.5	.4	851.6	1,229.1	.1	164	.3800	4.720
195	396.3125	379.7	.4	850.1	1,229.8	.2	160	.3888	4.832
200	407.5	381.7	.3	848.6	1,230.3	.1	157	.3973	4.945

SPECIFIC VOLUME OF STEAM.

The specific volume, v, of steam, in accordance with the experiments of Tate and Fairbairn, may be expressed by the formula—

$$v = 25.62 + \frac{49513}{P + 0.72}$$

VOLUME AND WEIGHT OF WATER.

The volume of water does not change in direct proportion with the temperature, its greatest density being at 39° F., at which one cubic foot weighs 62.425 pounds. At 32° it weighs 62.418, at 62° it weighs 62.355, and at the boiling point it weighs 59.640 pounds. One cubic foot of water is generally taken at 62.5 pounds = 7.48 U. S. gallons; one cubic inch of water = .036 pounds; one cubic foot of water = 6.2355 imp. gallons, or 7.48 U. S. gallons; one U. S. gallon of water = 8.34 pounds; one U. S. gallon of water = 231 cubic inches.

PRODUCTION OF STEAM.

The economical production of steam for industrial purposes is chiefly a question of fuel and the proper construction of boilers, grates, etc., and has been alluded to in the chapter on heat under the headings relating to fuel. For satisfactory arrangements as to boilers, etc., it may be assumed that one pound of fair average coal will produce about eight pounds of steam, more or less.

WORK DONE BY STEAM.

The theoretical ability of steam to do a certain amount of work is governed by the laws of thermodynamics above set forth, and the practical yield depends on a great many details in the mode of applying the force of steam practically, the consideration of which is beyond the limits of this treatise. For rough estimates, it is assumed that it requires from fifteen to thirty pounds of steam to produce a horse power, according to perfection of engine, per hour.

HEATING AREA OF BOILER.

If H is the nominal horse power of a boiler and A the effective heating area of the same, Box finds that—

$$A = 8\,(H + 2.5\,H)$$

A nominal horse power requires from 0.6 to 1.2 square feet of grate surface between the limits of sixty and three horse powers.

PRIMING.

The water which is mechanically drawn over from the boiler with the steam is called priming, and may be determined in the following manner given by Clark. Blow a quantity of the steam, the amount of priming in which it is desired to ascertain, into a vessel holding a given weight of cold water, noting the pressure and the weight of the steam blown in, and the initial and final temperatures of the mixture. An addition is to be made to the initial weight of water, to represent the weight of water equivalent to that of the vessel containing the water, in terms of their respective specific heats. A corresponding addition is to be made for such portion of the apparatus as is immersed in the water.

Let W= weight of condensing water, plus the equivalent weight of the receiver and apparatus immersed in the water.

w= weight of nominal steam discharged into the vessel under water.

$W+w$= gross weight of mixture of nominal steam and condensing water.

H= total heat of one pound of the steam, reckoned from the temperature of the condensing water.

Hw= total heat delivered by the gross weight of nominal steam discharged, taken as dry steam.

t= initial temperature of condensing water.

t'= final temperature of condensing water.

s= augmentation of specific heat of water due to rise of temperature.

L= latent heat of one pound of steam of the given initial pressure.

Lw= latent heat of steam discharged into the vessel, taking it as dry steam.

P= weight of priming or moisture in percentage of the gross weight of nominal steam.

$$P=100\frac{Hw-[(W+w)\times(t'-t+s)]}{Lw}$$

FLOW OF STEAM.

The flow of steam through pipes takes place according to Babcock after the following equation:

$$W=300\sqrt{\frac{D(p_1-p_2)d^5}{L\left(1+\frac{3.6}{d}\right)}}$$

In which formula W is the weight of steam in pounds which will flow per minute through a pipe of the length L in feet and the diameter d in inches, when p_1 is the initial pressure, p_2 the pressure at end of pipe, and D the density or weight per cubic foot of the steam.

Steam of a pressure of fifteen pounds per square inch (gauge pressure) flows into vacuum with a speed of 1,550 feet per second, and into air with a speed of 650 feet per second.

HYGROMETRY.

Hygrometry is the art of measuring the moisture contained in the atmosphere, or of ascertaining the hygrometric condition of the latter.

AIR SATURATED WITH MOISTURE.

The amount of aqueous vapor which can be held by a given volume of air increases with the temperature and decreases with the pressure. The air is called saturated with moisture when it contains all the moisture which it can contain at that temperature. The degree of saturation or hygrometric state of the atmosphere is expressed by the ratio of the aqueous vapor actually present in the air to that which it would contain if it were saturated. In accordance with Boyle's law the degree of saturation may also be expressed by the ratio of the elastic force of the aqueous vapor which the air actually contains to the elastic force of vapor which it would contain if saturated.

ABSOLUTE MOISTURE.

The absolute moisture is the quantity of aqueous vapor by weight contained in unit volume of air.

DEW POINT.

When the temperature of air containing moisture is lowered a point will be reached at which the air is saturated with moisture for that temperature, and a further lowering of temperature will result in the liquefaction of some of the moisture. This temperature is called the dew point.

DETERMINATION OF MOISTURE.

The moisture in the atmosphere may be determined by a wet bulb thermometer, which is an ordinary thermometer, the bulb of which is covered with muslin kept wet, and which is exposed to the air the moisture of which is to be ascertained. Owing to the evaporation of the water on the muslin the thermometer will shortly acquire a stationary temperature which is always lower than that of the surrounding air (except when the latter is actually saturated with moisture). If t is the temperature of the atmosphere and t_1 the temperature of the wet bulb thermometer in degrees Celsius, the tension, e, of the aqueous vapor in the atmosphere is found by the formula—

$$e = e_1 - 0.00077\,(t - t_1)\,h_1$$

e_1 being the maximum tension of aqueous vapor for the temperature t_1 as found in table, and h the barometric height in millimeters.

If e_2 is the maximum tension of aqueous vapor for the temperature t, the degree of saturation, H, is expressed by—

$$H = \frac{e}{e_2}$$

and the dew point is also readily found in the same table, it being the temperature corresponding to the tension e.

TABLE SHOWING THE TENSION OF AQUEOUS VAPOR IN MILLIMETERS OF MERCURY, FROM —30° C. TO 230° C.

Temp.	Tension.	Temp.	Tension.	Temp.	Tension.	Temp.	Tension.
—30	.39	21	18.5	94	610.4	105	907
—25	.61	22	19.7	94.5	622.2	107	972
—10	.9	23	20.9	95	633.8	110	1,077
—15	1.4	24	22.7	95.5	645.7	115	1,273
—10	2.1	25	25.6	96	657.5	120	1,491
— 5	3.1	26	25.0	96.5	669.7	125	1,744
— 2	4.0	27	26.6	97	682.0	130	2,030
— 1	4.3	28	28.1	97.5	694.6	135	2,354
0	4.6	29	29.8	98	707.3	140	2,717
1	4.95	30	31.6	98.5	721.2	145	3,125
2	5.3	35	41.9	99	732.2	150	3,581
3	5.7	40	55.0	99.1	735.9	155	4,088
4	6.1	45	71.5	99.2	738.5	160	4,551
5	6.5	50	92.0	99.3	741.2	165	5,274
6	7.0	55	117.5	99.4	743.8	170	5,961
7	7.5	60	148.0	99.5	746.5	175	6,717
8	8.0	65	186.0	99.6	749.2	180	7,547
9	8.6	70	232.0	99.7	751.9	185	8,453
10	9.1	75	287.0	99.8	754.6	190	9,443
11	9.7	80	354.0	99.9	757.3	195	10,520
12	10.4	85	432.0	100	760	200	11,689
13	11.1	90	525.4	100.1	762.7	205	12,956
14	11.9	90.5	535.5	100.2	765.5	210	14,325
15	12.7	91	545.8	100.4	772.0	215	15,801
16	13.5	91.5	556.2	100.6	776.5	220	17,39[illegible]
17	14.4	92	566.2	101	787.0	225	19,097
18	15.3	92.5	577.3	102	816	230	20,926
19	16.3	93	588.4	103	845	......	.....
20	17.4	93.5	599.5	104	876	......	....

Degrees C	120	134	144	152	159	171	180	199	213	225
Atmospheres	2	3	4	5	6	8	10	15	20	25

PSYCHROMETERS.

Instead of the wet bulb thermometer alone it is more convenient to use two exact thermometers combined (one with a wet bulb and the other with a dry bulb, to give the temperature of the air) to determine the hygrometric condition of the atmosphere or of the air in a room. Instruments on this principle can be readily bought, and are called psychrometers. If they are arranged with a handle, so that they can be whirled around, they are called "sling psychrometers." These permit a quicker correct reading of the wet bulb thermometer than the plain psychrometer, in which the thermometers are stationary and are impracticable at a temperature below 32° F., while the sling instrument can be read down to 27° F.

The following table can be used to ascertain the degree of saturation or the relative humidity:

RELATIVE HUMIDITY—PER CENT.

t (Dry Ther.)	Difference between the dry and wet thermometers ($t-t'$).												t (Dry Ther.)
	0°.5	1°.0	1°.5	2°.0	2°.5	3°.0	3°.5	4°.0	4°.5	5°.0	5°.5	6°.0	
28	94	88	82	77	71	65	60	54	49	43	38	33	28
29	94	89	83	77	72	66	61	56	50	45	40	35	29
30	94	89	84	78	73	67	62	57	52	47	41	36	30
31	95	89	84	79	74	68	63	58	53	48	43	38	31
32	95	90	84	79	74	69	64	59	54	50	45	40	32
33	95	90	85	80	75	70	65	60	56	51	47	42	33
34	95	91	86	81	75	72	67	62	57	53	48	44	34
35	95	91	86	82	76	73	69	65	59	54	50	45	35
36	96	91	86	82	77	73	70	66	61	56	51	47	36
37	96	91	87	82	78	74	70	66	62	57	52	48	37
38	96	92	87	83	79	75	71	67	63	58	54	50	38
39	96	92	88	83	79	75	72	68	63	59	55	52	39
40	96	92	88	84	80	76	72	68	64	60	56	53	40

The hygrometer of Marvin is a sling psychrometer of improved and approved construction.

HYGROMETERS.

While the term hygrometer applies to all instruments calculated to ascertain the amount of moisture in the air, it is specifically used to design instruments on which the degree of humidity can be read off directly on a scale without calculation and table. Their operation is based on the change of the length of a hair or similar hygroscopic substance under different conditions of humidity.

DRYING AIR.

To remove moisture from air more or less saturated with it, certain so called hygroscopic substances which have a great affinity for water may be applied. Chloride of calcium, dried at a dull red heat and powdered, may be used for this purpose, and when spread in a layer ⅝-inch thick and exposed to air at 48° F., with a humidity of 0.75, will absorb per square foot surface in each one of seven succeeding days the following amounts of moisture: 1,368, 1,017, 958, 918, 900, 802 and 703 grains respectively (Box).

VAPORIZATION.

The vaporization of water into the air depends on the hygrometric state of the atmosphere, and its amount in grains, R, per square foot and per hour with air perfectly calm, may be expressed according to Box by the following rule:

$$R = (e_2 - e)\,15$$

When the air into which the water evaporates is in motion the evaporation proceeds much faster, thus: For a fresh breeze—

$$R = (e_2 - e)\,66$$

for a strong wind—

$$R = (e_2 - e)\,132$$

and for a gale—

$$R = (e_2 - e)\,188.$$

The refrigeration which is produced by the vaporization of water into the air is about 900 B. T. units for each pound of water evaporated, or 0.117 units per grain of water evaporated.

PURITY OF WATER.

As natural water is never absolutely pure it is frequently of importance to ascertain the degree of purity of a water for certain purposes. The requirements to be made in regard to the purity of a water vary with the purposes for which it is to be used; water may be very good for drinking purposes, but at the same time it may be too hard for boiler feeding; and on the other hand a water may be good for boiler feeding, yet it may be too impure (bacteriologically) for drinking purposes. Similar distinctions obtain in other respects, so that it is impracticable to give general rules for the valuation of a water, unless they are based on an exact chemical analysis of the same. The crude chemical tests which are frequently recommended in this connection are of little or no value in most cases, and more frequently they are misleading. They generally only give qualitative indications, but in order to be able to judge a water correctly the relative quantities of its constituents must be known.

CHAPTER IV.—THE AMMONIA COMPRESSION SYSTEM.

GENERAL FEATURES.

The refrigeration in this system is brought about by the evaporation of liquid anhydrous ammonia, which takes place in coils of pipe termed the expander or refrigerating coils. These coils are either placed in the rooms to be refrigerated, or they are immersed in a bath of salt brine, which absorbs the cold. The salt brine is circulated in pipes through the rooms to be refrigerated by means of a pump. The ammonia, after having expanded, is compressed again by means of a compression pump called the compressor into another system of pipes called the condenser. The condenser is cooled off by running water, which takes away from the ammonia in the coils the heat which it has acquired through the compression, as well as the heat which it has absorbed while having evaporated in the expander. Owing to both pressure and withdrawal of heat, the ammonia assumes its liquid form again to pass again into the expander, thus repeating its circulation over and over again.

THE SYSTEM A CYCLE.

The refrigerating contrivance above described embodies a perfect cycle of operations. The working substance, ammonia in this case, returns periodically to its original condition. During each period a certain amount of heat, partly in the refrigerator and partly during compression (from work converted into heat), is added to the working substance and an exactly equivalent amount is abstracted from the working substance in the condenser by the cooling water.

THE COMPRESSOR.

The compressor is a strongly constructed cylinder in which a piston moves to and fro, having a valve through which the expanded ammonia from the refrigerating coils enters and another through which it is forced into the condenser. A double-acting compressor has two valves at each end of the compressor cylinder, and the packing for the piston rod must be made sufficiently long and tight to withstand the pressure of the ammonia. The compressor, like all other parts of the ammonia system. must be made of steel and iron, no copper or brass being admissible.

During the compression stage a certain amount of heat is evolved. If not otherwise stated, it is assumed in the following discussion, that enough heat is removed during compression to keep the vapor always in a saturated condition.

REFRIGERATING EFFECT OF CIRCULATING MEDIUM.

To arrive at numerical values of the quantities involved in the refrigerating process we may first determine the theoretical refrigerating effect, r, of the circulating medium.

If t be the temperature of the condenser, that is, the temperature of the cooling water leaving the condenser; if t_1 be the temperature of the refrigerator, that is, the temperature of the brine leaving the refrigerator; if s is the specific heat of the circulating liquid, and if h_1 is the latent heat of vaporization of one pound of the circulating medium in thermal units at the temperature t_1, we find the refrigerating effect, r, of one pound of the circulating fluid, expressed in thermal units after the following formula:

$$r = h_1 - (t - t_1)\, s$$

The term $(t - t_1)\, s$ represents the refrigeration required to reduce the temperature of the circulating fluid from the temperature t to the temperature t_1.

Practically speaking, the temperature of the ammonia in condenser will always be a few degrees higher than the water leaving the condenser, and the ammonia in refrigerating coil will always be a few degrees (5 to 10) lower than the outgoing brine.

WORK OF COMPRESSOR.

If the cycle of operation was a perfect reversible one, the work required from the compressor for every pound of the liquid circulating would be to lift the amount of heat, r, from the temperature t_1 to the temperature t. As explained already, this is not the case, and the whole amount of heat as represented by the latent heat of vaporization, namely, h_1, is to be lifted by the compressor through the range of temperature indicated. Hence the work theoretically required from the compressor expressed in thermal units, W, is therefore—

$$W = \frac{t - t_1}{T} h_1$$

T representing the temperature of the refrigerator expressed in degrees of absolute temperature ($t_1 + 460$).

HEAT TO BE REMOVED IN THE CONDENSER.

The theoretical number of heat units, D, which would have to be removed by the condenser water per pound of refrigerating fluid in circulation in the system, if the circulating fluid in compressor were always kept in a saturated condition from without by removing the surplus heat, could be expressed as follows:

$$D = h,$$

h being the latent heat of volatilization of one pound of the circulating liquid at the temperature of condenser (t).

The whole amount of heat, D_1, to be removed when including that which would cause superheating of the fluid in compressor, may be theoretically expressed as follows:

$$D_1 = \frac{t - t_1}{T} h_1 + h_1 - s(t - t_1).$$

AMOUNT OF SUPERHEATING.

The amount of heat, S, liable to cause superheating may therefore be expressed by the formula—

$$S = D_1 - D, \text{ or}$$

$$S = \frac{t - t_1}{T} h_1 + h_1 - h - s(t - t_1).$$

COUNTERACTING SUPERHEATING.

The surplus heat in compressor is removed in various ways: by injecting refrigerated oil, by surrounding the compressor with a cold water jacket, or by carrying liquid ammonia into the compressor, etc. While there is no doubt as to the advisability of preventing superheating as much as possible, the theoretical discussions regarding the relative merit of these expedients do not quite agree among themselves, nor with practical experience, and it would appear that besides theoretical considerations certain practical points have some bearing on this question, especially the degree to which the prevention of superheating is effected.

AMOUNT OF AMMONIA IN COMPRESSOR.

The additional amount of liquid ammonia that would have to be carried into the compressor with every pound of ammonia vapor entering the same, in order to keep the latter saturated during compression, may be expressed by the formula—

$$P = \frac{S}{h_1}$$

in which P stands for pounds of liquid ammonia so required.

NET THEORETICAL REFRIGERATING EFFECT.

The ammonia required to keep the vapor saturated in compressor has to be cooled down from the temperature t to the temperature t_1, and the refrigeration is reduced to that extent. Accordingly the net refrigerating effect, r_1, of every pound of circulating liquid volatilized in refrigerator, in case of wet compression is expressed by the formula:

$$r_1 = h_1 - (t - t_1)s - \frac{S}{h_1} s(t - t_1).$$

VOLUME OF THE COMPRESSOR.

The volume of the compressor is expressed by the amount of space through which the piston travels each stroke. If r be the radius of the compressor and b the length of stroke in feet, the active volume of the compressor, V, is—

$$V = r^2 \times b \times 3.145 \text{ cubic feet.}$$

If r and b are expressed in inches the formula would become—

$$V = \frac{r^2 b \times 3.145}{1728} \text{ cubic feet.}$$

CUBIC CAPACITY OF COMPRESSOR.

The cubic capacity of a compressor may be expressed by the amount of space which the piston travels through in one minute, only one way being counted in a single-acting, and both ways being counted for each revolution in a double-acting compressor. If m is the number of revolutions per minute, r the radius and b the length of stroke in feet of a compressor, the capacity of the same, C, if single-acting, is expressed by the formula:

$$C = r^2 \times 3.145 \times b \times m \text{ cubic feet per minute;}$$

if double-acting, it is twice that. If r and b are given in inches, the product must be divided by 1,728 to find C.

CLEARANCE.

As the piston does not exactly touch the cylinder ends, leaving always more or less dead space called *clearance*, the whole of the above capacity is not available on this account, and from 5 per cent to 7 per cent may be deducted from it for clearance. This may be called the reduced capacity of the compressor.

The exact percentage of clearance depends on a number of conditions, and may be approximately determined after the following equation:

$$C_1 = \frac{C\left[V - n\left(\frac{w}{w_1} - 1\right)\right]}{V}$$

In this equation C is the theoretical capacity of a compressor, and C_1 the corrected or reduced capacity in accordance with clearance. V is the volume traversed by piston in each stroke in cubic feet, n the actual clearance space left between piston and cylinder in cubic feet, w and w_1 the weights of equal volumes of ammonia at the pressure in condenser and refrigerator respectively.

REFRIGERATING CAPACITY OF COMPRESSOR.

The refrigerating capacity of a compressor does not alone depend on its cubic capacity, but also on surrounding circumstances, especially the temperature in condenser and refrigerator coils, and can, therefore, not be exactly determined without these data. For rough estimates it may be assumed, however, that under quite frequently prevailing conditions a cubic compressor capacity per minute of four feet will be equivalent to a capacity of one ton refrig. in twenty-four hours. (Fifty-six inches double-acting compressor capacity sixty revolutions.) If C_1 is the reduced compressor capacity per minute (that is, C less clearance) the corresponding refrigerating capacity, R, expressed in tons of refrigeration in twenty-four hours, may be found after the following formula:

$$R = \frac{C_1 \times 36 \times r}{v \times 7{,}100}$$

or approximately—

$$R = \frac{C_1 \times r}{200\, v} \text{ tons.}$$

In this formula v stands for the volume of one pound of ammonia vapor in cubic feet at the temperature of the refrigerator; the sign r stands for the maximum theoretical refrigerating capacity for each pound of ammonia passing the compressor.

The refrigerating capacity of a compressor, expressed in thermal units, R_1, per hour, is—

$$R_1 = \frac{C_1 \times 60 \times r}{v} \text{ units.}$$

AMMONIA PASSING THE COMPRESSOR.

The amount of ammonia, K, in pounds passing the compressor per minute is expressible thus:

$$K = C_1 \times w \text{ pounds,}$$

in which C_1 stands for the reduced compressor capacity per minute and w for the weight of one cubic foot of ammonia vapor at the temperature of the refrigerator or expansion coils.

NET REFRIGERATING CAPACITY.

As the last four formulæ allow for clearance, but not for other losses, it is more convenient and practically sufficiently correct in most cases to substitute in these formulæ C for C_1, and reduce the refrigerating capacity so found by 15 per cent, which should be ample for all losses, and give net refrigerating capacity.

HORSE POWER OF COMPRESSOR.

If $W = \frac{t - t_1}{T} h_1$ (in thermal units) is the power required by the compressor to lift the heat which became latent by the evaporation of one pound of ammonia in refrigerator, as shown before, and if K represents the amount of ammonia vapor entering the compressor per minute, the work to be done by the compressor per minute, W_1, expressed in thermal units, is—

$$W_1 = W \times K \text{ units.}$$

If expressed in foot-pounds, W_2, it is—

$$W_2 = 778\ W \times K \text{ foot-pounds.}$$

And if expressed in horse powers, W_3, it is—

$$W_3 = \frac{788}{33000}\ W \times K = 0.0234\ WK \text{ horse power.}$$

$$W_3 = 0.0234 \frac{t - t_1}{T} h_1 \times C \times w \text{ horse power.}$$

SIZE OF COMPRESSOR.

In order to determine the size of a compressor for a given refrigerating duty it is advisable to reduce the latter to an expression of heat units to be removed per hour; and if the same is understood to represent actual refrigerating capacity, some 15 per cent or more, according to circumstances, should be added for clearance and other losses, and in case the refrigerating capacity is required in the form of manufactured ice it should at least be doubled. The reduced refrigerating duty so obtained we will call r_2, v the volume of one pound of ammonia gas at the temperature of the outgoing brine, r_1 the refrigerating effect of one pound of ammonia for the temperatures employed, V the active volume swept over by the piston in each revolution (two times the volume of compressor if the same is double-acting), and m the number of revolutions per minute. Signs having this meaning, the following equations obtain:

$$Vm = \frac{r_2 v}{60 \times r_1} \text{ cubic feet.}$$

In this case Vm signifies the compressor capacity per minute. If m is given—

$$V = \frac{r_2 v}{60 \times r_1 m} \text{ cubic feet.}$$

If V is given—

$$m = \frac{r_2 v}{60 \times r_1 V} \text{ revolutions.}$$

NUMBER OF REVOLUTIONS AND PISTON AREA.

The number of revolutions of compressor varies with its size from forty to eighty revolutions per minute. When the compressor is worked directly by a steam engine, as is generally the case, the number of revolutions of the compressor is governed by those of the engine, and the area of the compressor piston must be in accordance with that of engine piston. The product of average pressure on engine piston with the area of the latter must always be greater than the product of the compressor piston area multiplied by the pressure in condenser coil if both the engine and compressor piston have the same length of stroke. If the stroke of compressor piston is shorter than that of engine piston its area can be made correspondingly larger.

USEFUL AND LOST WORK OF COMPRESSOR.

That part of the work of the compressor which is expressed by the foregoing equations for W_1, W_2 or W_3 may be considered as useful work of the compressor, while what work is done by the compressor in excess of that amount, due to superheating, friction and other causes, may be considered as lost work. The smaller the lost work the more perfect is the operation of the compressor.

DETERMINATION OF LOST WORK.

The lost work of a compressor may be determined in various ways, directly by interpretation of the indicator diagram and also indirectly in some cases. The lost work is the difference between the actual work done by the compressor and that theoretically required of the same, or expressed by formula, L standing for lost work in thermal units and W_6 for actual compressor work in thermal units:

$$L = W_6 - W_1$$

INDIRECT DETERMINATION OF ACTUAL WORK.

In a machine with submerged condenser, the actual work, W_6, of the compressor may be approximately determined in T. U. per hour after the following formula:

$$W_6 = (T - T_1)\,p - (t - t_1)\,g\,s_1$$

in which formula T is the temperature of outgoing, T_1 the temperature of incoming condenser water, t the temperature of cold brine, t_1 the temperature of returning brine, p the number of pounds of condensing water used per hour, g the number of pounds of brine circulated per hour, and s_1 the specific heat of the brine.

The actual compressor work found in this manner will be somewhat larger than that found from the indicator diagram, since it includes the lost work due to friction in the compressor. Allowance must also be made for amount of superheating neutralized otherwise than by condenser water.

HORSE POWER OF ENGINE.

The work required to operate the compressor, whether furnished by engine direct or by transmission and gearing, must be equal, or rather somewhat greater than the actual work of the compressor. It must exceed the work shown by the indicator by at least the amount due to friction of piston, etc. It is safe to assume that the indicated horse power of an engine, W_7, necessary to propel a compressor of a theoretical horse power, W_3, is at least about—

$$W_7 = 1.4\ W_3 \text{ horse power.}$$

In defective machines it may be more; seldom, however, it will be less.

WATER EVAPORATED IN BOILER.

The amount of water evaporated in boiler (for non-condensing engine) may be approximately estimated on the basis that twenty-five pounds of water are needed per hour per horse power in a well regulated boiler. The amount of water, A, evaporated for twenty-four hours is, therefore—

$$A = 25 \times 24 \times W_7 \text{ pounds.}$$

COAL REQUIRED.

If one pound of coal evaporates n pounds of water the amount of coal, F, required in twenty-four hours is approximately—

$$F = \frac{25 \times 24 \times W_7}{n} \text{ pounds.}$$

In a non-condensing engine about fifteen pounds of water are used per horse power per hour, and the foregoing formula in that case reads—

$$F = \frac{15 \times 24 \times W_7}{n} \text{ pounds.}$$

n differs for various kinds of fuel, but may be assumed equal to 8 for fair average coal.

EFFICIENCY OF COMPRESSOR.

The term efficiency covers a variety of meanings, and the meaning ought to be expressed clearly in each case. Generally efficiency is expressed by the number of units of heat removed from the refrigerator for every thermal unit of work done by compressor, which is also expressed by the quotient—

$$E = \frac{\text{Heat removed in refrigerator}}{\text{Work done by compressor in T. U.}}$$

This may be called the actual efficiency for a given case. As it varies not only with the machine, but also, and most decidedly so, with the local condition under which it works (temperature of refrigerator and condenser) it affords no criterion as to the lost work done by the compressor, *i. e.*, it is not an expression for the degree of perfection of the compressor.

In order to obtain an expression for this quality we must, according to Linde, compare the actual efficiency of a plant with the maximum theoretical efficiency of the plant when working under the same condition. The maximum theoretical efficiency, E_2, is expressed by Linde through the formula—

$$E_2 = \frac{T}{T_1 - T}$$

As we have seen above, this should more properly be substituted by the maximum theoretical efficiency, E_1, as explained in the above, at least if machines with the same circulating medium are to be compared, viz.:

$$E_1 = \frac{h_1-(t-t_1)s}{\frac{t-t_1}{T}h_1} \text{ or}$$

$$E_1 = \frac{T[h_1-s(t-t_1)]}{h_1(t-t_1)}$$

If R stands for the heat actually removed in refrigeration and Q for work actually performed by compressor, as ascertained by actual observation or test, we have for the actual efficiency, E, the expression—

$$E = \frac{R}{Q}$$

The ratio or proportion, n, between the actual and the theoretical capacity is therefore—

$$n = \frac{E}{E_1}$$

or if we insert the expressions found above—

$$n = \frac{Rh_1(t-t_1)}{QT[h_1-s(t-t_1)]}$$

DIFFERENT KINDS OF COMPRESSORS.

There are many constructive details in valves, etc., in the different makes of compressors which it is impracticable here to discuss. The principal difference, however, is due to the different methods in which superheating of the gas during compression is prevented or to whether the compressor is horizontal or vertical, double or single-acting, etc. By way of example we mention only a few typical ones.

THE LINDE COMPRESSOR.

This compressor is principally used for wet compression, the peculiarities of which have been mentioned above; it is a horizontal double-acting compressor with a deep packing, having a length of twelve inches or more in order to withstand the pressure of some 150 to 180 pounds. Since ammonia attacks India rubber, the best rubber packings for compressors are inlaid with cotton. Selden's, Garlock's and Common Sense packing are also used.

The Boyle compressor is vertical and single-acting, compressing only on the up stroke. The gas has free entrance to and exit from the cylinder below piston, calculated to keep cylinder and piston cool. The extreme lower portion of the pump forms an oil chamber to seal the stuffing box around piston.

THE DE LA VERGNE COMPRESSOR.

This compressor is also a vertical compressor, and superheating is counteracted by means of refrigerated oil, which is circulated through the compressor by means of a small pump. Another object of the oil is that its presence ahead and behind the piston abolishes the evil effects of clearance, or at least lessens the same materially. It furthermore affords excellent lubrication of the moving parts and helps to make the piston tight.

THE WATER JACKET COMPRESSOR.

This form of compressor is mostly vertical, its peculiarity being that the superheating is prevented by circulating cold water or brine through a water jacket which surrounds the compressor.

These compressors are frequently single-acting; in this case a shorter stuffing box (causing less friction) for piston rod may be used, since the pressure on the stuffing box is seldom more than thirty pounds.

TABLE SHOWING REFRIGERATING EFFECT OF ONE CUBIC FOOT OF AMMONIA GAS AT DIFFERENT CONDENSER AND SUCTION (BACK) PRESSURE IN B. T. UNITS.

Temperature of Gas in Degrees F.	Corresponding Suction Pressure. Lbs. per sq. in.	Temperature of the Liquid in Degrees F.								
		65°	70°	75°	80°	85°	90°	95°	100°	105°
		Correspg. Condenser Pressure (gauge) lbs. per sq. in.								
		103	115	127	139	153	168	184	200	218
	G. Pres									
—27°	1	27.30	27.01	26.73	26.44	26.16	25.87	25.59	25.30	25.02
—20°	4	33.74	33.40	33.04	32.70	32.34	31.99	31.64	31.30	30.94
—15°	6	36.86	36.48	36.10	35.72	35.34	34.96	34.58	34.20	33.82
—10°	9	42.28	41.84	41.41	40.97	40.54	40.10	39.67	39.23	38.80
— 5°	13	48.31	47.81	47.32	46.82	46.33	45.83	45.34	44.84	44.35
0°	16	54.88	54.32	53.76	53.20	52.64	52.08	51.52	50.96	50.40
5°	20	61.50	60.87	60.25	59.62	59.00	58.37	57.75	57.12	56.50
10°	24	68.66	67.97	67.27	66.58	65.88	65.19	64.49	63.80	63.10
15°	28	75.88	75.12	74.35	73.59	72.82	72.06	71.29	70.53	69.76
20°	33	85.15	84.30	83.44	82.59	81.73	80.88	80.02	79.17	78.31
25°	39	95.50	94.54	93.59	92.63	91.68	90.72	89.97	88.81	87.86
30°	45	106.21	105.15	104.09	103.03	101.97	100.91	99.85	98.79	97.73
35°	51	115.68	114.54	123.39	112.24	111.09	109.94	108.79	107.64	106.49

TABLE GIVING NUMBER OF CUBIC FEET OF GAS THAT MUST BE PUMPED PER MINUTE AT DIFFERENT CONDENSER AND SUCTION PRESSURES, TO PRODUCE ONE TON OF REFRIGERATION IN 24 HOURS.

Temperature of Gas in Degrees F.	Corresponding Suction Pressure. Lbs. per sq. in.	Temperature of the Gas in Degrees F.								
		65°	70°	75°	80°	85°	90°	95°	100°	105°
		Correspg. Condenser Pressure (gauge) lbs. per sq. in.								
		103	115	127	139	153	168	184	200	218
	G. Pres									
—27°	1	7.22	7.3	7.37	7.46	7.54	7.62	7.70	7.79	7.88
—20°	4	5.84	5.9	5.96	6.03	6.09	6.16	6.23	6.30	6.43
—15°	6	5.35	5.4	5.46	5.52	5.58	5.64	5.70	5.77	5.83
—10°	9	4.66	4.73	4.76	4.81	4.86	4.91	4.97	5.05	5.08
— 5°	13	4.09	4.12	4.17	4.21	4.25	4.30	4.35	4.40	4.44
0°	16	3.59	3.63	3.66	3.70	3.74	3.78	3.83	3.87	3.91
5°	20	3.20	3.24	3.27	3.30	3.34	3.38	3.41	3.45	3.49
10°	24	2.87	2.9	2.93	2.96	2.99	3.02	3.06	3.09	3.12
15°	28	2.59	2.61	2.65	2.68	2.71	2.73	2.76	2.80	2.82
20°	33	2.31	2.34	2.36	2.38	2.41	2.44	2.46	2.49	2.51
25°	39	2.06	2.08	2.10	2.12	2.15	2.17	2.20	2.22	2.24
30°	45	1.85	1.87	1.89	1.91	1.93	1.95	1.97	2.00	2.01
35°	51	1.70	1.72	1.74	1.76	1.77	1.79	1.81	1.83	1.85

THE ST. CLAIR COMPOUND COMPRESSOR.

This is a combination of two or more single-acting compressors after the principles of compound engines, in such a way that the ammonia is compressed part way at a lower pressure in one compressor and then transferred to another compressor, in which the higher compression is applied after the ammonia has passed an intermediate condenser.

WATER FOR COUNTERACTING SUPERHEATING.

The amount of refrigeration, U, required to counteract the superheating of ammonia in the case of dry compression may be expressed by—

$$U = S \times K \times 1440 \text{ units in twenty-four hours.}$$

In accordance with the above described devices, it is removed either by cooling the oil or by introducing water into the water jacket. The amount of water in gallons, g, used in the latter case per day may be approximated by the formula—

$$g = \frac{U}{8.33\,(t - t_1)} \text{ gallons.}$$

t being the temperature of the water leaving the water jacket, and t_1 being the temperature of the water entering the water jacket. The values for *S* and *K* have been given on pages 116 and 119.

THE BY PASS.

Most refrigerating machines are provided with a contrivance enabling the engineer to reverse the action of the compressor in such a way as to exhaust the condenser and compress into the refrigerator by the opening and the closing of appropriate valves, the combination of which constitutes what is called the by pass.

THE OIL TRAP.

This is a vessel placed between the compressor and condenser, through which the compressed vapor of ammonia is made to pass in order to deposit therein the oil drawn over with the ammonia from the lubricating materials used for oiling the stuffing boxes, etc. The inlet pipe should enter the trap sideways, so that the vapor may strike vertical surfaces and not the oil lying on the bottom of the trap. In some instances the oil trap is also surrounded by a water jacket.

CONDENSER.

The condenser consists of systems of pipes or coils into which the compressed ammonia is forced by the compressor. These coils are either immersed in the cooling water (submerged condenser) or the cooling water runs or trickles over them (open air, surface or atmospherical condenser). In passing through the condenser the ammonia yields to the cooling water the heat which it has acquired in doing refrigerating duty by its evaporation, and the heat which it has acquired during compression, the mechanical work done by compression having been converted into its equivalent of heat. This amount of heat is also equal to the latent heat of volatilization of the ammonia at the temperature of the condenser, and in addition to that the superheating which may have taken place.

SUBMERGED CONDENSER.

A submerged condenser consists of one or more sections of coils of 1¼ to 2-inch pipe. It is preferable to have a number of sections, connected by manifold inlets and outlets in such a way that one or more sections may be shut off for repairs or for other reasons. Instead of having the same size pipe all the way through, the pipe may be taken of larger size at the inlet for the vapor, and taper down, say, from 2-inch to 1-inch toward the outlet, where the ammonia is more or less liquid already; occupying a smaller space.

The hot ammonia vapors enter the condenser at the top, and the liquid ammonia leaves at the bottom where the cold water enters the condenser, which in turn leaves the condenser at the top. Special attention should be paid to an equal distribution of the water over the bottom of condenser, and a stirring apparatus should be provided to keep the water in motion around the condenser coils. The condenser should be more high and narrow, rather than short and wide, in order to assist the natural tendency for circulation.

AMOUNT OF CONDENSER SURFACE.

The efficiency of the condenser determines, in a great measure, the economical working of the machine, for which reason it is good policy to have as much condenser surface as practical considerations may permit. As to the actual amount of condenser surface to be employed, practice is the principal guide, and it has been found that for average conditions (incoming condenser water 70° and outgoing condenser water 80°, more or less) for each ton of refrigerating capacity (or for one-half ton ice making capacity) it will take forty square feet of condenser surface, which corresponds to sixty-four running feet of 2-inch pipe, and to ninety running feet of 1¼-inch pipe. Frequently 20 square feet of condenser surface, and even less, are allowed per ton of refrigeration (double that for actual ice making capacity), but this necessitates higher condenser pressure, etc., and is deemed poor economy by many engineers.

The number of square feet of cooling surface, F, required in a submerged condenser may be approximately calculated after the formula—

$$F = \frac{h.\,k}{m\,(t - t_1)} \text{ square feet.}$$

in which h is the heat of vaporization of one pound of ammonia at the temperature of the condenser, k the amount of ammonia passing the compressor per minute, and m the number of units of heat transferred per minute per square foot of surface of iron pipe having saturated ammonia vapor inside and water outside. t represents the temperature of the ammonia in the coils, and t_1 that of the cooling water outside of the coils, *i. e.*, mean temperature of the inflowing and outflowing cooling water.

Taking the above practical figures for condenser surface between 70° and 80° temperatures as a guide, the factor m is equal to 0.5, so that the formula reads:

$$F = \frac{h.\,k}{0.5\,(t - t_1)} \text{ square feet.}$$

This formula, like others which have been given on this subject, is an empirical one, but it has the advantage of simplicity, and yields results corresponding to the practical data given above.

The number of square feet of pipe surface can readily be converted into pipe lengths of any given size by referring to the table on dimensions of pipe.

AMOUNT OF COOLING WATER.

The heat which is transferred to the ammonia while producing the refrigeration, and also the heat equivalent to the work done upon the ammonia by the compressor (superheating being prevented), must be carried away by the cooling water, expressed in thermal units; and speaking theoretically, the sum of these two heat effects is equal to the heat of vaporization of the ammonia at the temperature of the condenser. On the basis of this consideration the amount of cooling water A, in pounds required per hour may be expressed by the formula—

$$A_1 = \frac{h.\,k \times 60}{t - t_1} \text{ pounds}$$

or in gallons after division by 8.33, the signs having the same significance as in the foregoing formula, with the exception of t, which represents the actual temperature of the outgoing, and t_1, which represents the actual temperature of the incoming cooling water.

Practically the amount of water used varies all the way from three to seven gallons per minute per ton, ice making capacity in twenty-four hours.

ECONOMIZING COOLING WATER.

Where cooling water is very scarce, and especially where atmospherical conditions, dryness of air, etc., are favorable, the cooling water may be re-used by subjecting the spent water to an artificial cooling process by running the same over large surfaces exposed to the air in a fine spray.

A device of this kind is described as being a chimney-like structure, built of boards, having a height of about twenty-six feet, the other dimensions being five by seven feet. Inside this structure are placed a number of partitions of thin boards, spaced four inches apart, extending to within six feet of the bottom of the structure; but the lower halves of these partitions are placed at right angles to those in the upper portion, this arrangement giving better results than unbroken partitions.

The water to be cooled enters the structure at the top, where by the use of funnel-shaped troughs it is spread evenly over the partitions and walls, and flows downward in thin sheets. At the base of the structure air is introduced in such quantity that the upward current has a velocity of about twenty feet per second. The air meeting the downward flow of water absorbs the heat by contact and also by vaporizing about 2 per cent of the water, reducing its temperature during the passage 27°, or from 83° to 56°. By this process the temperature of the water can be reduced from 5° to 15° below the temperature of the air, according to the amount of moisture in the latter. The chief expense to be considered in the process of re-cooling condenser water is the lifting of the water to the top of the structure. As a matter of course it is also good economy to use the hot condenser water for boiler feeding, as the equivalent of heat absorbed by the same is saved in the steam boiler.

OPEN AIR CONDENSER.

In the open air or atmospherical condenser the pipes through which the ammonia passes are arranged in the open air, exposed to a constant draft of air, if possible. The cooling water trickles over the pipes. The ammonia vapor flows in opposite direction, entering at the bottom of the condenser, the liquid passing off to the side into a vertical manifold as fast as it is condensed.

Other atmospherical condensers said to give excellent results are made in vertical sections of pipe, each section receiving the compressed vapor at the top from a common manifold, and discharging the liquid at the bottom into a common manifold, which leads to the liquid receiver.

PIPE REQUIRED.

The amount of condensing surface for an open air condenser is taken at the rate of forty square feet per ton of refrigerating capacity (or for one-half ton of ice making capacity). This is equivalent to 64 running feet of 2-inch pipe or 90 running feet of 1¼-inch pipe.

As in the case of the submerged condenser, much less pipe (twenty-five square feet per ton of refrigeration and less) is frequently used.

WATER REQUIRED.

The cooling water required for an atmospheric condenser is much less (upwards of 50 per cent and more) than for a submerged condenser, since the action of water is assisted by that of the air directly, and still more indirectly by causing some of the cooling water to evaporate, thus bringing about an extra absorption of latent heat.

It is claimed that where local conditions are favorable, the same cooling water may be used over and over again in an atmospherical condenser, if the same is built sufficiently high.

Another advantage of the open air condenser is due to the fact that all the water comes in direct contact with the surfaces to be cooled.

CONDENSER PRESSURE.

The pressure in the condenser depends on the temperature of the condensing water, and is always as high as or higher than the tension of ammonia vapor corresponding to the temperature of the water leaving the condenser (say about ten pounds higher).

LIQUID RECEIVER.

Generally a vessel, preferably a vertical cylinder holding about half a gallon for each ton of refrigerating capacity (in twenty-four hours) of the machine, is placed between the condenser and the expansion valve to receive and store the liquefied ammonia. It also serves as an additional oil trap, the oil being heavier than the ammonia settling on the bottom, where its presence is indicated by a gauge, and whence it can be withdrawn by opening a valve. A second gauge may be provided for on the liquid receiver, at about that point at which the pipe carrying the ammonia from the receiver to expander terminates within the receiver, in order to show that there is a sufficiency of liquid ammonia in the latter.

If the liquid receiver is to act as a storage room for all liquefiable ammonia in the plant in case of repairs, etc., it must be considerably greater than one-half gallon per ton of refrigerating capacity. In this case it is provided with valves, and they should never be closed unless the receiver is not over two-thirds filled with ammonia. To avoid explosions on this account the liquid receiver should be made big enough to contain the whole charge of ammonia twice over.

DIMENSIONS OF CONDENSERS.

The following tables, compiled by Skinkle, give the dimensions of both submerged and atmospheric condensers of some plants in actual operation, and allow much more pipe for the atmospheric than for the submerged condenser:

ATMOSPHERIC CONDENSERS.

Ice Making Capacity. In Tons.	Refrigerating Capacity, In Tons.	Condenser Pans.				Number of Pipes High.	Number of Pipes Wide.	Size of Pipe. Inches.	Length of Coil over bends. Feet.	Total feet of Pipe in Condenser.	Feet of Pipe per Ton, Ice Making Capacity.	Feet of Pipe per Ton, Refrigerating Capacity.
		Length of Pan. Feet.	Width of Pan. Feet.	Depth of Pan. Inches.	Thickness of Iron. Inches.							
12½	25	21	10⅔	8	3-16	40	5	1	17	3,680	294.4	147.2
20	35	24½	10⅔	8	3-16	40	5	1	21	4,440	222.	126.8
30	50	24½	14	8.	3-16	50	7	1	21	7,750	258.3	155.
40	75	24½	14	8	3-16	50	7	1¼	21	7,750	193.75	103.33
50	100	24½	14	8	3-16	90	7	1	21	13,950	279.	139.5
60	125	24½	14	12	3-16	80	7	1¼	21	12,400	206.6	99.2
80	150	27½	17	12	3-16	80	7	1¼	24	14,080	176.	93.86
				Aver	age f	or 1	in.	Pipe	per	ton,	263.42	142.12
				Aver	age f	or 1¼	in.	Pipe	per	ton,	192.12	98.79

SUBMERGED CONDENSERS.

Ice Making Capacity. In Tons.	Refrigerating Capacity. In Tons.	Tanks.				Number of Coils.	Pipes High.	Feet Long.	Size of Pipe. Inches.	Total Feet of Pipe in Condenser.	Feet of Pipe per Ton, Ice Making Capacity.	Feet of Pipe per Ton, Refrigerating Capacity.
		Length. Feet.	Width. Feet.	Depth. Feet.	Thickness of Iron. Inches.							
5	10	10	3⅓	6½	3-16	9	12	7½	1	855	171.	85.5
10	20	10	7½	6½	3-16	20	12	7½	1	1,900	190.	95.
12½	25	10	7½	6½	3-16	22	12	7½	1	2,090	167.	83.6
15	30	10	8⅓	6½	3-16	25	12	7½	1	2,375	151.6	79.16
20	35	10	10	6½	3-16	27	12	7½	1	2,565	128.25	73.28
30	50	10	10	12½	¼	27	24	7½	1	5,130	171.	102.6
40	75	14	10	13½	⅜	27	24	11½	1	7,695	191.1	102.6
60	110	14	13	13½	⅜	35	24	11½	1	9,975	166.25	90.68
									Aver	age,	167.	89.

THE FORECOOLER.

In order to save power and cooling water many plants are provided with supplementary condensers, or forecoolers, which consist of a coil or series of coils through which the compressed ammonia is made to pass before it enters the condenser proper. The forecooler is cooled by the spent or overflow water of the condenser.

If consisting of one coil, the forecooler should have the same size as the discharge pipe from the compressor; if consisting of a number of coils, the manifold pipe and the aggregate area openings of small pipes should equal that of the discharge pipe.

NOVEL CONDENSERS.

Condensers are now also built, in which the compressed gas, instead of entering a system of coils immersed in water, enters a cylinder or shell while the cooling water circulates through coils located within the cylinder.

Such a condenser is described by Hendrick as to consist of a heavy cast iron shell standing upright on a channel iron frame; it contains two or more spiral coils of 1¼-inch extra heavy pipe, the tails of which project through the heads of the shell and are united by manifolds. The ammonia gas, as discharged by the compressor, is delivered into the shell at the top, and as it becomes liquefied under the influence of pressure and by contact with the coils through which the condensing water is circulated (entering the lower ends of the coil), the liquid anhydrous ammonia collects in the bottom of the shell, which thus constitutes the liquid anhydrous receiver, and which is provided with suitable level and gauge. It will be seen that in this construction the water is subdivided into two or more separate and distinct streams, traveling through coils which vary in length from 100 to 175 feet, according to the size of the condenser. This is said to give a much better utilization of the cold in the water than the ordinary methods, where the condensing coils are submerged in a water tank, or where the coils are arranged so that the water trickles over them; in both cases the water simply traveling upward or downward ten to twenty feet. All coils are continuous from end to end.

On a similar principle brine coolers are made in which the brine circulates through systems of pipes, while the ammonia expands in a shell or cylinder surrounding the brine pipes.

PURGE VALVE.

At the highest point of the condenser, or on the discharge line next to the condenser, a purge valve should be provided for, to let off permanent gases.

DUPLEX OIL TRAP.

Frequently two oil traps are used, one of which, generally a larger one, is placed near the machine, and the other, the smaller one, near the condenser. When a forecooler is used the smaller trap is placed between it and the main condenser. The following table shows the sizes of traps that may be used:

Tons refrigeration	2 to 15	15 to 50	51 to 60	61 to 100
Small trap	8″×3′	10″×3′	12″×3′	12″×4′
Large trap	8″×5′	10″×6′	12″×6′	12″×8′

WET AND DRY COMPRESSION.

If superheating is prevented by carrying liquid ammonia into the compressor to keep the vapor always in a saturated condition, we say that we are working by wet compression; and if, on the other hand, the ammonia gas becomes superheated during compression, we are working by what is called dry compression. Some forms of compressors are specially adapted for wet compression; others for dry compression.

Opinions are much divided as to the relative merits of these two systems of compression. The theory shows a gain of economy in favor of wet compression, and the practical results do not contradict this, although the difference is not very great.

POWER TO OPERATE COMPRESSOR.

The power actually required to operate a compressor in order to produce a ton of refrigeration varies from one to two horse power, according to size of machine, other circumstances being equal. Very large machines may be operated with one horse power per ton of refrigerating capacity (in twenty-four hours), but generally one and one-third to one and one-half horse powers are required per ton for machines of over forty tons refrigerating capacity. Machines from ten to forty tons refrigerating capacity will require from one and one-half to two horse powers per ton, and still smaller machines will require up to two and one-half horse powers, and sometimes still more, per ton of refrigeration.

EXPANSION VALVE.

This valve is placed between the condenser, or rather, the liquid receiver, and the expansion or refrigerating coils. It is a peculiar valve, admitting of very fine adjustment, so as to enable the engineer to admit the required amount of liquid to the expander, and no more.

EXPANSION OF AMMONIA.

The expansion or volatilization of the liquid ammonia, by which the refrigeration is effected, takes place within series or coils of iron pipes. These pipes may be located in the rooms to be refrigerated (direct expansion system) or they may be placed in a bath of salt brine, which, after having been cooled in this way, is circulated in turn through the rooms to be refrigerated. (Indirect expansion, or brine system.)

SIZE OF EXPANSION COILS.

The surface or the size and length of expansion coils to be placed in the rooms to be refrigerated, or in the brine tank, like nearly all the pipe work in the refrigerating practice, is based on empirical rules.

There are no concise formulæ on these subjects, as exact experiments on the transmission of heat under circumstances obtaining in the refrigerating practice are almost entirely wanting.

Besides this, the conditions are very variable, owing to the change of pipe surface by atmospherical conditions, or by the deposit of ice and snow or by the deposit from the water, as in case of the condenser, difference in insulation, etc. For these reasons every manufacturer has his own rules; and whatever is said in this compend on this subject is abstracted from practical experience and subject to modifications in individual cases.

PIPING ROOMS.

The size of pipe usually employed for piping rooms varies from one to two inches, and the length required varies according to circumstances, more especially with the temperature or the back pressure of the expanding ammonia and the temperature at which the rooms are to be held. If a room is to be held at a temperature of 34°, and the temperature of the expanding ammonia is 10°, it will take only half as much pipe to convey a certain amount of refrigeration as it would take if the temperature of the expanding ammonia were at 22° F.

In the latter case, however, the machine works under conditions far more economical, and for this reason it is advisable to use the larger amount of pipe in order to be enabled to work with a higher back pressure.

TRANSMISSION PER SQUARE FOOT.

In allowing a difference of 8° to 15° between the temperatures inside and outside of the pipes it is variously assumed that one square foot of pipe surface will convey 2,500 to 4,000 units of refrigeration in twenty-four hours in direct expansion.

This figure nearly agrees with a transmission of heat at the rate of 10 B. T. units per hour, per square foot surface, for each degree F. difference between temperature inside and outside of pipe, in case of direct expansion. In the case of brine circulation the brine with the same back pressure has, of course, a much higher temperature than the ammonia, and for this reason the above difference will be much less, which explains the fact that from one and one-half to two times as much pipe is used with brine circulation as in direct expansion.

If the amount of piping is calculated on this basis, allowing a refrigeration of a certain number of B. T. units per cubic foot of space to be refrigerated, the result will generally fall short of the piping required after the rules laid down in the following paragraph. This is to be explained by the fact that the latter rules are given on a very liberal basis calculated to cover unfavorable cases as regards insulation, size of rooms, etc., it being understood that any possible surplus in piping will tend to increase the efficiency of machine. This remark applies not only to the rules for piping in following paragraph, but to rules on piping in most cases.

PRACTICAL RULE FOR PIPING.

Practically the matter, however, is not often calculated on this basis, but after a rule of thumb it is assumed (allowing for difference in insulation and size of rooms) that about one running foot of 2-inch pipe (direct expansion) will take care of ten cubic feet of space in houses which are to be kept below freezing down to a temperature of 10° F.

About one running foot of 2-inch pipe will take care of forty cubic feet of space in rooms to be kept at or above the freezing point, 32° F., or thereabouts.

About one running foot of 2-inch pipe will take care of sixty cubic feet of space in rooms to be kept at 50° F., and above, as in the case of ale storage.

In conformity with the remarks in preceding paragraph, we take it that these rules are intended to cover cases of rooms of 50,000 cubic feet capacity and less, poorly insulated, and operated with small differences in temperature. On a similar basis it is frequently assumed that one ton refrigerating capacity will take care of 4,500 cubic feet cold storage capacity to be held at 32° to 35° F., and that from 260 to 300 feet of 1¼-inch pipe will properly distribute one ton of refrigeration.

Relating to the question of piping rooms, condensers and brine tanks, it may be understood once for all that there are two sides to this also. One contemplates a less expensive plant by reducing piping to a minimum frequently at the expense of economical working. The other side aims at increasing the capacity by ample pipe surface, and therefore the first outlay for a plant will be greater, but probably will pay better in the end.

DIMENSIONS OF PIPE.

One running foot of 2-inch pipe is equal to 1.44 feet of 1¼-inch pipe, and 1.8 feet of 1-inch pipe, as regards surface. For similar comparisons and calculations the following tables will be found convenient:

DIMENSIONS OF STANDARD PIPE.

Nominal Inside Diameter.	Actual Inside Diameter.	Actual Outside Diameter.	Thickness.	Internal Circumference.	External Circumference.	Length of Pipe per Sq. Ft. of Outside Surface	Internal Area.	External Area.	Length of Pipe Containing One Cubic Foot.	Weight per Foot of Length.
In.	In.	In.	In.	In.	In.	Ft.	In.	In.	Ft.	Lbs.
⅛	0.270	0.405	0.068	0.848	1.272	9.434	0.0572	0.129	2500	0.243
¼	0.364	0.54	0.068	1.144	1.696	7.075	0.1041	0.229	1385	0.421
⅜	0.494	0.675	0.091	1.552	2.121	5.657	0.1916	0.358	751.5	0.562
½	0.623	0.84	0.109	1.957	2.652	4.502	0.3048	0.554	472.4	0.845
¾	0.824	1.05	0.113	2.589	3.299	3.637	0.5333	0.866	270.0	1.126
1	1.048	1.315	0.134	3.292	4.134	2.903	0.8627	1.357	166.0	1.670
1¼	1.380	1.66	0.140	4.335	5.215	2.301	1.496	2.164	96.25	2.258
1½	1.611	1.90	0.145	5.061	5.969	2.201	2.038	2.835	70.65	2.694
2	2.067	2.375	0.154	6.494	7.461	1.611	3.355	4.430	42.36	3.667
2½	2.468	2.875	0.204	7.754	9.032	1.328	4.783	6.491	30.11	5.773
3	3.067	3.50	0.217	9.636	10.966	1.091	7.388	9.621	19.49	7.547
3½	3.568	4.0	0.226	11.146	12.566	0.955	9.837	12.566	14.56	9.055
4	4.026	4.5	0.237	12.648	14.137	0.849	12.730	15.904	11.31	10.728
4½	4.508	5.0	0.247	14.153	15.708	0.765	15.939	19.635	9.03	12.492
5	5.045	5.563	0.259	15.845	17.475	0.629	19.990	24.299	7.20	14.567
6	6.065	6.625	0.280	19.054	20.813	0.577	28.889	34.471	4.98	18.764
7	7.023	7.625	0.301	22.063	23.954	0.505	38.737	45.663	3.72	28.410
8	7.982	8.625	0.322	25.076	27.096	0.444	50.039	58.426	2.88	28.348
9	9.001	9.688	0.344	28.277	30.433	0.398	63.633	73.715	2.26	34.677
10	10.019	10.65	0.366	31.475	33.772	0.355	78.838	90.792	1.80	40.041

DIMENSIONS OF EXTRA STRONG PIPE.

A table giving dimensions of extra strong pipe will be found in the Appendix.

BRINE SYSTEM.

In the brine system the expansion coils, as stated, are placed in separate vessels containing salt brine, which is cooled down to the desired degree. The brine so cooled is then conducted through pipes located in the rooms to be refrigerated by means of force pumps. In ice making the cells or boxes containing the water for ice making are suspended in the brine tank.

SIZE OF PIPE IN BRINE TANK.

The amount of piping allowed in brine tank is also a matter of practical experience. Generally 120 to 150 running feet of 1¼-inch pipe are allowed per ton of refrigerating capacity (in 24 hours) in brine tank for general refrigeration.

In case of ice making 250 to 300 running feet of 1¼-inch pipe are allowed in brine tank per ton of ice to be manufactured in twenty-four hours.

TABLE OF BRINE TANKS AND COILS.

The following table shows the dimensions of some brine tanks and coils for different capacities, expressed in tons of refrigerating capacity (not ice making capacity).

Capacity in Tons Refrigeration.	Length of Tank in Feet.	Width of Tank in Feet.	Height of Tank in Feet.	Thickness of Iron, Inches.	Cubic Feet of Brine when Filled within Six In. of Top.	Number of Coils.	Pipes High.	Feet Long.	Size of Pipe, Inches.	Total Feet of Collage in Brine Tank.	Feet of Collage per Ton of Refrigeration, ½ Ton of Ice.
25 tons....	16	13	8½	¼	1,664	38	12	11	1	5,016	200
35 "	20	13	8½	¼	2,080	46	12	11	1	6,072	173.5
50 "	20	13	11	¼	2,730	46	16	11	1	8,188	163.4
75 " ..	22	15	15	⅜	4,785	52	20	13	1	13,963	186
										4)	722.9
Average per ton.	....	....	....	....	....	....	....	....	....	....	180.7

PIPES FOR BRINE CIRCULATION.

In the case of brine circulation there must be another series of coils in rooms to be refrigerated, through which the brine circulates, as the brine does not circulate as fast as the ammonia vapor, and for other reasons the surface of brine coils in storage rooms must be much larger than in case of direct expansion under conditions otherwise similar.

In round figures it is generally assumed that the area of pipe surface in case of brine circulation should be from one and one-half to two times as large as in case of direct expansion.

RULES FOR LAYING PIPES.

The pipes in storage rooms should be placed where they are least in the way.

They should be arranged in independent sections connected by manifolds in such a way that each section can be shut out to throw off the frost.

TABLE FOR EQUALIZING PIPES.

The size of main pipe is given in the column at the left. The number of branches is given in the line on top, and the proper size of branches is given in the body of the table on the line of each main and beneath the desired number of branches.

In commercial sizes the normal 1¼-inch pipe is generally over size; often as large as 1⅜. It is safe to call it 1.3 inches, and it is so figured in the table. Exact sizes are given for branch pipes. The designer of the pipe system can thus better select the commercial sizes to be used.

Size of Main Pipe.	Number of Branches.								
	2	3	4	5	6	7	8	9	10
1 in.	.758	.644	.574	.525	.488	.459	.435	.415	.398
1¼ "	.985	.838	.747	.683	.635	.597	.556	.540	.518
1½ "	1.14	.967	.861	.788	.733	.689	.653	.623	.597
2 "	1.52	1.29	1.15	1.05	.977	.918	.870	.830	.796
2½ "	1.89	1.61	1.44	1.31	1.22	1.15	1.09	1.09	.995
3 "	2.27	1.92	1.72	1.58	1.47	1.38	1.31	1.25	1.19
3½ "	2.65	2.26	2.01	1.84	1.71	1.61	1.52	1.45	1.39
4 "	3.03	2.58	2.30	2.10	1.95	1.84	1.74	1.66	1.59
4½ "	3.41	2.90	2.58	2.36	2.20	2.07	1.96	1.87	1.79
5 "	3.79	3.22	2.87	2.63	2.44	2.30	2.18	2.08	1.99
6 "	4.55	3.87	3.45	3.15	2.93	2.75	2.61	2.49	2.39
7 "	5.30	4.51	4.02	3.68	3.42	3.21	3.05	2.91	2.79
8 "	6.06	5.16	4.59	4.20	3.91	3.67	3.48	3.32	3.18
9 "	6.82	5.80	5.17	4.73	4.40	4.13	3.92	3.74	3.58
10 "	7.58	6.44	5.74	5.25	4.88	4.59	4.35	4.15	3.98
12 "	9.08	7.73	6.89	6.30	5.86	5.51	5.22	4.98	4.78

In brine circulation the brine should also be pumped through series of pipes running in the same direction, and connected by manifolds to decrease friction.

Further information in regard to piping rooms, etc., will be found in the chapters on Cold Storage, Brewery Refrigeration, etc.

TABLE SHOWING THE NUMBER OF GALLONS DISCHARGED PER MINUTE BY A SINGLE-ACTING PUMP OF A GIVEN DIAMETER AND STROKE AT 10 STROKES PER MINUTE.

LENGTH OF STROKE IN INCHES.

Diameter of Pump Barrel, in Inches.

In.	1	2	3	4	5	6	7	8	9	10	12	14	15	16	18	20	24	In.
1	.034	.068	.102	.136	.170	.204	.238	.272	.306	.340	.408	.476	.510	.544	.612	.680	.816	1
1¼	.053	.106	.159	.212	.266	.319	.372	.425	.478	.531	.637	.744	.797	.850	.956	1.062	1.275	1¼
1½	.076	.153	.229	.306	.382	.459	.535	.612	.688	.765	.918	1.071	1.147	1.224	1.378	1.530	1.836	1½
1¾	.104	.208	.312	.416	.521	.625	.729	.833	.937	1.041	1.249	1.457	1.562	1.666	1.874	2.082	2.499	1¾
2	.136	.272	.408	.544	.680	.816	.952	1.088	1.224	1.360	1.632	1.904	2.040	2.176	2.488	2.720	3.264	2
2½	212	.425	.637	.850	1.062	1.275	1.487	1.700	1.912	2.125	2.550	2.975	3.187	3.400	3.825	4.250	5.100	2½
3	.306	.612	.918	1.224	1.530	1.836	2.142	2.448	2.754	3.060	3.672	4.284	4.590	4.896	5.508	6.120	7.344	3
3½	.416	.833	1.249	1.666	2.082	2.499	2.915	3.332	3.748	4.165	4.998	5.831	6.247	6.664	7.497	8.330	9.996	3½
4	.544	1.088	1.632	2.176	2.720	3.264	3.808	4.352	4.896	5.440	6.528	7.616	8.160	8.704	9.792	10.880	13.056	4
4½	688	1.377	2.065	2.754	3.442	4.121	4.819	5.508	6.196	6.885	8.262	9.639	10.327	11.016	12.393	13.770	16.524	4½
5	.850	1.700	2.550	3.400	4.250	5.100	5.950	6.800	7.650	8.500	10.200	11.900	12.750	13.600	15.300	17.000	20.400	5
5½	1.028	2.057	3.085	4.114	5.142	6.171	7.199	8.228	9.256	10.285	12.342	14.399	15.427	16.456	18.513	20.570	24.684	5½
6	1.224	2.448	3.672	4.896	6.120	7.344	8.568	9.792	11.016	12.240	14.688	17.136	18.360	19.584	22.032	24.480	29.376	6
7	1.666	3.332	4.998	6.664	8.330	9.996	11.662	13.328	14.994	16.660	19.992	23.324	24.990	26.656	29.998	33.320	39.984	7
8	2.176	4.352	6.528	8.704	10.880	13.056	15.232	17.408	19.584	21.760	26.112	30.464	32.640	34.816	39.168	43.520	52.224	8
9	2.754	5.508	8.262	11.016	13.770	16.524	19.278	22.032	24.786	27.540	33.048	38.556	41.310	44.064	50.572	55.080	66.096	9
10	3.400	6.800	10.200	13.600	17.000	20.400	23.800	27.200	30.600	34.000	40.800	47.600	51.000	54.400	61.200	68.000	81.600	10
12	4.896	9.792	14.688	19.584	24.480	29.376	34.272	39.168	44.064	48.960	58.752	68.544	73.440	78.336	88.128	97.920	107.504	12
15	7.650	15.300	22.950	30.600	38.250	45.960	53.550	61.200	68.850	76.500	91.800	107.100	114.750	122.400	137.700	153.000	183.600	15
18	11.016	22.032	33.048	44.064	55.080	66.096	77.112	88.128	99.144	110.160	132.192	154.224	165.240	176.256	198.288	220.320	264.384	18
20	13.600	27.200	40.800	54.400	68.000	81.600	95.200	108.800	122.400	136.000	163.200	190.400	204.000	217.600	244.800	272.000	326.400	20
24	19.584	39.168	58.752	78.336	97.920	117.504	137.088	156.672	176.256	195.840	235.008	274.176	293.760	313.344	352.512	391.680	470.016	24
In.	1	2	3	4	5	6	7	8	9	10	12	14	15	16	18	20	24	In.

The quantities given in the table are in gallons, and are calculated for single-acting pumps, at ten strokes per minute. The quantity for any other number of strokes per minute may be found by multiplying the quantity noted in the table by the ratio of the factor 10 to the given number of strokes. For double-acting pumps the quantity noted in the table should be doubled.

THE BRINE PUMP.

The circulation of the refrigerated brine through the refrigerating coils in storage rooms, etc., is accomplished by the brine pump. The size of the brine pump may be estimated on the basis that the brine should not travel faster than sixty feet per minute. The table on opposite page will be found convenient in this connection.

PREPARING BRINE.

The brine is a solution of some saline matter in water, in order to depress the freezing point of the latter. Generally chloride of sodium or common salt is used for this purpose. To make the brine it is well to use a water tight box, 4 × 8 feet, with perforated false bottom and compartment at end, with overflow pipe for brine to pass off through a strainer. The salt is spread on false bottom, and the water fed in below the false bottom as fast as a solution of the proper strength will form. A wooden hoe or shovel may be used for stirring to accelerate solution.

TABLE SHOWING PROPERTIES OF SOLUTION OF SALT.
(Chloride of Sodium.)

Percentage of Salt by Weight.	Pounds of Salt per Gallon of Solution.	Degrees on Salometer at 60° F.	Weight per Gallon at 39° F.—4° C.	Specific Gravity at 39° F. — 4° C.	Specific Heat.	Freezing Point, Fahrenheit.	Freezing Point, Celsius.
1	0.084	4	8.40	1.007	0.992	30.5	—0.8
2	0.169	8	8.46	1.015	0.984	29.3	—1.5
2.5	0.212	10	8.50	1.019	0.980	28.6	—1.9
3	0.256	12	8.53	1.023	0.976	27.8	—2.3
3.5	0.300	14	8.56	1.026	0.972	27.1	—2.7
4	0.344	16	8.59	1.030	0.968	26.6	—3.0
5	0.433	20	8.65	1.037	0.960	25.2	—3.8
6	0.523	24	8.72	1.045	0.946	23.9	—4.5
7	0.617	28	8.78	1.053	0.932	22.5	—5.3
8	0.708	32	8.85	1.061	0.919	21.2	—6.0
9	0.802	36	8.91	1.068	0.905	19.9	—6.7
10	0.897	40	8.97	1.076	0.892	18.7	—7.4
12	1.092	48	9.10	1.091	0.874	16.0	—8.9
15	1.389	60	9.26	1.115	0.855	12.2	—11.0
20	1.928	80	9.64	1.155	0.829	6.1	—14.4
24	2.376	96	9.90	1.187	0.795	1.2	—17.1
25	2.488	100	9.97	1.196	0.783	0.5	—17.8
26	2.610	104	10.04	1.204	0.771	—1.1	—18.4

STRENGTH OF BRINE.

Generally speaking, the brine must contain sufficient salt to prevent its freezing at the lowest temperature in freezing tank, and by referring to the accompanying table one can answer the question for himself on this basis very readily.

To determine the weight of one cubic foot of brine multiply the values given in column 4 by 7.48.

To determine the weight of salt to one cubic foot of brine multiply the values given in column 2 by 7.48.

POINTS GOVERNING STRENGTH OF BRINE.

Therefore if the temperature in the freezing tank does not go below 15° F., it would be quite sufficient to use a brine containing 15 per cent of salt (salometer degrees 60), as from the above table it appears that such a solution does freeze below that temperature. On the other hand, if the temperature of freezing does not go below 20° F., a brine containing only 10 per cent salt would be sufficient for the same reason, etc. This table also explains why it would be irrational to use stronger solutions of salt than these, for, as we see from the column showing specific heat, the same grows smaller as the concentration of the brine increases, and consequently the stronger the brine the less heat a given amount of brine will be able to convey between certain definite temperatures. There is another danger connected with the use of too strong, especially of concentrated, brine in refrigeration. Such brine may cause clogging of pipes, etc., on account of depositing salt. This danger, however, is not so great as that of having the solution too thin, for while it may be concentrated enough not to freeze in the brine tank, it may be still too weak to withstand the temperature obtaining in the expansion coil, so that a layer of ice will form around the latter which interferes with the prompt absorption of heat from the brine. For this reason the surface of the expansion coils in brine tank should be inspected from time to time to see if any ice has formed on them.

SIMPLE DEVICE FOR MAKING BRINE.

An ordinary barrel with a false bottom three inches above the real bottom, perforated with ¼-inch holes, is a practical contrivance for making brine. The space above the false bottom is filled with salt nearly to the top of the barrel. Ordinary water is admitted below the false bottom, and the ready brine runs out at the top through a pipe, which is best inclosed in a wire screen filled with sponges. The pipe carrying off the brine should be about ½-inch larger than the pipe admitting the water.

SUBSTITUTE FOR SALOMETER.

In case one is unable to readily obtain a salometer, a Beaume hydrometer, or a Beck hydrometer scale, both of which are in quite general use for taking the strength of acids, etc., can be used as well. Their degrees compared with specific gravity and percentage of salt are shown in the following table, and, as will be seen, do not differ so very much from the degrees of the salometer scale:

Percentage of Salt by Weight.	Specific Gravity.	Degrees on Beaume's scale, 60° F.	Degrees on Beck's scale, 60° F.
0	1.0000	0	0
1	1.0072	1	1.2
5	1.0362	5	6
10	1.0733	10	12
15	1.1114	15	17
20	1.1511	19	23
25	1.1923	23	28

CHLORIDE OF CALCIUM.

Some engineers prefer to use chloride of calcium for the preparation of brine in preference to common salt. It is higher in price than the latter, but is said to keep the pipes cleaner, causing less wear and a better conduction of heat.

The physical properties of the chloride of calcium solution, as appears from the subjoined table, are quite similar to those of common salt. The freezing point, however, can be depressed several degrees lower by the use of the former, and for this reason the use of chloride of calcium may be advisable in such extreme cases. Otherwise the preparation of the solution of chloride of calcium is the same as that of ordinary brine.

PROPERTIES OF SOLUTION OF CHLORIDE OF CALCIUM.

Percentage by Weight.	Specific Heat.	Spec. Grav. at 60° F.	Freezg. Pt. degrees F.	Freezg. Pt. degs. Cels.
1	0.996	1.009	31	— 0.5
5	0.964	1.043	27.5	— 2.5
10	0.896	1.087	22	— 5.6
15	0.860	1.134	15	— 9.6
20	0.834	1.182	5	—14.8
25	0.790	1.234	—8	—22.1

BRINE CIRCULATION VS. DIRECT EXPANSION.

The principal reason why brine circulation is still preferred by many to direct expansion, is to be sought in fear entertained with regard to the escaping ammonia in case the pipes should leak. The danger from this source, however, seems to have been greatly exaggerated, as but few accidents of this kind have been known, the pressure in the ammonia pipe being generally not much higher than in the brine coils.

Another advantage frequently quoted in favor of brine circulation is the fact that comparatively great quantities of refrigerated brine are made and stored ahead, a supply which can be drawn on in case the machinery should have to be stopped for one reason or another.

In case of a prolonged stoppage, refrigerating brine made by dissolving ice and salt together can be circulated through the brine pipes, which is also impracticable in case of direct expansion.

It is also claimed that in small plants, in case of brine circulation, the general machinery might be stopped and only the brine pump be kept going to dispense the surplus refrigeration which had been accumulated in the brine during the day.

THE DRYER.

The dryer is an attachment of more recent coinage with which many compression plants are provided, its purpose being the drying of ammonia gas. It is a kind of trap on the suction pipe connected in such a manner (by means of a by-pass) that the gas can be passed through it when necessary.

This trap is provided with removable heads for the introduction of some moisture absorbing substance (freshly burnt unslaked lime, as a rule) and for the withdrawal of the spent absorbent.

LIQUID TRAP.

It is also recommended to have an additional trap between the expansion valve and the expanding coils. The vaporization then takes place within the chamber or trap, and oil and other undesirable foreign matter will be deposited in this trap, and will not be carried over into the expansion coils. The trap is provided with a by-pass, so that it can be cleaned without stoppage.

If such a trap can be placed within the rooms to be refrigerated it may be of some advantage; but if it has to be placed outside, as in the case of brine circulation, much refrigeration is wasted.

CHAPTER V.—ICE MAKING AND STORING.

SYSTEMS OF ICE MAKING.

One of the principal uses of mechanical refrigeration is the production of artificial ice, which is carried out after different methods or systems. The two methods which are most generally used are the so-called can system and the plate system.

ICE MAKING CAPACITY OF PLANT.

From the temperature of brine tank respectively, the temperature in expansion coils (which will be from 5° to 10° lower), the temperature in condenser coil from the size of compressor, etc., the theoretical refrigerating capacity of the plant can be calculated as above shown, making allowance for clearance, etc., as mentioned.

The ice making capacity of the plant is, of course, much below this theoretical refrigerating capacity. An allowance from 6 to 12 per cent loss due to radiation in brine tank, pipes, etc., must be made in the start, and in addition to that a further allowance for the refrigeration of the water from the ordinary temperature to that of freezing, and for the refrigeration of the ice from 32° to the temperature of the brine. For that and other reasons it may be assumed that the ice making capacity of a machine is from 40 to 60 per cent of its theoretical refrigerating capacity.

CAN SYSTEM.

In making ice by what is called the can system, the water is placed in cans or molds made of galvanized iron of convenient shape, which are inserted in a tank filled with brine, the latter being kept cool by coils of pipe in which the expanding ammonia circulates. Temperature of brine varies from 10° to 25° F., 15° F. being considered favorable.

SIZE OF CANS.

The cans or molds for freezing vary in size and shape. The sizes of cans in most common use are shown in the following table:

No. 1 can,	8½×15×32,	weight of cake,	100 lbs.,	No. 18 Iron.
" 2 "	8½×16×44,	" "	150 "	" 18 "
" 3 "	11 ×11×32,	" "	100 "	" 18 "
" 4 "	11 ×22×32,	" "	200 "	" 16 "
" 5 "	11 ×22×44,	" "	300 "	" 15 "

The weight is net. Allowance is made for about 5 per cent more to allow for loss in thawing, etc.

DIMENSIONS OF ICE MAKING TANKS.

Table compiled by Skinkle, giving sizes of some freezing tanks, piping and molds in actual operation.

Tons Ice Making Capacity.	No. of Tanks.	Sizes of Tanks. Length of Tank Feet & Inches.	Sizes of Tanks. Width of Tank Feet & Inches.	Sizes of Tanks. Depth of Tank in Inches.	Sizes of Tanks. Thickness of Plates, Inches.	No. of Coils.	Size of Pipe, Inches.	No. of Pipes High.	Length of Coils.	Total Feet of Pipe in Tank.	Feet of Pipe per Ton Ice Making Capacity.	Number of Ice Molds in Tank.	Size of Molds in Inches.	Net Weight of Ice from Each Mold.	Number of Molds per Ton Ice Making Capacity.	Number Hours for Freezing Each Mold.	Remarks.
½	1	17—0	6—2	33	3-16	7	1	6	15—4	644	322	60	8×15×33	100 lbs.	30	36	
3	1	17—0	9—0	33	3-16	10	1	6	15—0	900	300	90	8×15×33	100 "	30	36	
5	1	17—0	14—9	33	3-16	16	1	6	15—0	1440	288	150	8×15×33	100 "	30	36	
10	1	29—0	19—0	33	¼	25	1	8	17—0	3400	340	192 24	11×22×33 11×11×33	200 " 100 "	20.4	48	
12½	1	37—6	19—0	33	¼	33	1	8	17—0	4488	329	256 32	11×22×33 11×11×22	200 " 100 "	21.36	52.2	Special
15	1	43—0	19—0	33	¼	37	1	8	17—0	5032	335	288 36	11×22×33 11×11×33	200 " 100 "	20.4	48	
20	2	29—0	19—0	33	¼	25	1	8	17—0	3400	340	192 24	11×22×33 11×11×33	200 " 100 "	20.4	48	*
30	2	43—0	19—0	33	¼	37	1	8	17—0	5032	335	288 36	11×22×33 11×11×83	200 " 100 "	20.4	48	*
30	1	43—0	30—0	48	⅜	35	1¼	8	28—0	7840	261	480	11×22×45	300 "	16	57.6	
60	2	56—0	20—0	48	⅜	49	1¼	10	18—0	8820	294	432	11×22×45	300 "	14.4	51.8	
60	2	43—0	30—0	48	¼	35	1¼	8	28—0	7840	261	480	11×22×45	300 "	16	57.6	*

Average of 1-inch Pipe per Ton, 327 Feet. Average of 1¼-inch Pipe per Ton, 272 Feet.

* Twenty-ton Tanks are duplicate 10-ton tanks }
Thirty-ton " " " 15 " " } Dimensions of one tank only are given in each instance.
Sixty-ton " " " 30 " " }

TIME FOR FREEZING.

At about 14° to 15° F. of the brine, 11-inch ice will take about forty-five to fifty hours to close, and 10-inch ice about thirty-eight to forty-four hours, and 8-inch ice about twenty-eight to thirty-two hours. If the temperature is 10° F. it will take about 20 to 25 per cent less time, but ice will be more brittle. These figures relating to the time of freezing are given on the basis of first rate conditions all around, which are seldom if ever attained in practical working. For this reason estimates on the size of brine tank, number of cans, etc., are generally made on the basis of the following freezing times: Fifty to seventy-two hours for ice eleven inches thick; forty to sixty hours for ice ten inches thick, and thirty-six to fifty hours for ice eight inches thick. On this basis twenty cans are required (300-pound cans, seventy-two hours' freezing time, 11-inch ice) for each ton of ice making capacity per day, and the room in freezing tank must be in accordance therewith.

Siebert, after a formula of his own, gives the following freezing time table:

FREEZING TIMES FOR DIFFERENT TEMPERATURES AND THICKNESSES OF CAN ICE.

Thickn'ss.	1 in.	2 in.	3 in.	4 in.	5 in.	6 in.	7 in.	8 in.	9 in.	10 in.	11 in.	12 in.
Temp. 10°	0.32	1.28	2.86	5.10	8.00	11.5	15.6	20.4	25.8	31.8	38.5	45.8
" 12°	0.35	1.40	3.15	5.60	8.75	12.6	17.3	22.4	28.4	35.0	42.3	50.4
" 14°	0.39	1.56	3.50	6.22	9.70	14.0	19.0	25.0	31.5	39.0	47.0	56.0
" 16°	0.44	1.75	3.94	7.00	11.0	15.8	21.5	28.0	35.6	43.7	53.0	63.0
" 18°	0.50	2.00	4.50	8.00	12.5	18.0	24.5	32.0	40.5	50.0	60.5	72.0
" 20°	0.58	2.32	5.25	9.30	14.6	21.0	28.5	37.3	47.2	58.3	70.5	84.0
" 22°	0.70	2.80	6.30	11.2	17.5	25.2	34.3	44.8	56.7	70.0	84.7	100.0
" 24°	0.88	3.50	7.88	14.0	21.0	31.5	42.8	56.0	71.0	87.5	106.0	120.0

It will be noticed on closer inspection that in this table the time for freezing different thicknesses of ice is proportional to the square of the thickness. Thus, to freeze a block ten inches thick takes 100 times as long as to freeze a block of one inch thickness, and four times as long to freeze a block of four inches thickness, than it takes to freeze one of two inches thickness, and so on.

PIPE IN BRINE TANK.

About 250 feet of 2-inch or 350 feet of 1¼-inch pipe, or its equivalent according to the temperature of brine and capacity of machine, are generally used per ton of ice per twenty-four hours.

Less pipe is frequently used, even as low as 150 feet of 2-inch pipe, and 200 feet of 1¼-inch pipe per ton of ice making capacity (in twenty-four hours), but in that case the back pressure must be carried excessively low, which duly increases the consumption of coal and the wear and tear of machinery. It is also claimed that when the agitation in brine tank is very perfect and the ammonia expansion pipes have short runs (from header) eighty-five to 100 square feet of pipe in brine tank per ton of actual daily ice making capacity will be sufficient. These figures agree somewhat with the ones given in the foregoing paragraph.

ARRANGEMENT OF FREEZING TANK.

The size and length of pipe in brine tank should be arranged in such a manner that each row of molds is passed by an ammonia pipe on each side, preferably on the wide side of mold. The series of pipes in freezing tank are connected by manifold, the liquid ammonia entering the manifold at the lower extremity, and the vapor leaving by the suction manifold placed at the higher extremity of the refrigeration coils.

When working with wet vapor of ammonia, the liquid should enter at the upper extremity, and leave for compressor at lower extremity of refrigeration coils.

The refrigerating tank should be well insulated by wainscoting made of matched boards. The space between wainscoting and tank (about ten to eighteen inches) should be filled in with sawdust, cork or other insulating material. It is recommended that brine tank insulation should be twelve to eighteen inches thick on sides of tank, and at least twelve inches under the bottom.

Brine tanks are made of sheet iron or steel, wood and also of cement. Each kind has its admirers, according to circumstances, local and otherwise. Tank steel plate is said to make the best job, if properly built, and will last from ten to twelve years.

Wooden tanks are built of 2×4 or 2×6-inch planks, according to size of tank, and when built that way lined with ⅞-inch matched flooring. All the 2×4 or the matched flooring is laid and bedded in pure hot asphaltum before being nailed together. Cedar or cypress and hard yellow pine wood are recommended for brine tanks.

Cement tanks must be made of the best cement, and thoroughly hardened and dried and coated with hot asphaltum before being used.

SIZE OF BRINE TANK.

The brine tank should be no larger than is required to receive the molds, the refrigeration coils and the agitator. Generally two inches space are left between molds and three inches space where the pipes pass between them. Three feet additional length for tank are allowed for agitator. Otherwise the size of the brine tank depends on the size of the mold, *i. e.*, the time which it will take to freeze the contents solid. If it takes forty-eight hours to close the cans, the freezing tank must hold twice as much as is expected to be turned out in twenty-four hours.

THE BRINE AGITATOR.

The brine agitator is a little contrivance calculated to keep up a steady motion of the brine; it generally consists of a small propeller, driven by belt, which keeps up a constant motion of the brine from one side of the tank to the other.

HARVESTING CAN ICE.

The molds containing the ice are withdrawn from the freezing tank in small plants by "hand tackle," in larger plants by the power crane. The cans are removed by the crane to the dipping tank containing hot water, called the hot well, in which the cans are suspended for a short time, hoisted up again and turned over on an inclined plane or similar contrivance when the blocks of ice drop out and slide into the storage room. In some factories a sprinkling device takes the place of the hot well.

PLATE SYSTEM.

In making ice after the so called plate system, hollow plates through which cold brine or ammonia can be made to circulate are immersed vertically into tanks filled with water, and the ice forms gradually on both sides of the plates, thus purifying itself of any air or other impurities on its surface, which in the can system concentrate themselves toward the center, forming an impure core. For this reason it is not necessary to distill or boil water which is otherwise pure for ice making after the plate system as it is required in the can system, and hence a saving of coal by the plate system. On the other hand, the latter system requires more skill to manipulate it successfully in all its details, and the plant is more expensive to install and keep in repair. The comparisons between the two systems as to cost depend largely on the size of plants and local conditions. The following table of comparison showing the cost of production per ton of the two systems in first-rate plants will meet average conditions. It is derived from Denton, and corrected after the experiences of St. Clair and others.

	Can System.	Plate Syst.m.
Harvesting and storing, Denton.	.11	.06
Engineers and firemen	.13	.12
Coal at $3.50 per gross ton	.42	.24
Water pumped at 5c. per 1,000 cubic feet	.013	.026
Interest and depreciation at 10 per cent	.246	.327
Repairs	.027	.034
	.946	.807

SIZE OF PLATES.

The plates vary in size; generally they are 10×14 feet in area, and may be made by welding pipe into continuous coil. The spaces between the pipes are filled out by metal strips, the whole forming a solid plate.

TIME FOR FREEZING.

The freezing on the plates to form ice of a thickness of about twelve to fourteen inches takes from nine to fourteen days, forming cakes of ice weighing several tons.

HARVESTING PLATE ICE.

When the ice on the plate has become thick enough, hot ammonia taken from the system before it enters the condenser is let into the plate coil, where it loosens the ice from the metal in a few minutes. The cake is then split, and grooves cut by circular saws or hand plows enable the splitting of the whole cake into pieces of desired size, ready for market.

STORAGE OF MANUFACTURED ICE.

The question whether it is more economical to shut down the ice plant during the winter and have a plant of sufficient size to supply the summer demand, or to store ice during the winter months and get along with a smaller plant, appears now to be decided in favor of the latter system, at least under generally prevailing conditions.

ICE FOR STORAGE.

The best, clear, solid ice, without any core of any kind, is also the best for storage. Some insist that ice for storage should not be made at temperatures higher than 10° to 14° in brine tank, but where the storage or ante-room is kept cool, this is hardly required.

CONSTRUCTION OF STORAGE HOUSES.

Storage houses for manufactured ice are built on the same principle as storage houses for natural ice. Efficient insulation is the principal consideration. The house should be built as nearly square as possible, the roof should have a good pitch, and both gable ends, as well as the top, should be ventilated. The escape of cold air, as well as the ingress of warm air at the bottom should be well guarded against. A plain house may be built of frame with 2×8 studdings, lined inside with P. & B. building paper and 1-inch boards. The outside to be lined with one thickness of boards and two-ply paper, the 8-inch space between being filled with tan bark. The outside has a 4-inch air space; is then lined outside with tongued and grooved weather boarding. The roof is covered with paper, and has an 8-foot ventilator on top.

ANTE-ROOM.

Storage houses for manufactured ice should be provided with an ante-room holding some fifty tons of ice and over, so as to obviate the frequent opening of the storehouse proper. This ante-room should be kept cool by pipes supplied with refrigerated brine or ammonia from the machine.

Fifty cubic feet of ice as usually stored will equal about one ton of ice.

REFRIGERATING ICE HOUSES.

In order to keep the ice intact in storage rooms, etc., the same must be refrigerated by artificial means. Generally a brine or direct expansion coil is used for that purpose.

The refrigeration and size of coils required may be calculated after the rules given above and further on under "Cold Storage." For rough estimations it is assumed that such rooms require about ten to sixteen B. T. U. refrigeration per cubic feet contents for twenty-four hours.

About one foot of 2-inch pipe (or its equivalent in other size pipe) per fourteen to twenty cubic feet of space are frequently allowed in ice storage houses for direct expansion, and about one-half to one-third more for brine circulation.

The pipes should be located on the ceiling of the ice storage house. It is also important that the house is well ventilated from the highest point, and thoroughly drained to prevent any accumulation of moisture below the bed of ice. A foundation bed of one and a half to two feet of cinders greatly assists the drainage of the house. Ice storage houses should be painted white, but not with white lead or zinc, as a mineral paint, like barytes or patent white, will emit less heat.

PACKING ICE.

Different methods obtain in packing ice into storage houses.

Some place the blocks on edge, and as closely together as possible, and place the other blocks on top exactly over each other (no breaking of joints). Between the times of storing the ice is covered with dry sawdust or soft (not hard) wood planer shavings. The top layer is always covered with dry sawdust or shavings.

Others recommend strongly the use of ½-inch strips between layers of manufactured ice in the storehouse, the cakes being separated, top, side and bottom, from all others in the house.

Instead of sawdust, etc., rice chaff is used in the south, and it can be dried and re-used. Straw or hay is also used in places. When sawdust is used in packing ice the layer must not be too thick, as this would create heat in itself.

It is also recommended to store the ice with alternate ends touching and alternately from one and a half to two inches apart, somewhat similar to a collapsed worm fence, alternating on each row. This prevents the ice from freezing together solidly, so that it may be easily separated. The cakes should not be parallel with each other, and should never be stored unless the temperature is at, or below, the freezing point. Prairie hay is the best for covering; oat or wheat is next best, with sawdust last. Six inches of hay should be used between the ice and the wall, well packed. There should be no covering used until the house is filled. Use hay first, secondly straw, and last sawdust if no hay can be got. In warmer climates ice should be stored and covered immediately on coming from the tank at a very low temperature, say 12° or 15°.

SHRINKAGE OF ICE.

In an ice storage house without artificial refrigeration the average shrinkage from January to July will be about one-tenth pound of ice for every twenty-four hours for every square foot of wall surface. In round numbers it may amount to from 6 to 10 per cent of the ice stored in the six months mentioned.

HEAT CONDUCTING POWER OF ICE.

From an interpretation of practical data, it appears that about ten B. T. U. of heat will pass through a square foot of ice one inch thick in one hour for every degree Fahrenheit difference between the temperatures on either side of the ice sheet.

WITHDRAWING AND SHIPPING ICE.

In withdrawing ice from storage care should be taken that the water from the top does not get down to the ice below. Where there is an ante-room the same is filled from time to time from the main storage room, to withdraw from as occasion requires. For the shipment of ice in large quantities, in cars, boats, etc., it is packed the same as for storage. Small quantities of ice are frequently shipped by express, etc., in bags well packed with sawdust or the like.

In withdrawing ice from storage houses ("breaking out") skilled labor is required, and besides this the proper tools, viz.: Two breaking out bars, one for bottom and one for side breaking; otherwise much ice will be broken and wasted.

The small pieces of ice remaining on top layer, as well as any wet shavings or other material, should be removed each time when ice is taken from the house.

SELLING OF ICE.

The selling and delivery of ice is generally done by the coupon system.

It is a system of keeping an accurate account with each customer of the delivery of and the payment for ice by means of a small book containing coupons, which in the aggregate equal 500 or 1,000 or more pounds of ice, each coupon representing the number of pounds of ice taken by the customer every time ice is delivered.

These books are used in the delivery of ice in like manner as mileage books or tickets are used on the railroad. A certain number of coupons are printed on each page, each coupon being separated from the others by perforation, so that they are easily detached and taken up by the driver when ice is delivered.

Such books are each supplied with a receipt or due bill, so that if the customer purchases his ice on credit all that is necessary for the dealer to do is to have the customer sign the receipt or due bill and hand him the book containing coupons equal in the aggregate to the number of pounds of ice set forth in the receipt or due bill. The dealer then has the receipt or due bill, and the customer has the book of coupons. The only entry which the dealer has to enter against such purchaser in his books is to charge him with coupon book number, as per number on book, to the amount of 500, 1,000 or more pounds of ice, as the value of the book so delivered may be. The driver then takes up the coupons as he delivers the ice from day to day.

WEIGHT AND VOLUME OF ICE.

One cubic foot of ice weighs fifty-seven and one-half pounds at 32°.

One cubic foot of water frozen at 32° makes 1.0855 cubic feet of ice, the expansion being 8½ per cent by freezing.

One cubic foot of pure water at the point of its greatest density, 39° F., weighs 62.43 pounds.

HANDLING OF ICE.

The handling of ice during transit and delivery to the retail customer is a matter to which all possible attention should be given, especially by the dealers in manufactured ice, in order to reap the full benefit for the expense and care bestowed by them on the making of a pure article. The wagons in which the water is delivered should be in a clean, sanitary condition in fact as well as in appearance. The men in charge of them should not walk around in the wagons with muddy boots. The ice should not be slid on dirty sidewalks, and then be washed off with water from the same bucket with which the horses are watered. These things, although they may seem to be of little consequence, are nevertheless watched and commented on, and go far to discredit the just claims made by the manufacturer of ice in favor of his product. The same remarks hold good for the shipment of ice in railroad cars. They should also be properly cleaned, and in case any covering material is needed, it should be selected with the same care as that for the covering of ice in storage at the factory.

COST OF ICE.

The cost to manufacture and to keep in readiness for shipping a ton of ice varies greatly with circumstances, notably the price of fuel, the kind of water, the regularity with which the plant is operated, etc. The cost, therefore, is all the way from $1 to $2.50 per ton.

It is also found that one pound of average coal will make from five to ten pounds of ice, according to circumstances, and that from three to seven gallons of water are required per minute to make one ton of ice in twenty-four hours.

COST OF MAKING ICE.

The cost of making ice varies also considerably with the size of plant. Of a model plant producing about 100 tons of ice per twenty-four hours the following data of daily expense are recorded, and we consider them very low:

Chief engineer	$ 5.00
Assistant engineer	6.00
Firemen	4.00
Helpers	5.00
Ice pullers	9.00
Expenses	12.00
Coal, at about $1.10 per ton	18.00
Delivering at 50c per ton (wholesale delivery)	50.00
Repairs, etc.	3.00
Insurance, taxes, etc.	6.00
Interest on capital	20.55
Total for 100 tons of ice	$138.55

Calculating on the smaller production of twenty tons in twelve or twenty-four hours we obtain the following figures:

	Twenty tons in 12 hours.	Twenty tons in 24 hours.
Engineer	$ 2.50	$ 5.00
Fireman	1.50	3.00
Watchman	1.00	....
Coal	3.00	3.00
Repairs	.50	.50
Total for 20 tons of ice	$8.50	$11.50
Average per ton	.42.5 cts	57.5 cts.

APPROXIMATE COST OF OPERATING ICE FACTORIES.

The last figures on the cost of ice given on the preceding page are too low, for the reason that they leave out delivery, helpers, interest and a number of other things. The same may be said of the figures in the accompanying table, taken from the Frick Co.'s catalogue; and therefore the necessary additions and alterations, for price of coal, etc., must be made in each individual case to avoid gross errors which otherwise are unavoidable.

Tons ice per day.	Engineers $1.50 to $5.00 per day.		Oilers $1.25 per day.		Firemen $1.25 per day.		Tankmen and Laborers $1.00 per day.		General Helpers $1.25 per day.		Coal 15 cts. per Cwt., or $3.00 per ton.		Oil Waste, Lights and Sundries.	Daily Operating Expenses.	Ice per ton.
1	2	$3 00		. . .		. . .		. . .		. . .	500	$0 75	$0 50	$4 25	$4 25
2	2	3 00		. . .		. . .		. .		. . .	1,000	1 50	50	5 00	2 50
4	2	3 00		. . .		. . .	1	$1 00		. . .	2,000	3 00	50	7 50	1 88
6	2	3 50		. . .		. . .	1	1 00		. . .	2,700	4 05	50	9 05	1 51
10	2	3 50		. . .		. . .	1	1 00		. . .	4,500	6 75	75	12 00	1 20
15	2	3 75		. . .		. . .	2	2 00		. . .	6,700	10 05	1 00	16 80	1 12
20	2	3 75		. . .		. . .	2	2 00		. . .	8,000	12 00	1 00	18 75	94
25	2	4 00		. . .	2	$2 50	2	2 00		. . .	8,400	12 60	1 10	22 20	89
30	2	4 00		. . .	2	2 50	2	2 00		. . .	9,500	14 25	1 25	24 00	80
35	2	4 00		. . .	2	2 50	2	2 00		. . .	10,800	16 20	1 50	26 20	75
40	2	4 50		. . .	2	2 50	3	3 00		. . .	11,500	17 25	1 75	29 00	73
50	1	2 75		. . .	2	2 50	4	4 00		. . .	14,300	21 45	2 00	34 45	69
60	2	5 00	1	$1 25	3	3 75	4	4 00		. . .	16,000	24 00	2 25	40 25	67
75	2	5 00	1	1 25	3	3 75	4	4 00	1	$1 25	20,000	30 00	3 00	48 25	65
100	2	5 50	2	2 50	4	5 00	6	6 00	1	1 25	25,000	37 50	4 00	61 75	62

SKATING RINKS.

Artificial ice is also used for skating rinks to be operated all the year round. The amount of refrigeration, piping, etc., required for such installations depends largely on local conditions and other circumstances.

A skating rink in Paris 7,700 square feet has 15,000 feet 1-inch pipe, and the refrigerating machine requires a 100-horse power engine.

A skating rink in San Francisco, 10,000 square feet, is operated by machine of sixty tons refrigerating capacity.

The skating floor at the Shenley Park Casino in Pittsburg is constructed as follows: It consists of a 3-inch plank floor covered with two thicknesses of impervious paper; the second floor likewise covered, leaving an air space below. About 80,000 pounds of coke breeze, or about ten inches in thickness, was placed on the last named floor, the whole surmounted by 3 × 6-inch yellow pine decking, carefully spiked down and joints calked, the whole finished with a heavy coat of brewer's pitch, this preventing any dampness from reaching the insulation. Nearly 300,000 feet of lumber were used for this structure, the rink being 70 × 225 feet, or about 16,000 square feet. On the top of the floor, with the ends extending through the two ends of tank, which are rendered water tight, are 72,000 feet of 1-inch extra heavy pipe, and they are simply straight pipes 228 feet long, connected at each end by a manifold. They are operated by direct expansion. This rink will, in case of a rush, accommodate 1,100 people, and one having one-quarter of its surface would probably suffice for a patronage of 200 people. The refrigerating machine used to operate this plant has a refrigerating capacity of about 160 tons.

QUALITY OF ICE.

The keen competition between manufactured and natural ice has brought up a number of questions touching the relative merits of these articles. Although it is quite generally conceded that ice made from distilled water is in every respect purer and more healthful than natural ice, still there are claims to the contrary, some claiming that natural ice will last longer, others that distillation takes the life out of the water and ice, etc. As far as the keeping is concerned, there is no difference if the blocks are wholly frozen without holes or cracks in them; and as to the life in manufactured ice, it is certainly one of its advantages that all bacterial life is killed in the same.

WATER FOR ICE MAKING.

Expressed broadly, water that is fit for drinking purposes is fit for ice making, but while for drinking purposes a moderate amount of air and mineral matter in the water is more or less desirable, for ice making the absence of both is necessary if the ice is to be clear

But even if a natural ice from a certain source is apparently or temporarily free from pathogenic (disease) bacteria, it may nevertheless be suspected of possible or future contamination if its analysis indicates contamination with sewage or other waste matter. This is to be suspected when the ice or the water melted from the same contains an excess of ammonia, especially albuminoid ammonia, of nitrates and of chlorides. In order to give expression to this condition of things, many municipalities have special laws defining the purity required for marketable ice. The corresponding ordinance in the city of Chicago demands that: "All ice to be delivered within the city of Chicago for domestic use shall be pure and healthful ice, and is hereby defined to be ice which, upon chemical and bacteriological examination, shall be found to be free from nitrates and pathogenic bacteria, and to contain no more than nine-thousandths of one part of free ammonia and nine-thousandths of one part of albuminoid ammonia in 100,000 parts of water."

CLEAR ICE.

Although ice that is impure may be clear, and ice which is practically pure may be cloudy or milky, clear ice is nevertheless desirable, and generally called for. While many natural waters will furnish clear ice after the plate system, the can system always requires boiling, and generally previous distillation and reboiling of the water in order to furnish clear ice. (It sometimes happens that the ice of some cans is white and milky, while that of others is clear. This is generally due to a leak in the cans yielding the milky ice, whereby brine enters the same. It may be readily detected by the taste of the ice.)

BOILING OF WATER.

In case a natural water is almost free from mineral matter (or if the same consists chiefly of carbonates of lime and magnesia), and contains only suspended matter and air in solution, it may be rendered fit for clear ice making by vigorous boiling, either with or without the assistance of a vacuum, and with or without subsequent filtration, as the case may require.

DISTILLED WATER.

In order to save a vast amount of fuel, 40 per cent and upward, the exhaust steam from the engine is generally used to supply the distilled water as far as it goes, and a deficiency is supplied directly from steam boiler.

The impurities, such as grease, etc., carried by the exhaust steam, are removed by a so-called steam filter, and then the vapors are passed through a condenser constructed on the same principles as the ammonia condenser. The condenser may be submerged in water or be an atmospherical or open air condenser. For cooling, the overflow water from the ammonia condenser is used in all cases.

AMOUNT OF COOLING WATER.

If 960 B. T. U. is the latent heat of steam, and the temperature of the cooling water when it reaches the condenser is t_1, and when it leaves the condenser is t, the theoretical amount of cooling water, P, in pounds required per ton of distilled water is—

$$P = \frac{2000 \times 960}{t - t_1}$$

To this from 2 to 20 per cent should be added for loss, etc., according to size of plant.

SIZE OF CONDENSER.

If t is the mean temperature of the cooling water, that is, the average between the temperature of the water entering and leaving the condenser, and if t_1 is the average temperature in the condenser (presumably about 210° F.), then the number of square feet condenser surface, S, per ton of water in twenty-four hours is found after the rule—

$$S = \frac{2000 \times 960}{(t_1 - t)\, n \times 24}$$

n being the number of B. T. U. transferred by one square foot surface of iron pipe for each degree F. difference in one hour, steam being on one side and water on the other. For practical calculations fifty feet of 1¼-inch pipe are allowed in a steam condenser for each ton of ice produced in twenty-four hours.

This is equivalent to about twenty-two square feet of condenser surface per ton of ice made in twenty-four hours. If we assume that this amount of surface is calculated to prove fully sufficient even if the cooling water has a temperature of 130° F. (the range temperature in this case being $212 - \frac{212 + 130}{2} = 40$) we find $n = 100$ or very nearly that.

From experiments quoted on page 27 it appears that n varies from 200 to 500 units, and is still more, nearly twice that, in case of brass or copper pipe which is frequently used for steam condensers in distilling apparatus. We may assume, therefore, that $n = 100$ will give ample condensing in extreme cases, and also allow for decrease in the heat transmitting power of iron pipes on account of oxidation, incrustation and the like.

The condenser should be provided with an efficient gas and air collector.

In case the natural water is very impure, a filtration of the same before it enters the steam boiler is very advisable and frequently resorted to.

Various kinds of filters are used, sponge, charcoal and sand filters most generally; in exceptional cases boneblack filters are used also. In case the water contains much dissolved organic matter, filtering with addition of alum is found very advantageous in many cases.

REBOILING AND FILTERING WATER.

The condensed distilled water contains air in solution, and sometimes also certain other volatile substances, possessing more or less objectionable flavors. To free it from both, the water is subjected to vigorous reboiling in a separate tank. Impurities thrown to the surface are skimmed off.

The reboiling of the water must not be done by live steam (no perforated steam coil) if the water has naturally a bad smell.

As a still further means of purification a charcoal or other filter is used, through which the water passes after reboiling.

COOLING THE DISTILLED WATER.

The filtered and boiled distilled water is now passed through a condenser coil over which cold water (water which is afterward used on ammonia condenser) passes, and after it is cooled down here as much as practicable, it runs to the storage tank which is generally provided with a direct expansion ammonia coil to reduce the temperature of the water as near to the freezing point as possible. From the storage tank the freezing cans are filled as required.

INTERMEDIATE FILTER.

Frequently another water filter is placed between the water cooler and the storage tank.

DIMENSIONS OF DISTILLING PLANT.

As is the case with most other appliances in the refrigerating practice, the dimensions of the different parts of a distilling plant vary considerable with different manufacturers. For superficial guidance we will quote one or two examples.

TEN-TON DISTILLING PLANT.

Open air condenser consisting of ninety-six pipes, each five feet long and one and one-quarter inch diameter.

Reboiler four feet diameter, three feet high, containing steam coil of about sixty feet ½-inch pipe.

Intermediate cooler to bring temperature of reboiled water to about 80° consists of eighteen pieces 1¼-inch pipe, each twelve feet long.

Charcoal filter thirty inches in diameter and seven feet high. Layer of charcoal five feet high.

Cooling and storage tank three feet diameter and seven feet high, contains 250 feet 1¼-inch pipe for direct ammonia circulation.

In the installation of a plant it is generally prudent to expect an increase in the production, and on this basis the above dimensions might well apply to smaller plants, say downward to five tons.

THIRTY-TON PLANT.

Steam filter three feet diameter, seven feet high with five consecutive wire screens, sixteen meshes per inch.

Surface condenser containing 100 pieces 1-inch brass pipe, each four and a half feet long. (*On this basis and on the assumption made in the discussion of the last formula n would be equal to about 400 units for brass, which nearly agrees with the experiments quoted on page 26.*)

Reboiler twenty-four inches diameter, six feet high, containing four feet steam pipe ten inches in diameter.

Intermediate cooler, thirty-two pieces of 2-inch pipe, each seventeen and a half feet long.

Two charcoal filters, each three feet diameter, seven feet high. Layer of charcoal five and a half feet high.

Cooling and storage tank six and a half feet diameter and eight feet high, containing 750 feet 1¼-inch pipe for direct ammonia expansion.

Sand filter two feet diameter, four feet high.

THE SKIMMER.

The skimmer is a contrivance which is arranged in many plants between the reboiler and the intermediate cooler, to skim off oil or any other light impurities which may float on the water. It is a small cylindrical vessel with an overflow at the top and connected to the reboiler with a straight pipe on the one side, and on the other with the intermediate cooling coil. The flow of the distilled water to the latter coil is so regulated that a small amount of water will always overflow from the skimmer, taking with it the impurities. Sometimes the skimmer is provided with a steam coil to keep the water boiling, thus facilitating the rising to the surface of impurities.

BRINE CIRCULATION.

Among the other devices used for brine circulation besides propeller wheels, paddles, etc., we mention the pump—preferably a centrifugal pump with a system of brine suction and discharge pipes located inside of the freezing tank, to take out the suction and return the discharge brine at regular intervals of space throughout the length and breadth of the tank, so that every spot between the cans is drawn into the circulation.

ECONOMIZING FUEL.

As much of the overflow water from the steam condenser as may be needed for boiler feeding should be made to pass through a feed water heater located between the steam filter or oil separator and the condenser. Through this heater the hot steam passes first, to make the feed water as hot as possible.

ARRANGEMENT OF PLANT.

It is essential that the whole of the distilling apparatus is kept clean, sweet and free from iron rust; for these reasons the plant should be so arranged that all tanks, pipes, etc., which contain or conduct the distilled water are constantly filled with the same.

The plant should be cleaned as often as necessary by steaming the same out.

DEFECTS OF ICE.

Water which has gone through the process of distillation, condensation, reboiling, skimming, etc., does not always make unobjectionable ice, perfectly clear, without core or without taste and flavor. Ice may be practically pure, wholesome and palatable while containing these defects, and although most successful manufacturers know how, as a rule, to avoid these defects, still, occasionally they turn up and often prove to be a great annoyance.

WHITE OR MILKY ICE.

White or cloudy or milky ice is generally due to the presence of air in the distilled water; it is caused by deficient reboiling or by overworking the reboiler, by a deficient supply of steam to the distilled water condenser. In the latter case, a vacuum is formed through the rapid condensation of the steam, and more air is drawn in and mixed with the steam than can be driven away by the usual extent of reboiling. If in this case the supply of steam cannot be increased, the amount of cooling water running over the condenser must be reduced, in order to keep the pressure up in the condenser. Otherwise the distilled water must be more thoroughly reboiled. Air is also drawn in sometimes during the filling of the cans, through leaks in the distilled water pipes, etc.

Frequently milky or streaky ice is also due to leaks in the freezing can, through which brine may be allowed to mix with the water in the can, which will then show as white or milky ice or as white spots or streaks. The salty taste of these parts readily shows their cause, which may be remedied by mending the cans.

ICE WITH WHITE CORE.

The white core which forms in the ice from the last portion of the freezing water is due to mineral water (generally carbonate of lime and magnesia) derived from the natural water, from which it has not been successfully separated, this separation being the principal object of the distilling process. In most cases the core is caused by the priming of the boiler, by carrying too much water, or by overworking the boiler and also not blowing off the boiler often enough, in which case the mineral constituents of the water accumulate and increase the danger of priming. The most rational remedy in this case is boilers large enough to make overworking, high water and priming entirely impossible. Another important remedy is the purification of the water before it enters the boilers.

ICE WITH RED CORE.

The red core in ice is brought about by a separation of oxide of iron in the ice, which was kept in solution in the water in the form of carbonate of iron. This sediment is nearly always derived from the iron of the plant, more especially the coils. It frequently sets in during the second season of the working of a plant, and then is directly traceable to the rust which has formed within during the idle months or during shorter stoppages. To prevent this, the pipes and tanks might be kept filled with distilled and thoroughly reboiled water. If the water supply carries much carbonic acid, this substance may contaminate the distilled water in such a manner as to dissolve iron from coils, etc., which is afterward deposited in the ice, as set forth.

It has also been proposed to use pipe tinned inside for the distilled water condenser, and, if possible, tinned surface throughout the distilled water plant, to avoid the possibility of contamination with iron from this source. If the water supply carries carbonate of iron in solution, this may also become the cause of a red core, but only in case the boiler primes or is overworked and foul, and if the filters do not do their duty.

The formation of this red core will doubtless be avoided in the future by proper treatment of the water and more careful management of boilers and plant. For the present, in cases where prevention is impossible, a cure may be effected by cooling the distilled water down to about 36° to 38°, at which temperature the iron will separate and may be separated by means of ordinary small sand or sponge filter.

A radical, but rather expensive and troublesome, means to prevent the formation of a core of any kind, consists in the removal of the water still remaining unfrozen in the nearly complete ice block, just before the core begins to form, by means of a syringe and refilling the space with clear distilled water.

DANGERS OF FILTRATION.

A core in the ice may also be caused by mineral matter, which has been imparted to the distilled water by the very process of filtration. When, as sometimes happens, the distilled water is charged with carbonic acid gas, and boneblack (not previously chemically treated) is used as filtering material, the water will take up a certain amount of carbonate of lime from the boneblack, and cause a white core. Other impurities in filtering material will cause similar cores.

COLOR, TASTE AND FLAVOR OF ICE.

Regarding the odor and taste possessed by some distilled water, or by ice made therefrom, and also the greenish color shown by some ice, they are due to the presence of minute quantities of volatile matter (belonging to the hydrocarbon class), which are derived from the natural water supply or from the lubricating materials. If their presence is due to the water, these defects, as in fact also most other defects in ice, will become more apparent if the boilers are allowed to become foul; and, on the other hand, if the boilers are cleaned and blown off with sufficient frequency (in the case of vile water as often as once in twenty-four hours) these defects, like others, may be so reduced as to become almost unnoticeable. Priming of the boilers, of course, also increases these as well as other defects.

If odor, taste or color of the ice are derived from the lubricating oil (which also sometimes causes cloudy ice) efficient oil or steam filters, kept in proper order, are the best remedy. An improper and excessive use of cylinder oil should also be carefully guarded against.

BEST USE OF BONEBLACK.

Where these preventive remedies do not apply, the distilled water may be freed from these defects (taste and odor) by filtering it through granulated boneblack; and where this is found too expensive, as in the absence of means for revivifying the spent boneblack, the latter may be used, after having been reduced to an impalpable powder. In this shape a pound or two of boneblack will go a long way, and will suffice to withdraw any smell or taste from a ton of water. To this end this powder should be intimately mixed with the distilled water in the said proportion before the last filtration, which will retain the boneblack, together with the impurities which it has absorbed from the water.

Blood charcoal will act even more efficiently in this respect, but it is very doubtful whether its superiority to boneblack, powdered equally fine, is sufficient to overcome its high price (eighty-five cents per pound for the best imported article). With this material it is also important to make sure that it has been freed from all soluble constituents before using.

NUMBER OF FILTERS REQUIRED.

Regarding the number and kind of filters required, it would appear from the foregoing that this question must be settled separately for individual cases. When the distilled water supply is charged with much oily matter, with odoriferous volatile products, and also with mineral substances held in solution, we shall doubtless stand in need, at least for the time being, of an oil or steam filter, of a charcoal or boneblack filter (or boneblack powder) and of a filter between the freezing can and the distilled water or cooling tank.

If mineral matters were entirely absent the last filter would not be needed, and if volatile products are absent the charcoal or boneblack treatment may be dispensed with, and *vice versa;* and in case where the vapor from which the water is to be condensed is absolutely pure, and the coils and tanks of the condensing apparatus likewise, no filters, skimmers and the like will be required at all.

It is to be hoped that within the near future the natural water supplies will be so improved, and the management of boilers, engines, lubricators, condensing coils, reboilers, etc., will be manipulated universally in such a manner that the purity of the ice can be insured without so much attention. In this respect, frequent cleaning of boilers, blowing out of coils by steam when stopping and starting, and careful lubricating are among the first points to be considered.

Under all circumstances, however, a simple but efficient filter between distilled water storage tank and the freezing cans will always be found a valuable help and safeguard. The filtering apparatus recently introduced for this purpose, consisting of two perforated disks with special filtering cloths between, is a neat and compact apparatus which seems to satisfy all demands as regards easy application, simple operation, economy of space, little attention and efficiency.

ROTTEN ICE.

When complaints are made about the "quick melting away" of manufactured ice, it will be found that it is generally caused by incomplete cakes, or cakes which have not completely closed in the center. The increased surface thus given to a cake causes it to melt away quicker, in increasing proportion as the surface of the whole increases by this procedure. For these reasons holes in ice must be avoided, and every piece of ice should be frozen solid all over.

So called rotten ice also melts away quickly; it is ice, the surface of which is also increased by cracks proceeding from the outside to the center. Such ice is frequently withdrawn from the outside layers of stored manufactured ice not protected by mechanical refrigeration during the storage, and the application of such refrigeration is the best remedy for it.

TEST FOR WATER AND ICE.

Water if properly distilled (and of course ice made from such water, likewise) if slowly evaporated on a piece of platinum foil on a spirit lamp or a Bunsen gas burner, should leave no solid residue. If care is used in performing the operation a piece of thin glass plate may be used instead of the platinum foil.

PURE WATER.

The opinions on the requirements to be made of a water supply vary considerably; the following may stand for a sample of what some authorities demand of a water fit for drinking and other domestic purposes, and in some measure it may also be applied to ice.

1. Such water should be clear, temperature not above 15° C.
2. It should contain some air.
3. It should contain in 1,000,000 parts:
   Not more than 20 parts of organic matter.
   Not more than 0.1 part of albuminoid ammonia.
   Not more than 0.5 part of free ammonia.
4. It should contain no nitrites, no sulphurated hydrogen, and only traces of iron, aluminum and magnesium. Besides the mentioned substances it should not contain anything that is precipitable by sulphureted ammonia.
5. It must not contract any odor in closed vessels.
6. It must contain no saprophites and leptothrix and no bacteria and infusoria in notable quantities.

7. Addition of sugar must cause no development of fungoid growth.

8. On gelatine it must not generate any liquefying colonies of bacteria.

DEVICES FOR MAKING CLEAR ICE.

Besides the plate system and the use of distilled water, a number of contrivances have been devised for the manufacture of clear ice from natural water. The efficiency of these devices is based upon the motion which they keep up in the water in various ways. Their detailed description cannot be attempted here; moreover it seems that they have not given much satisfaction generally; probably they are too cumbersome and too uncertain in their performance.

THE CELL SYSTEM.

From the other methods in use for ice making, we may yet mention the cell system, which is in use on the continent to some extent. It consists of a series of walls of cast or wrought iron placed from twelve to eighteen inches apart, the space between each pair of walls being filled with the water to be frozen. The cooled brine circulates within a number of spaces left in the walls, and the ice forms on the walls, increasing in thickness until the two opposite layers meet. If thinner blocks are required, freezing may be stopped at any time, and the ice removed. In order to detach the ice from the walls warmer brine may be circulated through the cell walls to loosen the ice. It stands to reason that impurities of the water will be separated from the same on the ice, if the two opposite layers are not allowed to meet. It will take, however, nearly double the time to freeze a block of a given thickness if the two layers are not allowed to meet to form one solid block.

COST OF REFRIGERATION.

In order to arrive at the possible remunerability of a refrigerating plant calculated to turn out artificial ice, it is but fair to compare the cost of the latter with the price of pure natural ice in the available market. If, however, on the other hand, a refrigerating plant is calculated to replace natural ice in the cooling of storage room, ice boxes, etc., the above calculation must be changed as a matter of course.

CHAPTER VI.—COLD STORAGE.

COLD STORAGE.

Cold storage in general comprises the preservation of perishable articles by means of low temperature, and is one of the principal cases to which artificial refrigeration is applied.

STORAGE ROOMS.

Cold storage rooms, like ice houses, are built to be as perfectly insulated and protected as possible against the egress of cold and ingress of heat. They are kept cold by systems of pipe lines through which circulates either refrigerated ammonia (direct expansion) or cooled brine (brine system). The size of the house depends on the storage requirements; they should be built as nearly square as possible, be properly ventilated, have double doors and windows, and all other protections that will insure the best insulation possible. The size of cold storage rooms varies from that of a small ice box of a few cubic feet capacity to that of gigantic storehouses of several million cubic feet space.

CONSTRUCTION OF COLD STORAGE HOUSES.

It is not within the scope of this treatise to go into details on this subject; nevertheless the descriptions of two specimens of walls for insulated buildings for storage and other purposes, which have given excellent satisfaction, may find a place here.

CONSTRUCTION OF WOOD.

A strong and well insulated wall of wood may be constructed by placing 2×6-inch studs twenty-four inches apart; and in order to form outside of wall nail on them first a layer of 1-inch matched boards, then a layer of two-ply paper, and again a layer of 1-inch matched boards. On the inside a layer of 1-inch matched boards is nailed on the studs, and against these boards 2×2-inch studs are placed twenty-four inches apart. In order to form the inside of wall one layer of 1-inch matched boards is nailed on the 2×2-inch studs, then a layer of two-ply paper, and lastly another layer of 1-inch matched boards on top of this paper. The spaces left between the 2×2-inch studs are left as air spaces, while the spaces between the 2×6-inch studs are filled in with sawdust crushed cork or the like.

CONSTRUCTION OF BRICK AND TILES.

For brick and tile construction the outside of the walls is formed of a brick wall sixteen inches or more in thickness, according to size and height of building. On the inside the wall is plastered. Again, a wall built of 4-inch hollow tiles is placed at a distance of three inches from the plaster coating of the brick wall, and a coat of plaster or cement on tiles on the inside finishes the whole wall. The space between the tiles and brick wall may be filled in with cork, sawdust or some other insulating material.

If the space between tiles and brick is filled with mineral wool, the wall represents a fire-proof structure.

OTHER CONSTRUCTIONS.

The following materials and dimensions have been recommended for walls of cold chambers by Taylor:

Fourteen-inch brick wall, 3½-inch air space, 9-inch brick wall, 1-inch layer of cement, 1-inch layer of pitch, 2×3-inch studding, layer of tar paper, 1-inch tongued and grooved boarding, 2×4-inch studding, 1-inch tongued and grooved board, layer of tar paper, and, finally, 1-inch tongued and grooved boarding, the total thickness of these layers or skins being 3 feet 3 inches.

Thirty-six-inch brick wall, 1-inch layer of pitch, 1-inch sheathing, 4-inch air space, 2×4-inch studding, 1-inch sheathing, 3 inch layer of mineral or slag wool, 2×4-inch studding, and, finally, 1-inch sheathing; total thickness 4 feet 7 inches.

Fourteen-inch brick wall, 4-inch pitch and ashes, 4-inch brick wall, 4-inch air space, 14-inch brick wall; total thickness 3 feet 4 inches.

Fourteen-inch brick wall, 6-inch air space, double thickness of 1-inch tongued and grooved boards, with a layer of water-proof paper between them, 2-inch layer of best quality of hair felt, second double thickness of 1-inch tongued and grooved boards, with a similar layer of paper between them; total thickness, 2 feet 2 inches.

Fourteen-inch brick wall, 8-inch layer of sawdust, double thickness of 1-inch tongued and grooved boards, with a layer of tarred water-proof paper between them, 2-inch layer of hair felt, second double thickness of 1-inch tongued and grooved boards with similar layer of paper between them; total thickness, 2 feet 4½ inches.

The cold storage chambers built at the St. Katherine dock, London, are constructed as follows:

On the concrete floor of the vault, as it stood originally, a covering of rough boards 1¼ inches in thickness was laid longitudinally. On this layer of boards were then placed transversely, bearers formed of joist 4½ inches in depth by 3 inches in width, and spaced 21 inches apart. These bearers supported the floor of the storage chamber, which consisted of 2½-inch battens tongued and grooved. The 4½-inch wide space or clearance between this floor and the layer or covering of rough boards upon the lower concrete floor was filled with well dried wood charcoal. The walls and roof were formed of uprights 5½×3 inches fixed upon the floor joists or bearers, and having an outer and inner skin attached thereto; the former consisting of 2-inch boards, and the latter of two thicknesses or layers of 1¼-inch boards with an intermediate layer of especially prepared brown paper. The 5½-inch clearance or space between the said inner and outer skeins of the walls and roof was likewise filled with wood charcoal, carefully dried.

CONSTRUCTION OF SMALL ROOMS.

Small storage rooms, down to ice boxes, are always built of wood, paper, cork, etc., on lines similar to those given for wooden walls, but with endless variations.

CONSTRUCTIONS AND THEIR HEAT LEAKAGE.

The following construction of walls for cold storage buildings, taken from the catalogue of the Fred W. Wolf Co., have also been practically tested, and the approximate heat leakage through them per square foot and per degree of difference in temperature between inside and outside of the room, is also given in British thermal units in twenty-four hours.

FIREPROOF WALL AND CEILING.

Brick wall of thickness to suit height of building, 3-inch scratched hollow tiles against brick wall, 4-inch space filled with mineral wool, 3-inch scratched hollow tiles, cement plaster. Heat leakage 0.70 B. T. U.

The ceiling to match this wall consists of the following layers: Concrete floor, 3-inch book tiles, 6-inch dry underfilling, double space hollow tile arches, cement plaster. Heat leakage 0.80 B. T. U.

WOOD INSULATION AGAINST BRICK WALL.

The following wood insulation against a brick wall has a leakage of 1.74 B. T. U., and consists of the following layers:

Brick wall, against which are nailed wooden strips 1 × 2 inches. On these are nailed two layers of 1-inch sheathing with two layers of paper between; next we have 2 × 4-inch studs sixteen inches apart, filled in between with mineral wool, 1-inch matched sheathing, two layers of paper; 1 × 2-inch strips, sixteen inches apart from centers; double 1-inch flooring with two layers of paper between.

CONSTRUCTIONS OF WOOD.

The following constructions of wall, ceiling and floor may be followed for cold storage rooms when built of wood:

The wall is constructed as follows: Outside siding, two layers of paper, 1-inch matched sheathing, 2 × 6-inch studs, sixteen inches apart from centers, two layers of 1-inch sheathing, with two layers of paper between, 2 × 4-inch studs, sixteen inches apart from centers, filled in between with mineral wool, 1-inch sheathing, two layers of paper, 2 × 2-inch strips, sixteen inches from center to center, two layers 1-inch flooring, with two layers of paper between. The heat leakage through this wall is 2.90 B. T. U.

The ceiling has the following details:

A double 1-inch floor with two layers of paper between, 2 × 2-inch strips, sixteen inches apart from center, filled in between with mineral wool, two layers of paper, 1-inch matched sheathing, 2 × 2-inch strips, sixteen inches apart, filled between with mineral wool, two layers of paper, 1 inch matched sheathing, joists, double 1-inch flooring, with two layers of paper between. The heat leakage through this ceiling amounts to 2.17 B. T. U.

The details of the floor are as follows:

Two-inch matched flooring, two layers of paper, 1-inch matched sheathing, 4 × 4-inch sleepers, sixteen inches apart from centers, filled between with mineral wool, double 1-inch matched sheathing, with twelve layers of paper between, 4 × 4-inch sleepers sixteen inches apart from centers imbedded in 12-inch dry underfilling.

The heat leakage through this floor is given at 1.92 B. T. U.

PIPING.

All ammonia brine and heating pipes, headers and mains ought to be in the corridors, well insulated.

CONSTRUCTIONS WITH AIR INSULATIONS.

In the following constructions, taken from the catalogue of the De La Vergne Refrigerating Machine Co., the insulating spaces are made by confined bodies of air, it being claimed by some that any filling of these spaces with loose non-conducting material will settle in places. The penetration of air and moisture is specially guarded against by the use of pitch in connection with brick or stone, or by paper where wood is used. Joints between boards should be laid in white lead and corners should be protected by triangular pieces of wood with paper placed carefully behind.

CONSTRUCTIONS OF WOOD.

The main walls of buildings (for refrigerators of hotels, restaurants and cold storage in general) built on the foregoing principles, have the following details, commencing inside: ⅞-inch spruce, insulating paper, ⅞-inch spruce, 1-inch air space, twelve inches square, ⅞-inch spruce, insulating paper, ⅞-inch spruce, 1-inch air space, ⅞-inch spruce, insulating paper, ⅞-inch hard wood.

The ceiling or floor, when the room above or below is not cooled, has the following details, commencing below the joists: ⅞-inch board, insulating paper, ⅞-inch board, floor beams, ⅞-inch board, insulating paper, ⅞-inch board (two inches air space, ⅞-inch board, insulating paper, ⅞-inch board). If room above is cooled, the parts in parenthesis may be omitted.

Partitions between two cooled rooms, where difference of temperature does not exceed 20°, may be constructed as follows: ⅞-inch board, insulating paper, ⅞-inch board, 1½-inch air space, ⅞-inch board, insulating paper, ⅞-inch board.

For main inside walls between two rooms, of which one is not cooled, the following construction may be followed: ⅞-inch board, insulating paper, ⅞-inch board, two inches air space, ⅞-inch board, insulating paper, ⅞-inch board, two inches air space, ⅞-inch board, insulating paper, ⅞-inch board.

CONSTRUCTION IN BRICK.

The outer walls in buildings of brick may be constructed as follows, commencing outside: Brick wall of proper strength, two coats of pitch, two inches air space, ⅞-inch board, insulating paper, ⅞-inch board, two inches air space, ⅞-inch board, insulating paper, ⅞-inch board.

The ceiling may be constructed as follows, when room above is not cooled (commencing at the top layer): One inch asphalt, two inches concrete, brick, wooden strips, ⅞-inch board, insulating paper, ⅞-inch board, two inches air space, ⅞-inch board, insulating paper, ⅞-inch board.

If the difference in temperature between the lower and upper room does not exceed 20° F. the following construction for ceiling may be used: One inch asphalt, two inches concrete, brick.

SURFACE OF INTERIOR WALLS.

It is claimed that the porosity of the surfaces of walls in cold storage rooms is in a measure responsible for the spoiling of provisions. Such walls, if made of cement, plaster and similar semi-porous material, possess sufficient moisture to give rise to all sorts of putrefactive and bacterial growths, allowing them to thrive under favorable conditions. A further objection to this kind of walls is the quicker radiation of heat through them. For these reasons it has been urged that the walls in cold storage houses for cold and especially meat storage, should be made from porcelain, and that they should be cleaned several times during the year.

REFRIGERATION REQUIRED.

The amount of refrigeration required in a given case depends on a number of circumstances and conditions, the size of the room, the frequency with which the articles are brought in and removed, their temperature, specific heat of produce, etc. For these reasons it is impossible to give a simple general rule, and the following figures, which are frequently used in rough calculations, must be considered as approximations only:

For storage rooms of 1,000,000 cubic feet and over, 20 to 40 B. T. U. per cubic foot per twenty-four hours.

For storage rooms 50,000 cubic feet and over, 40 to 70 B. T. U. per cubic foot per twenty-four hours.

For boxes or rooms 1,000 cubic feet and over, 50 to 100 B. T. U. per cubic foot per twenty-four hours.

For boxes less than 100 cubic feet, 100 to 300 B. T. U. per twenty-four hours.

For rooms in which provisions are to be chilled, about 50 per cent additional refrigeration may be allowed in approximate estimations. For actual freezing the amount should be doubled (see also Meat Storage).

PIPING AND REFRIGERATION.

The foregoing rules on refrigerating capacity, as well as those given elsewhere, and including also the rules for piping given on pages 134 to 138, and elsewhere, have in common one vital defect, in that they fit only one given temperature or rooms of one certain size. This condition of things necessarily gives rise to numerous misunderstandings and many errors, and for this reason I have endeavored to outline some tables which would do equal justice to all the elements involved, or at least indicate how this could be done. The desire of the author to supply such much needed tables without further delay must be an excuse for their imperfections, as so far only comparatively few of the values given therein could be verified by data taken from actual experience.

TABULATED REFRIGERATING CAPACITY.

The amount of refrigeration required for cold storage buildings for provisions, beer, meat, ice, etc., depends, as has been mentioned repeatedly, principally on the size of the rooms, their insulation, the maximal outside temperature and the minimal inside temperature (leaving openings, opening of doors and refrigeration of contents, etc., out of the question). The chief variants among these quantities are the degree of insulation, the size of rooms or houses and the minimal temperature within (the latter depending on the objects of storage); while for the maximal outside temperature we may agree upon a certain fixed quantity, which for approximate calculations will apply for a large territory of the United States, at least.

We may safely take this maximal temperature for most of the United States at 80° to 90° F., so it will amply cover 85° F.

Doing this, we can readily outline a table which will show the amount of refrigeration required for rooms of different sizes and of different insulation for any given temperature, as, for instance, the following table, which gives the number of cubic feet in cold storage buildings which can be covered by one ton of refrigerating capacity for rooms of different sizes, for different temperatures and for different (excellent and poor) insulation during a period of twenty-four hours:

NUMBER OF CUBIC FEET COVERED BY ONE TON REFRIGERATING CAPACITY FOR TWENTY-FOUR HOURS.

Size of building in cub. ft. more or less.	Insulation.	Temperature ° F.					
		0°	10°	20°	30°	40°	50°
100	excellent	150	600	800	1,000	1,600	3,000
	poor	70	300	400	600	900	2,000
1,000	excellent	500	2,500	3,000	4,000	6,000	12,000
	poor	250	1,500	1,800	2,500	5,000	10,000
10,000	excellent	700	3,000	4,000	6,000	9,000	18,000
	poor	300	1,800	2,500	3,500	7,000	14,000
30,000	excellent	1,000	5,000	6,000	8,000	13,000	25,000
	poor	500	3,000	3,500	5,000	11,000	20,000
100,000	excellent	1,500	7,500	9,000	14,000	20,000	40,000
	poor	800	4,500	5,000	8,000	16,000	35,000

The next table is constructed on the same basis, giving the amount of refrigeration required per cubic foot of space for storage rooms of different sizes for different temperatures, expressed in British thermal units, and for a period of twenty-four hours.

REFRIGERATING CAPACITY IN B. T. U. REQUIRED PER CUBIC FOOT OF STORAGE ROOM IN TWENTY-FOUR HOURS.

Size of building in cub. ft. more or less.	Insulation.	Temperature ° F.					
		0°	10°	20°	30°	40°	50°
100	excellent	1,800	480	360	284	180	95
	poor	4,000	960	480	470	330	140
1,000	excellent	550	110	95	70	47	24
	poor	1,100	190	165	110	55	28
10,000	excellent	400	95	70	47	30	16
	poor	900	160	110	81	40	20
30,000	excellent	280	55	47	35	22	11
	poor	550	95	81	55	26	14
100,000	excellent	190	38	30	20	14	7
	poor	350	63	55	35	18	4

The expression "excellent insulation" in the above and following tables may be taken to refer to walls, ceilings, etc., the heat leakage of which does not exceed two B. T. U. for each degree F. difference in temperature per square foot in twenty-four hours; and the expression "poor insulation" may be taken to refer to walls, etc., the heat leakage in which amounts to four B. T. U. and more. The average of the amounts of refrigeration, space and pipes given in the tables may be taken for average good insulation, other circumstances being equal.

TABULATED AMOUNTS OF PIPING.

The amount of piping required for cold storage buildings depends, in the first place, on the amount of refrigeration to be distributed thereby, and therefore indirectly on the same conditions as does the amount of refrigeration required. In addition thereto the amount of piping also depends on the difference between the temperature within the refrigerating or direct expansion pipes, and without. As this difference may be varied arbitrarily by the operator, and necessarily differs for different storage temperatures, it would be veiy difficult to arrange a table fitting all possible conditions.

However, it stands to reason that for each storage temperature there is one preferable brine or expansion temperature, and the accompanying tables on piping are expected to fit these temperatures for practical calculations.

LINEAL FEET OF 1-INCH PIPE REQUIRED PER CUBIC FOOT OF COLD STORAGE SPACE.

Size of building in cub. ft. more or less.	Insulation.	Temperature ° F.					
		0°	10°	20°	30°	40°	50°
100	excellent	3.0	0.78	0.48	0.36	0.24	0.15
	poor	6.0	1.50	0.90	0.66	0.48	0.30
1,000	excellent	1.0	0.26	0.16	0.12	0.08	0.05
	poor	2.0	0.50	0.30	0.22	0.16	0.10
10,000	excellent	0.61	0.16	0.10	0.075	0.055	0.035
	poor	1.2	0.33	0.20	1.15	0.11	0.07
30,000	excellent	0.5	0.13	0.08	0.06	0.040	0.025
	poor	1.	0.25	0.15	0.11	0.03	0.05
100,000	excellent	0.38	0.10	0.06	0.045	0.03	0.009
	poor	0.75	0.20	0.12	0.09	0.06	0.018

The quantities of pipe given in the foregoing table refer to direct expansion, and should be made one and one-half times to twice that long for brine circulation. They also refer to 1-inch pipe, and by dividing the lengths given by 1.25, or multiplying them by 0.8, the corresponding amount of 1¼-inch pipe is found. To find the corresponding amount of 2-inch pipe, the length given in the table must be divided by 1.8, or multiplied by 0.55.

The next table is for the same purpose as the one preceding, but it shows the number of cubic feet of storage building which will be covered by one foot of 1-inch pipe during a period of twenty-four hours for different sized rooms and different storage temperatures.

NUMBER OF CUBIC FEET COVERED BY ONE FOOT OF ONE-INCH IRON PIPE.

Size of building in cub. ft. more or less.	Insulation.	Temperature ° F.					
		0°	10°	20°	30°	40°	50°
100	excellent	0.3	1.3	2.1	2.8	4.2	7.0
	poor	0.15	0.7	1.1	1.5	2.1	3.5
1,000	excellent	1.0	4.	6.0	8.4	12.4	20.
	poor	0.5	2.	3.2	4.5	6.2	10.
10,000	excellent	1.7	6.	10.	13.	18.	28.
	poor	0.85	3.	5.	6.5	9.	14.
30,000	excellent	2.0	8.	14.	18.	25.	40.
	poor	1.0	4.	7.	9.	13.	20.
100,000	excellent	2.6	10.	17.	22.	33.	110.
	poor	1.3	5.	8.5	11.	17.	55.

The number of cubic feet of space given in the last table as being covered by one lineal foot of pipe refers to direct expansion, and only one-half to two-thirds of that space would be covered by the same amount of pipe in case of brine circulation.

The figures in this table also refer to 1-inch pipe; and to find the corresponding amounts of cubic feet of space which would be covered by one lineal foot of 1¼-inch pipe, the numbers given in the table have to be multiplied by 1.25 or be divided by 0.8. To find the corresponding amount of space which will be covered by one lineal foot of 2-inch pipe, the numbers given in the table must be multiplied by 1.8 or divided by 0.55.

The foregoing tables are calculated for a maximum outside temperature of 80° to 90° F. If the same is materially more or less about 10 per cent of refrigeration and piping should be added or deducted for every 5° F. more or less, as the case may be.

TABLES FOR REFRIGERATING CAPACITY.

The accompanying table designed by Criswell is calculated on the lines laid out in the foregoing paragraphs, on the assumption that the walls, ceiling and floor of the cold storage building have an average heat leakage of three B. T. U. per square foot in each twenty-four hours for each degree Fahrenheit difference in temperature outside and inside of building. The maximum temperature is taken at 82° F. Accordingly the total refrigeration for such a building is found by multiplying its total surface in square feet (third column of table) by 3, and the difference between the temperature in degrees Fahrenheit within the storage building and 82° F. It is then divided by 284,000 to reduce the refrigerating capacity to tons of refrigeration.

We will take for an example the building, 25×40×10. Its surface is 3,300 square feet, and the total refrigeration required for a temperature of 32° within the cold storage house is therefore $3{,}300 \times 3 \times (82-32) = \frac{495{,}000}{284{,}000} =$ 1.53 tons, or, in round numbers, 1.5 tons.

The building here referred to contains 10,000 feet, consequently one ton of refrigeration would cover $\frac{10{,}000}{1.51}$ =6,600 cubic feet of such a building. This figure should agree with the corresponding figure, given in the accompanying table (at least, approximately so), some of the figures in the table being obtained by interpolation or averaging. If we compare this table with the table given on page 175 we will note several apparent discrepancies. They are explained by the desire to give a very liberal estimate in the tables on page 175, and to make allowance not only for the refrigerating of the contents, but also for the opening of doors. These are doubtless the reasons why the refrigerating capacity for smaller rooms in table on page 175 appears so large, especially at lower temperatures, as in these cases the opening of doors, etc., acts most wastefully.

TABLE FOR REFRIGERATING CAPACITY.

SIZE OF BUILDING.				NUMBER OF CUBIC FEET PER TON OF REFRIGERATION AT TEMPERATURES GIVEN.						
				Temperatures.						
Dimensions building.	Contents, cubic feet.	Surface in square feet.	Ratio cubic feet to square feet.	0°	8°	16°	24°	32°	40°	48°
5x 4x 5	100	130	1.3	900	1,100	1,300	1,500	1,700	1,900	2,100
8x 10x 10	800	520	.65	1,800	2,200	2,600	3,000	3,400	3,800	4,200
10x 10x 10	1,000	600	.6	1,940	2,376	2,808	3,240	3,670	4,104	4,530
25x 40x 10	10,000	3,300	.33	3,600	4,400	5,200	6,000	6,700	7,600	8,400
20x 50x 20	20,000	4,800	.24	4,860	5,940	7,020	8,100	9,180	10,260	11,340
30x 50x 20	30,000	6,200	.206	5,670	6,930	8,190	9,650	10,710	11,970	13,230
40x 50x 20	40,000	7.600	.19	6,300	7,700	9,100	10,500	11,900	13,300	14,700
50x 50x 20	50,000	9,000	.18	6,480	7,920	9,360	10,800	12,240	13,680	15,120
60x 50x 20	60,000	10.400	.17	6,840	8,360	9,880	11,400	12,920	14,440	15,960
80x 50x 20	80,000	15.200	.165	7,200	8,800	10,700	12,000	13,600	15,200	16,800
100x 50x 20	100,000	16,000	.16	7,200	8,800	10,400	12,000	13,600	15,200	16,800
100x100x 20	200,000	28,000	.14	8,100	9,900	11,700	13,000	15,300	17,100	18,900
100x100x 30	300,000	32,000	.106	11,030	13,486	15,938	18,390	20,840	23,290	25,740
100x100x 40	400,000	36,000	.09	13,050	15,950	18,850	21,750	24,650	27,550	30,450
100x100x 50	500,000	40,000	.08	14,400	17,600	20,800	24,000	27,200	30,400	33,600
100x100x 60	600,000	44,000	.073	16,200	19,800	23,400	27,000	30,600	34,200	37,800
100x100x 70	700,000	48,000	.07	16,650	20,350	24,050	27,750	81,450	35,150	38,850
100x100x 80	800,000	52,000	.065	18,000	22,000	26,000	30,000	34,000	38,000	42,000
100x100x 90	900,000	56,000	.062	18,900	23,100	27,300	31,500	85,700	39,900	44,100
100x100x100	1,000,000	60,000	.06	19,350	23,650	27,950	32,250	36,550	40,850	45,150
Lineal feet.	Cubic feet.	Square feet.	Ratio.	Cubic feet.	Cubic feet.	Cubic feet.	Cubic feet.	Cubic feet.	Cubic feet.	Cubic feet.

DOORS IN COLD STORAGE.

It may not be amiss on this occasion to state that the doors of cold storage buildings and rooms and ice boxes play a most important rôle in the economy of a plant; and therefore their construction, which is frequently left to the discretion of an ordinary carpenter, is a matter of the greatest importance. Not only should they be constructed on the basis of the least heat transmission, but so framed and hung as to be tight and remain so for the longest possible time, as well as open freely at all times. Readjustments long neglected involve financial losses in many directions, often expensive repairs, when a proper construction would avoid both by rendering the first needless. Facility for easily and quickly opening and closing, fastening and unfastening is most important. Workmen persistently leave doors open while going in and out if these points be neglected, with a consequent great ingress of heat and moisture. For this reason it is but fair to recognize the laudable exertion of those firms who make the rational construction of doors used in cold storage buildings, rooms, etc., a special feature.

CALCULATED REFRIGERATION.

For more exact estimates the refrigeration required in a given case may be calculated by allowing first for the refrigeration required to keep the storage at a certain given temperature in consequence of the radiation through walls; and second for the refrigeration required to cool the articles or provisions from the temperature at which they enter the storage room down to the temperature of the latter.

RADIATION THROUGH WALLS.

If the number of square feet contained in a wall, ceiling, floor or window be f, the number of units of refrigeration, R, that must be supplied in twenty-four hours to offset the radiation of such wall, ceiling or floor, may be found after the formula:

$$R = f\,n\,(t - t_1) \text{ B. T. units,}$$

or expressed in tons of refrigeration

$$R = \frac{f\,n\,(t - t_1)}{284000} \text{ tons.}$$

In these formulæ t and t_1 are the temperatures on each side of the wall, and n the number of B. T. units of heat transmitted per square foot of such surface for a difference of 1° F. between temperature on each side of wall in twenty-four hours. The factor n varies with the construction of the wall, ceiling or flooring, from 1 to 5.

For single windows the factor n may be taken at 12, and for double windows at 7 (*Box*).

For different materials one foot thick we find the following values for n:

For pine wood.......	2.0 B. T. U.	For sawdust.........	1.1 B. T. U.
" mineral wool...	1.6 " " "	" charcoal, pow'd	1.3 " " "
" granulated cork	1.3 " " "	" cotton	0.7 " " "
" wood ashes.....	1.0 " " "	" soft paper felt .	0.5 " " "

For brick walls of different thicknesses the factor n may be taken as follows after *Box:*

½ brick	4½ inches thick	n = 5.5	B. T. Units.
1 "	9 " "	" = 4.5	" " "
1½ "	14 " "	" = 3.6	" " "
2 "	18 " "	" = 3.0	" " "
3 "	27 " "	" = 2.6	" " "
4 "	36 " "	" = 2.2	" " "

For walls of masonry of different thicknesses the factor n may be taken as follows after *Box:*

Stone walls	6 inches thick,	n = 6.2	B. T. U.
" "	12 " "	" = 5.5	" " "
" "	18 " "	" = 5.0	" " "
" "	24 " "	" = 4.5	" " "
" "	30 " "	" = 4.3	" " "
" "	36 " "	" = 4.1	" " "

German authorities give values for n which are less than one-half of the values here quoted.

For air tight double floors of wood properly filled underneath so that the atmosphere is excluded, and for ceilings of like construction, n is equal to about 2 B. T. U. An air space sealed off hermetically between two walls has the average temperature of the outside and inside air, hence its great additional insulating capacity. If the air space is hermetically sealed inside and outside, it appears that its thickness is immaterial; half an inch is as good as three inches.

If a wall is constructed of different materials having different known values for n, viz., n_1, n_2, n_3, etc., and the respective thicknesses in feet d_1, d_2, d_3, the value, n, for such a compound wall may be found after the formula of Wolpert, viz.:

$$n = \frac{1}{\frac{d_1}{n_1} + \frac{d_2}{n_2} + \frac{d_3}{n_3}}$$

In case of an air space perfectly sealed off the factor n may be determined for that portion of the wall between the air space and the outside, which value is then inserted into the formula—

$$R = f\,n\,(t - t_1)$$

But in this case while t_1 stands for the maximum outside temperature t stands for the temperature of the air space, which may be averaged from the inside and outside temperature, taking into consideration the conductibility and thickness of the component parts of the wall.

In the selection of insulating substances, their power to withstand moisture plays an important part in most cases. In this respect cork is a very desirable material, likewise pitch and mixtures of asphalt; lamp black and a mixture of lamp black with mica scales is also used with great success, especially in portable refrigerating chambers, refrigerator cars and the like, as it will not pack from jolting, owing to its lightness and elasticity, and it also withstands moisture very well.

REFRIGERATING CONTENTS.

If the amount of refrigeration required to replace the cold lost by the transmission of walls, windows, ceilings, etc., has been determined upon, the refrigeration required to reduce the temperature of the goods placed in storage to that of the storage room is next to be ascertained.

If p, p_1, p_2, etc., be the number of pounds of different produce introduced daily into the storage room and s, s_1, s_2, etc., their respective specific heat, t their temperature and t_1 the temperature of the storage room, we find the amount of refrigeration, R, in B.T. units required daily to cool the ingoing product after the formula:

$$R = (p\,s + p_1\,s_1 + p_2\,s_2)\,(t - t_1) \text{ B. T. units.}$$

or, expressed in tons of refrigeration:

$$R = \frac{(p\,s + p_1\,s_1 + p_2\,s_2)\,(t - t_1)}{284000} \text{ tons.}$$

The specific heat of some of the articles frequently placed in cold storage may be found in the following table:

SPECIFIC HEAT AND COMPOSITION OF VICTUALS.

	Water.	Solids.	Specific Heat above Freezing Calc.	Specific Heat below Freezing Calc.	Latent Heat of Freezing Calc.
Lean beef	72.00	28.00	0.77	0.41	102
Fat beef	51.00	49.00	0.60	0.34	72
Veal	63.00	37.00	0.70	0.39	90
Fat pork	39.00	61.00	0.51	0.30	55
Eggs	70.00	30.00	0.76	0.40	100
Potato	74.00	26.00	0.80	0.42	105
Cabbage	91.00	9.00	0.93	0.48	129
Carrots	83.00	17.00	0.87	0.45	118
Cream	59.25	30.75	0.68	0.38	84
Milk	87.50	12.50	0.90	0.47	124
Oyster	80.38	19.62	0.84	0.44	114
Whitefish	78.00	22.00	0.82	0.43	111
Eels	62.07	37.93	0.69	0.38	88
Lobster	76.62	23.38	0.81	0.42	108
Pigeon	72.40	27.60	0.78	0.41	
Chicken	73.70	26.30	0.80	0.42	

CALCULATION OF SPECIFIC HEATS OF VICTUALS.

The specific heats in the fifth column of the foregoing table is calculated after the formula

$$s = \frac{a + 0.2b}{100} = 0.008\,a + 0.20$$

in which formula s signifies the specific heat of a substance containing "a" per cent of water and "b" per cent of solid matter; 0.2 is the value which has been uniformly assumed to represent the specific heat of the solid constituents of the different articles in question. If the articles are cooled below freezing, which takes place below 32° F., the specific heat changes, owing to the fact that the specific heat of frozen water is only about half of that of liquid water. In conformity with this fact, and considering that the specific heat of the solid matter is not apt to change under these circumstances, we find the specific heat, s', of the same articles in a frozen condition after the following formula:

$$s' = \frac{0.5a + 0.2b}{100} = 0.003a + 0.20$$

and in this way I have obtained the figures in the sixth column of the above table.

The figures in the last column, showing the latent heat of freezing, have been obtained by multiplying the latent heat of freezing water, which is 142 B. T. U. by the percentage of water contained in the different materials considered. In this manner the specific heat for other articles may be readily calculated.

For still more approximate determination we may assume that the specific heat of all kinds of produce is about 0.8. On this basis the amount of refrigeration, R, required to reduce the temperature of the produce to that of the refrigerating room is—

$$R = P\,(t - t_1)\ 0.8 \text{ units.}$$

And expressed in tons=

$$R = \frac{P\,(t - t_1)}{355000} \text{ tons of refrigeration.}$$

P being the total weight of the produce introduced daily.

FREEZING GOODS IN COLD STORAGE.

If, in addition to the refrigeration of the goods to be stored the same have to be actually frozen and cooled down to a certain temperature below freezing, the refrigeration as calculated in the foregoing paragraph must be corrected, for the water contained in the goods must be frozen, which requires an additional amount of refrigeration. On the other hand, the specific heat of the frozen water being one-half of that of water, this circumstance lessens somewhat the amount of refrigeration required below freezing point. Therefore if p represents the number of pounds of water contained in a daily charge for cold storage to be chilled and reduced to a temperature, t_1, the amount, R, found by the foregoing rules must be corrected by adding to it an amount of refrigeration equivalent to—

$$p\,(126 + 0.5\,t_1) \text{ units.}$$

CONDITIONS FOR COLD STORAGE.

For the preservation of perishable goods by cold storage the temperature is the main factor, although other conditions, such as clean, dry, well ventilated rooms and pure air, are of paramount importance. Humidity is almost as important as temperature. Extreme cold temperature will react on certain goods like eggs, fruits, etc , so that when taken out the change of temperature will deteriorate their quality quickly. Hence the conditions under which articles must pass from cold storage to consumption are often of as vital importance as the cold storage itself, for which reason special rules must be followed in special cases.

MOISTURE IN COLD STORAGE.

Besides the temperature in a cold storage room the degree of moisture is of considerable importance.

It is neither necessary nor desirable that the storage room should be absolutely dry; on the contrary, it may be too dry as well as it may too damp. If the room is too dry it will favor the shrinkage and drying out of certain goods. If the room is too damp goods are liable to spoil and become moldy, etc. For this reason the moisture should always be kept below the saturation point. This condition can be ascertained by the hygrometic methods described in the chapter treating on water and steam.

There is little danger that the rooms will ever be too dry; on the other hand, they are not required to be absolutely dry, and as to chemical dryers, such as chloride of calcium, oatmeal, etc., they are probably superfluous, with proper ventilation and refrigerating machinery properly applied.

Generally the artificial drying of air is considered superfluous in cold storage, as the air is kept sufficiently dry by the condensation that forms on the refrigerating pipes. In this way the moisture exhaled by fruits, etc., is also deposited. Special care, however, is to be taken to remove the ice from the coils from day to day as it forms, in which case it is readily removable. Chemical dryers are seldom used in storage houses refrigerated by artificial refrigeration. Freshly burnt lime is sometimes used in egg rooms.

In cold storage houses operated by natural ice, chemical or physical absorbents, such as oatmeal, slacked lime, chloride of calcium and chloride of magnesium are frequently used. The latter substance is the principal constituent of the waste bittern of salt works, which is sometimes used for drying air in the cold storage of fruit.

The waste bittern is spread out on the entire surface of the floor, and, if needed, on additional surfaces above it. One square foot of well exposed bittern, either in the dry state or state of inspissated brine, will be enough to take up the moisture arising from two to six bushels of fruit, varying according to its condition of greenness or ripeness. The floors of the preserving room should be level, so that the thick brine running from the dry chloride may not collect in basins, but spread over the largest surface. The moisture from the fruit taken up by the absorbent varies from about three to ten gallons for every 1,000 bushels of fruit weekly. The spent chlorides or the spent waste bittern may be revived by evaporation, by which they are boiled down to a solid mass again.

The waste bittern is also used as a crude hydrometer by dissolving one ounce of the same in two ounces of water and by balancing the shallow tin dish containing this mixture on a scale placed in the cold storage room. If the scale keeps balanced, it indicates the proper state of dryness, but if the weight of the mixture increases, the moisture in the room is increasing and the means for keeping the air dry should be put in operation.

DRY AIR FOR REFRIGERATING PURPOSES.

To produce a dry air by mechanical means St. Clair considers the entire absence of any condensing or refrigerating surface in the space to be refrigerated absolutely necessary. The rapid circulation of the air in the room is also of vital importance; and in such circulation no contact of the incoming cold air with the outgoing warm air to cause condensation is the result aimed at. To insure these conditions he places the *refrigerator* at the highest point, and has communicating air shafts from the bottom of the same to the rooms to be cooled. Like shafts ascend from the top of the rooms cooled to top of the refrigerator. The refrigerating coils in the refrigerator are kept at a temperature of zero to 15° below, and a small stream of strong brine is allowed to drip over the coils to a pan underneath, being pumped back to the upper drips as fast as deposited. This brine will have a temperature ranging from zero to 4° below. The action is said to be simple and effective; *all* moisture is either condensed or frozen instantly as it comes in contact with such low temperature, and an absolutely dry air descends in the air shafts to the rooms to be cooled.

VENTILATION OF COLD STORAGE ROOMS.

The foul air in storage rooms is removed by ventilation, which is effected in various ways. Frequently the change of air brought about by opening doors, etc., is considered sufficient; in some cases windows are opened from time to time. Ventilating shafts located in the ceiling of storage rooms are also often used as means to effect a change of air. A small rotary fan, located in the engine room and connected with the storage rooms by galvanized iron pipes, provided with gates or valves, is a very efficient device to remove foul air.

Where fans cannot be applied for want of motive power or other reasons a ventilating shaft, if properly constructed, will answer every purpose, and is much less expensive to operate. The air ducts, or pipes, should be located in the hallways, and connection made thence to each room through the side wall near the ceiling, and some suitable device should be arranged on the end of the pipe extending into the cooling room to regulate the amount of ventilation. The several air ducts leading from the various hallways should have a common ending, and connection made thence to the smoke stack. The strong up draft from the furnace insures ample ventilation from rooms at all times, provided that the pipes are made air tight and large enough for the purpose.

The simple expedient of a ventilating shaft extending just outside of the building without being raised to a considerable height, or some provision made to artificially produce a draft, often proves inoperative as a means of ventilating refrigerating rooms, because the air in the rooms, becoming cold, settles to the floor and escapes through crevices about the doors or when the doors are opened, causing a down draft, and in many cases overbalancing the uptake of the ventilating pipe.

FORCED CIRCULATION.

Of the various recent devices for forced circulation and the drying of air in cold storage, most are based on the principle of St. Clair delineated in the foregoing paragraph. It may also be combined with any system of artificial ventilation which may be brought about by fans, ventilators, etc. The introduction of air cooled a few degrees below the temperature of the storage room (by drawing the air over refrigerated surface, as is done in the St. Clair and similar systems) insures dry ventilation.

VELOCITY OF AIR.

If, as in the St. Clair system of forced circulation, the air after having been cooled (and dried) by being passed over the refrigerating coils located in the top part of the storage rooms, falls down from the bottom of the coil through a shaft or shafts to the bottom of the room, while the hot air from the top of the room ascends to the top of the coil by shafts or a shaft, the velocity of the air current thus produced by a difference in temperature, or rather by a difference in gravity due thereto, may be expressed by the following formula:

$$V = 1346 \sqrt{\frac{T - T_0}{T_0} (1 \times 0.0021\ T_0)}$$

In this formula T and T_0 are the temperatures (in degrees absolute Fahrenheit) of the air in the hot and cold air shafts respectively, which are supposed to have the same sectional area, and V is the velocity with which the air moves through the shafts in feet per second.

NUMERICAL RULES FOR MOISTURE.

The proper degree of humidity in cold storage rooms, especially also for the storage of eggs (to avoid mold and shrinkage at the same time) is of the utmost importance, and Cooper finds that the relative humidity should differ with the temperature at which the rooms are kept. Thus a room kept at 28° F. should have a relative humidity of 80 per cent, while a room kept at 40° F. should have a humidity of only 53 per cent, and intermediate degrees of humidity for intermediate temperatures. At least one correct normal thermometer (to correct the others by) should be kept in each cold storage plant.

DISINFECTING COLD STORAGE ROOMS.

Meat rooms and other cold storage rooms may be disinfected if necessary by formaldehyde vapors, which are produced by burning wood spirit in an ordinary spirit lamp, the wick of which is covered by a platinum wire screen, in the form and size of a thimble, to make it only glow, and not burn with a flame. Special lamps are made also for this purpose.

COLD STORAGE TEMPERATURES.

Generally speaking, the temperature of cold storage rooms is about 34° F. For chilling the temperature of the room it is generally brought down to 30° F., and in the case of freezing goods from 10° F. to 0° F.

The temperatures and other conditions considered best adapted for the cold storage of different articles of food, provisions, etc., have been compiled in the following paragraphs, which reflect the views of practical and successful cold storage men as expressed by them in *Ice and Refrigeration:*

STORING FRUITS.

The temperatures for storing fruits are given in the following table:

Fruit.	Remarks.	°F
Apples		30-40
Bananas		34-36
Berries, fresh	For three or four days	34-36
Canteloupes	Carry only about three weeks	32
Cranberries		33-34
Dates, figs, etc		34
Fruits, dried		35-40
Grapes		32-40
Lemons		36-45
Oranges		36
Peaches		35-45
Pears		33-36
Watermelons	Carry only about three weeks	32

In general, green fruits and vegetables should not be allowed to wither. Citrus fruits should be kept dry until the skin yields its moisture, then the drying process should be immediately checked. For bananas no rule can be made; the exigencies of the market must govern the ripening process, which can be manipulated almost at will.

Fruits, especially tender fruits, should be placed in cold storage, just when they are ripe. They will keep better than if put in when they are not fully ripe.

Pears will stand as low a temperature as 33°. Sour fruit will not bear as much cold as sweet fruit. Catawba grapes will suffer no harm at 26°, while 36° will be as cold as is safe for a lemon.

The spoiling of fruit at temperatures below 40° F. is due to moisture.

ONIONS.

Onions, if sound when placed in cold storage, can be carried several months and come out in good condition. It is important that the onions be as dry as possible when put into cold storage. If they can be exposed to a cool, dry wind, they will lose much of their moisture. They are usually packed in ventilated packages or crates. It is claimed, however, that they will keep all right in sacks, if the sacking is not too closely woven, and stored in a special way, being arranged in tiers so the air has free access. Authorities differ as to the best temperature at which to keep the onions, the range being from 30° to 35° F. But 32° to 33° seems to be generally preferred. The rooms should be ventilated and have a free circulation of dry air. Onions should not, of course, be stored in rooms with other goods. When the onions are removed the rooms should be well aired, thoroughly scrubbed and, after the walls, ceiling and floor are free from moisture, should be further purified and sweetened by the free use of lime or whitewash; and a good coat of paint or enamel paint would be advantageous, after which the rooms can be used for the storage of other goods, though some practical cold storage men are of the opinion that such rooms should not afterward be used for the storage of eggs, butter or other articles so sensitive and susceptible to odors, but should be set aside for the storage of such goods as would not be injured by foreign odors.

Attempts have been made to kiln dry onions, but this was found impracticable, owing to the fact that the extreme heat required to penetrate the tough outer skin of the onion caused it to soon decay. Experiments have also been made with evaporating onions after removing the outer skin, but this was also unsuccessful. There is no difficulty, however, in keeping onions in cold storage for six or seven months and having them come out in perfect condition, if the above suggestions are followed.

PEARS.

Pears, like other tender fruit, should be placed in cold storage when still firm, and before the chemical changes which cause the ripening have set in; and they must be handled very carefully. The temperature at which to store them is from 33° to 40° F. The pears after having been kept in cold storage will spoil very rapidly after coming out, and should be consumed as short a time thereafter as may be.

Pears should be picked as soon as the stem will readily part from the twig, and before any indications of ripeness appear; and, as in the case of apples, should immediately be placed in storage, but the temperature should not be as low as for apples.

Few kinds of pears can be kept as late as April and May; even after January there is considerable risk. The temperature should be between 33° and 40°, but, as for all winter storage goods, must be constant and uniform, for which reason the rooms should have heating as well as chilling pipe. The paper wrapper will best protect them from touching each other in storage.

LEMONS.

The best storage temperature for lemons is allowed to be 45° and below, but below 36° F. they are liable to be injured, if kept at that temperature for any length of time. The acid, which is the principal ingredient of lemons, is decomposed, and those containing the least acid will stand the least cold. Lemons should not be expected to keep good in cold storage over four months. Lemons stored during the first three months of the year are said to hold good for at least five months, but if stored later it is more difficult to preserve them.

GRAPES.

Grapes for cold storage must be well selected and very carefully packed. No crushed or bruised or partly decayed berries are allowable; a whole lot may be tainted by a single berry. Grapes lose much in flavor and taste in cold storage. Malagas hold their flavor best, and will last till Christmas and even longer, but the Concord and other softer grapes will not hold out after Thanksgiving day, as a rule. The best temperature is from 32° to 40°.

At the latter temperature the flavor appears to suffer less, especially with the Concord, and the lower temperature has more effect on the Concord than on the Malaga, it appears, generally speaking.

APPLES.

Apples may be kept either in barrels or boxes or in bulk, it is said, with equally good results. The barrels, etc., if kept in storage for any length of time, must be refilled to make up for shrinkage, before being put on the market. Opinions as to best temperature for apples vary all the way from 30° to 40°. The latter temperature should not be exceeded in any case. If the air in cold storage is too dry it wilts the apples, and if it is too damp it bursts and scalds apples, especially if the temperature is not low enough. The so called "Rhode Island Greening" seems to be most susceptible to scalds. Apples should be picked early and put in cold storage with the least possible delay. Apples when stored in barrels should not be stored on ends, but preferably on their sides. A temperature of 33° is considered most favorable by some.

In storing apples eight to ten cubic feet storage room space is allowed per barrel, and twenty to twenty-five tons daily refrigerating capacity per 10,000 barrels.

STORING VEGETABLES.

ARTICLES.	° F.
Asparagus	34
Cabbage	32-34
Carrots	33-34
Celery	33-35
Dried beans	32-40
Dried corn	35
Dried peas	40
Onions	32-34
Parsnips	33-34
Potatoes	34-36
Sauerkraut	35-38
Sweet corn	35
Tomatoes	34-35

Asparagus, cabbage, carrots, celery, are carried with little humidity; parsnips and salsify, same as onions and potates, except that they may be frozen without detriment.

FERMENTED LIQUORS.

ARTICLES.	° F.
Beer, ale, porter, etc	33-42
Beer, bottled	45
Cider	30-40
Ginger ale	36
Wines	40-45
Clarets	45-50

The temperatures at which these articles are to be kept in storage is of course not the temperature at which they should be dealt out for consumption. Beer, ale and porter should not be offered for consumption at a temperature below 52° F., and temperatures between 57° and 61° are even preferable on sanitary grounds, which, however, are often disregarded to insure a temporarily refreshing palate sensation.

STORING FISH AND OYSTERS.

Fish if previously frozen should be kept at 25° after being frozen. Oysters should not be frozen. The following temperatures are given:

ARTICLES.	° F.
Dried fish	35
Fresh fish	25-30
Oysters	33-40
Oysters in shell	40
Oysters in tubs	35

A successful firm describes the freezing of fish as follows:

When the fish are unloaded from the boats they are first sorted and graded as to size and quality. These are placed in galvanized iron pans twenty-two inches long, eight inches wide and two and a half inches deep, covered with loosely fitting lids, each pan containing about twelve pounds. The pans are then taken to the freezers. These are solidly built vaults with heavy iron doors, resembling strong rooms, and filled with coils of pipes so arranged as to form shelves. On these shelves the pans are placed, and as one feature of the fixtures is economy of space, not an inch is lost. The pans are kept here for twenty-four hours in a temperature at times as low as 16° below zero. Each vault or chamber has a capacity of two and a half tons, and there are sixteen of them, giving a total capacity of forty tons, which is the amount of fish that can be frozen daily if required.

On being taken out of the sharp freezers the pans are sent through a bath of cold water, and when the fish are removed they are frozen in a solid cake. These cakes are then taken to the cold storage warehouse, which is divided into chambers built in two stories, almost the same as the sharp freezers. The cakes of fish, as hard as stone, are packed in tiers and remain in good condition ready for sale. It is possible to preserve them for an indefinite time, but as a rule frozen fish are only kept for a season of from six to eight months. They are frozen in the spring and fall when there is a surplus of fish, and sold generally in the winter or in the close season when fresh fish cannot be obtained.

For shipment, fish may be packed in barrels after the following directions: Put in a shovelful of ice at the bottom of the barrel, and be always careful to see that auger holes are bored into the bottom of the barrels, to let the water leak out as fast as it is produced by the melting ice. After putting in a shovelful of fine ice, crushed by an ice mill, put in about fifty pounds of fish; then another shovelful of ice on top of the fish, etc., until the barrel is full, always leaving space enough on the top of the barrel to hold about three shovelsful of ice. By shovels, scoop shovels are meant.

Oysters are said to keep six weeks safe at 40°. In one instance they have been kept ten weeks at this temperature for an experiment.

STORING BUTTER.

Butter is preserved both ways: by keeping the same at the ordinary cold storage temperatures, and also by freezing. Both processes have given satisfactory results, but it appears that those obtained by actual freezing are quite superior, the flavor and other qualities of the butter being perfectly preserved by the freezing. To obtain the best results butter should be frozen at a temperature of 20° and the variation should not be over 2° to 3°. For long storage, however, butter, like fish, should be frozen quickly at a temperature of from 5° to 10°, and subsequently it should be kept at about 20° F. Ash and spruce tubs make the best packages for butter.

As regards thawing it, it is simply taken from the freezer, as in the case of ordinary cold storage goods, without paying any attention to the thawing out process. The thawing comes naturally, and the effect that it has upon the butter is to give it a higher and quicker flavor when thawed out than when frozen. When selling frozen goods it is sometimes necessary to let them stand out a little time in order to get the frost out of the butter; particularly so in the case of high grade goods, for the thawing develops the flavor. June butter is considered the best for packing and storage. It is essential to exclude the air from butter while being held in cold storage, hence cooperage must be the best, and soaked in brine for twenty-four hours. If the top of the butter is well covered with brine, a temperature of 33° to 35° will answer.

For ordinary cold storage of butter and similar articles, the following temperatures are given:

ARTICLES.	° F.
Butter	32–35
Butterine	35
Oleomargarine	35

STORING CHEESE.

The best temperature for the storage of cheese is generally considered 32° to 33°, and should not vary more than 1°. Cheese should not have been subjected to any high temperature before being placed in cold storage.

Cheese should be well advanced in ripening before it is placed in cold storage, to avoid bad smell in the house. It generally enters the cold storage room in June and July, and leaves by the end of January, sooner or later when needed. It will keep much longer, however, over a year when needed. It must be kept from freezing. If frozen, it must be thawed gradually, and consumed thereafter as soon as possible, or otherwise it will spoil internally. The humidity of the room must keep the cheese from shrinking and cracking, but the room must not be damp either, otherwise mold will set in.

MILK.

Milk is not as a rule kept in cold storage except for a short period. It has been proposed, however, to concentrate milk by a freezing process, by which part of the water in the ice is converted into ice. The ice is allowed to form on the surface of the pans, which are placed in cold rooms, and the surface of the ice is broken frequently, to present a fresh surface for freezing.

EGGS.

Eggs should be carefully selected before being placed in cold storage, and every bad one picked out by candling. The best temperature for storing eggs is between 32° and 33° F. As eggs are very sensitive and will absorb bad odors, etc., it is not advisable to store them together with cheese or other products exhaling odors.

For some purposes the contents of eggs may be stored in bulk. In this case the eggs are emptied into tin cans containing about fifty pounds and stored for any length of time at 30° F. They must be used quickly after thawing.

Eggs are generally placed in cold storage in April and early May; later arrivals will not keep as well. They are seldom kept longer than February. The temperature best suited for eggs is supposed to be between 31° and 34° by American packers, but English dealers claim that 40° to 45° is equally good. The humidity of the air in the cold storage room has doubtless a great bearing on this question.

Eggs which have been stored at 30° must be used soon after leaving storage, while eggs kept at 35° to 40° will keep nice for a longer time, as the germ has not been killed in the latter, and consequently they taste fresh. Eggs for the market, especially those to go in cold storage, must not have been washed. Washed eggs have a dead and lusterless looking shell, looking like burned bone through a magnifying glass.

It is also recommended that eggs in cold storage should be reversed at least twice weekly.

The age of eggs may be approximately determined by the following method, based upon the decrease in the density (through loss of moisture) of the eggs as they grow old: Dissolve two ounces of salt in a pint of water, and when a fresh egg is placed in the solution it will immediately sink to the bottom of the vessel. An egg twenty-four hours old will sink below the surface of the water, but not to the bottom of the vessel. An egg three days old will swim in the liquid, and when more than three days old will float on the surface. The older the egg the more it projects above the surface, an egg two weeks old floating on the surface with but very little of the shell beneath the water.

Experiments have been made for the preservation of eggs by dipping them in chemicals, but with no notable success. It is reported that when preserved in lime water, or in a solution of waterglass or by coating with vaseline they will keep for eight months, but doubtless not without some detrimental alteration in taste and flavor.

DRYING OF EGG ROOMS, ETC.

For the drying of egg rooms, etc., Mr. Cooper recommends supporting a quantity of chloride of calcium above the cooling coils, over which the air is circulated by mechanical means. The brine formed by the absorption of moisture by the chloride of calcium will then trickle down over the pipes and thereby effectually prevent any formation of frost on the pipes, and therefore keep them at their maximum efficiency at all times, The air, in passing over the brine moistened surface of the coils, is purified, and the brine, after falling to the floor of the cooling room, goes to the sewer, and no further contamination takes place. The re-use of the salt after redrying is objected to by some on account of these contaminations; but it seems to us that they will be rendered entirely harmless if the salt is dried at a sufficiently high temperature, and this can hardly be avoided if the water is all driven off, to do which requires calcination at a tolerably high temperature, a temperature which is far above that at which all germs are destroyed.

STORAGE OF MISCELLANEOUS GOODS.

ARTICLES.	REMARKS.	° F.
Canned Goods:		
Fruits		35
Meats		35
Sardines		35
Flour and Meal:		
Buckwheat flour		40
Corn meal		40
Oat meal		40
Wheat flour		40
Miscellaneous:		
Apple and peach butter		40
Chestnuts		33
Cigars		35
Furs, woolens, etc		25-32
Furs, undressed		35
Game to freeze	Long storage	0- 5
Game, after frozen	Short storage	25-28
Hops		33-36
Honey		36-40
Nuts in shell		35-38
Maple syrup, sugar, etc		40-45
Oil		35
Poultry, after frozen	Short storage	28-30
Poultry, to freeze	Long storage	5-10
Syrup		35
Tobacco		35

LOWEST COLD STORAGE TEMPERATURES.

Temperatures below zero Fahrenheit are hardly of any utility in cold storage, although in some instances even lower temperatures are produced. A room piped about four cubic feet of space to one lineal foot 1-inch pipe, direct ammonia expansion, could be brought to 8° F. below zero. Theoretically a temperature of —28° F. can be produced with ammonia refrigeration at a back pressure equal to that of the atmosphere (and even lower at lower pressures), but practically it is not likely that temperatures lower than —20° F. can be obtained with ammonia, although it may be done by carbonic acid; but as stated before, it is to no purpose as far as cold storage is concerned.

CHAPTER VII.—BREWERY REFRIGERATION.

PRINCIPAL OBJECTS OF BREWERY REFRIGERATION.

The principal uses for refrigeration in a brewery are as follows:

First.—Cooling of the wort from the temperature of the water as it can be obtained at the brewery to the temperature of the fermenting tuns (about 40° F.).

Second.—Withdrawal of the heat developed by the fermentation of the wort.

Third.—Keeping cellars and store rooms at a uniform low temperature of about 32° to 38° F.

Fourth.—Cooling brine or water to supply attemperators in fermenting tubs.

Fifth.—For the storage of hops and prospectively in the malting process.

ROUGH ESTIMATE OF REFRIGERATION.

Frequently the amount of refrigeration required for breweries is roughly estimated (in tons) by dividing the capacity of the brewery in barrels made per day by the figure (4). As a matter of course, this can answer only for very crude estimates. For closer estimates the different purposes for which refrigeration is required must be considered separately.

SPECIFIC HEAT OF WORT.

The wort by the fermentation of which the beer is produced consists chiefly of saccharine and dextrinous matter dissolved in water. Its specific heat, which is the chief quality that concerns us now, varies with the

Strength of Wort in Per Cent after Balling.	Corresponding Specific Gravity.	Corresponding Specific Heat.
8	1.0320	.944
9	1.0363	.937
10	1.0404	.930
11	1.0446	.923
12	1.0488	.916
13	1.0530	.909
14	1.0572	.902
15	1.0614	.895
16	1.0657	.888
17	1.0700	.881
18	1.0744	.874
19	1.0788	.867
20	1.0832	.861

amount of solid matter which it contains; this may be ascertained by finding its specific gravity by means of a saccharometer or other hydrometer. The specific heat of wort of different strength or specific gravity may be found from the accompanying table.

These figures are calculated for a temperature of 60° F. For every degree Fahrenheit that the temperature of the wort is below 60°, the number 0.00015 must be added to the specific gravity given in above table, and for every degree above the number 0.00015 must be subtracted. Thus the specific gravity of a wort of 13 per cent being acccording to the table 1.0530 at 60°, at 50° it would be $60-50=10\times0.00015=0.0015$ more, or 1.0545.

PROCESS OF COOLING WORT.

The wort as prepared in the brewery is boiling hot, and has to be cooled to the temperature of the fermenting tuns. It is first cooled—at least, generally so—by exposing it to the atmosphere in the cooling vat, in which, however, it should not remain over two to three hours, nor at a temperature below 110° F. After this the wort is allowed to trickle over a system of coils through which ordinary cold water circulates by which the temperature of the wort is reduced to that of the water, about 60° F. or thereabouts. A system of coils, generally placed below the one mentioned already, finishes the cooling process by reducing the temperature of the wort to about 40° F. or below—in ale breweries to about 55° F. This is done by circulating either cooled (sweet) water or refrigerated brine or refrigerated ammonia through the latter coils while the wort trickles over the same.

REFRIGERATION REQUIRED FOR COOLING WORT.

The amount of cooling required in this latter operation must be furnished by artificial refrigeration, and its amount expressed in B. T. units, U, may be calculated exactly if we know the number of barrels, B, of wort to be cooled, its specific heat, s, and its specific gravity, g, after the following formula:

$$U=B\times259\times g\times s\,(t-40)\text{ units,}$$

in which t stands for the temperature to which the wort can be cooled by the water to be had at the brewery.

To reduce this amount of refrigeration to tons of refrigeration it must be divided by 284,000.

SIMPLE RULE FOR CALCULATION.

Assuming that the average temperature of the wort after it has been cooled by the water as it is obtainable at the brewery, is about 70° F., and that the average strength of wort in breweries is between 13 and 15 per cent of extract, corresponding to a specific weight of about 1.05, and to a specific heat of 0.9, the above formula may be simplified and the refrigeration required daily for the cooling of the wort of a brewery of a daily capacity of B barrels, expressed as follows:

$$U = B \times 7400 \text{ units.}$$

Or, expressed in tons of refrigeration, U_1

$$U_1 = \frac{B \times 7400}{284000} = \frac{B}{38.4} \text{ tons.}$$

In other words, about one ton of refrigeration is required for about thirty-eight barrels of wort under the conditions mentioned. If the water of the brewery cools the wort to 60°, one ton of refrigeration would answer for about fifty-two barrels of wort.

The former figure on one ton of refrigeration for forty barrels of wort is generally adapted for preliminary estimates.

SIZE OF MACHINE FOR WORT COOLING.

The capacity of an ice machine is generally expressed in tons of refrigeration produced in twenty-four hours. However, the wort in a brewery must be cooled in a few hours; therefore, in order to find the capacity of the ice machine required to do the above duty the number of tons of refrigeration found to be required to do the cooling of the wort must be multiplied by the quotient $\frac{24}{h}$ in which h means the time expressed in hours in which the cooling of the wort must be accomplished. This of course applies to cases in which a separate machine is used for wort cooling, as is done in large breweries.

Frequently the cooling of the wort is accomplished by employing nearly the whole refrigerating capacity of the brewery for this purpose for a comparatively short time.

INCREASED EFFICIENCY IN WORT COOLING.

In these cases, therefore, the total refrigerating capacity of a brewery must never be less than that required to do the wort cooling in the desired time when all other refrigerating activity is suspended during that time. In this connection it should, however, be mentioned that the brine system, as well as the direct expansion system, may be made to work with increased efficiency when applied to wort cooling. In the former case this may be accomplished by storing up cooled brine ahead, and in the latter case by allowing the ammonia to re-enter the compressor at a much higher temperature after having been used for wort cooling than in other cases.

HEAT PRODUCED BY FERMENTATION.

The cooled wort is now pitched with yeast and allowed to ferment, by which process the saccharine constituents of the wort are decomposed into alcohol and carbonic acid with the generation of heat after the following formula:

$$\underset{\text{Maltose.}}{C_{12}H_{22}O_{11}, H_2O} = \underset{\text{Alcohol.}}{4\,C_2H_5OH} + \underset{\text{Carbonic Acid.}}{4\,CO_2} + \underset{\text{Heat.}}{66{,}000 \text{ units.}}$$

In other words, this means that 360 pounds of maltose during fermentation will generate 66,000 pounds Celsius units of heat, or that one pound of maltose while decomposed by fermentation will generate about 330 B.T. units of heat.

CALCULATING HEAT OF FERMENTATION IN BREWERIES.

If the weights of the wort and that of the ready beer are determined by means of a Balling saccharometer, and are b and b_1 respectively, the heat, H, in B. T. units generated during the fermentation of B barrels of such wort, may be determined after the formula—

$$H = \frac{B \times 0.91\,(b - b_1)\,(259 + b)\,330}{100} \text{ units.}$$

And the refrigeration required to withdraw this heat from the fermenting rooms, expressed in tons, U, of refrigerating capacity is—

$$U = \frac{H}{284{,}000} \text{ tons.}$$

SIMPLE RULE FOR SAME PURPOSE.

Again, if we assume that the wort on an average shows 14 per cent on the saccharometer, and after fermentation it shows 4 per cent, the above formula, giving the refrigeration in tons, U_1, in tons required in twenty-four hours to withdraw the heat generated by the fermentation of B barrels of wort turned in on an average daily, may be simplified as follows:

$$U_1 = \frac{B}{34} \text{ tons.}$$

In other words, one ton of refrigerating capacity is required for every thirty-four barrels of beer produced on an average per day of above strength. This rule will apply to pretty strong beers; for weaker beer it may become much less, so that one ton of refrigeration will answer for fifty barrels, and even more. This shows the importance of this branch of the calculation, which is frequently passed over in a "rule of thumb" way.

For preliminary estimates one ton of net refrigerating capacity is allowed to neutralize the heat generated by the fermentation of twenty-five barrels of beer.

DIFFERENT SACCHAROMETERS.

If in the above determinations of the strength of wort of beer any other kind of saccharometer has been used its readings can be readily transformed into readings of the Balling scale, by using the table on the following page, which may also be used in connection with the other tables on hydrometer scales in this book. In this way any hydrometer may be made available for the purpose contemplated in the above formula.

REFRIGERATION FOR STORAGE ROOMS.

Besides the heat generated by fermentation, the heat entering the fermenting and storage rooms from without must be carried away by artificial refrigeration, so as to keep them at a uniform temperature of 32° to 38° F. The amount of refrigeration required on this account is also frequently estimated by a "rule of thumb," allowing all the way from twenty to seventy units of refrigeration for every cubic foot of room to be kept cool during twenty-four hours. The difference in refrigeration is due to the size of the buildings and to the manner in which the walls and roofs are built.

Generally thirty units are allowed per cubic foot of space, in rough preliminary estimates, for capacities over 100,000 cubic feet.

For capacities between 5,000 and 100,000 cubic feet from forty to seventy units are allowed, and above 100,000 from twenty to forty units per cubic foot of space. Sometimes, after another way of approximate figuring, about 20 to 100 units of refrigeration (generally 50) are allowed per square foot of surrounding masonry ceiling and flooring.

TABLES FOR THE COMPARISON OF DIFFERENT SACCHAROMETERS AMONG THEMSELVES AND WITH SPECIFIC GRAVITY.

Kaiser, Balling or Extract per cent.	Long's Saccharometer.	Gendar's Saccharometer.	Specific Gravity.	Weight per Barrel of 31½ Gallons.	Kaiser, Balling or Extract per cent.	Long's Saccharometer.	Gendar's Saccharometer.	Specific Gravity.	Weight per Barrel of 31½ Gallons
0.00	0.00	0.00	1.000	262.41	12.00	17.45	14.64	1.0488	275.21
.25	.36	.30	1.001	262.66	.25	.83	.96	1.0498	275.49
.50	.72	.60	1.002	262.92	.50	18.21	15.28	1.0509	275.76
.75	1.08	.90	1.003	263.18	.75	.60	.60	1.0520	276.04
1.00	.44	1.20	1.004	263.45	13.00	.99	.92	1.0530	276.32
.25	.80	.50	1.005	263.71	.25	19.38	16.24	1.0540	276.60
.50	2.16	.80	1.006	263.97	.50	.77	.55	1.0551	276.88
.75	.52	2.10	1.007	264.23	.75	20.16	.86	1.0562	277.15
2.00	.88	.40	1.008	264.50	14.00	.55	17.17	1.0572	277.42
.25	3.24	.70	1.009	264.76	.25	.94	.48	1.0582	277.68
.50	.60	3.00	1.010	265.02	.50	21.33	.80	1.0593	277.96
.75	.96	.30	1.011	265.28	.75	.72	18.12	1.0604	278.25
3.00	4.32	.60	1.012	265.55	15.00	22.11	.43	1.0614	278.52
.25	.68	.90	1.013	265.81	.25	.50	.75	1.0625	278.80
.50	5.04	4.20	1.014	266.07	.50	.89	19.07	1.0636	279.09
.75	.40	.50	1.015	266.33	.75	23.27	.39	1.0646	279.35
4.00	.76	.80	1.016	266.60	16.00	.66	.71	1.0657	279.63
.25	6.12	5.10	1.017	266.86	.25	24.05	20.03	1.0668	279.92
.50	.48	.40	1.018	267.12	.50	.44	.35	1.0679	280.21
.75	.84	.70	1.019	267.38	.75	.83	.67	1.0690	280.50
5.00	7.20	6.00	1.020	267.65	17.00	25.22	21.00	1.0700	280.77
.25	.56	.30	1.021	267.91	.25	.61	.33	1.0711	281.06
.50	.92	.60	1.022	268.17	.50	26.00	.66	1.0722	281.34
.75	8.28	.90	1.023	268.43	.75	.39	.99	1.0733	281.63
6.00	.64	7.20	1.024	268.69	18.00	.78	22.32	1.0744	281.92
.25	9.00	.50	1.025	268.96	.25	27.17	.65	1.0755	282.21
.50	.36	80	1.026	269.22	.50	.56	.98	1.0766	282.50
.75	.72	8.10	1.027	269.48	.75	.96	23.31	1.0777	282.78
7.00	10.08	.40	1.028	269.74	19.00	28.36	.64	1.0788	283.08
.25	.44	.70	1.029	270.00	.25	.76	.97	1.0799	283.37
.50	.80	9.00	1.030	270.27	.50	29.16	24.30	1.0810	283.65
.75	11.16	.30	1.031	270.53	.75	.56	.63	1.0821	283.93
8.00	52	.60	1.032.	270.79	20.00	.95	.96	1.0832	284.21
.25	.96	.96	1.0332	271.11	.25	30.34	25.29	1.0843	284.49
.50	12.32	10.26	1.0342	271.37	50	.73	.62	1.0854	284.77
.75	.68	.57	1.0352	271.64	.75	31.12	.95	1.0865	265.05
9.00	13.04	.88	1.0363	271.91	21.00	.50	26.27	1.0876	285.33
.25	.40	11.19	1.0374	272.19	.25	.87	.60	1.0887	285.62
.50	.76	.50	1.0384	272.47	.50	32.25	.93	1.0898	285.91
.75	14.12	.81	1.0394	272.74	75	.64	27.26	1.0909	286.19
10.00	.48	12.11	1.0404	273.00	22.00	33.04	.59	1.0920	286.47
.25	.84	42	1.0415	273.28	.25	.44	.92	1.0931	286.77
.50	15.21	.73	1.0425	273.56	.50	.84	28.25	1.0942	287.06
75	.58	13.06	1.0436	273.84	.75	34.23	.58	1.0953	287.36
11.00	.95	.37	1.0446	274.11	23.00	.63	.91	1.0964	287.66
.25	16.32	.68	1.0457	274.39	.25	35.03	29.24	1.0975	288.96
.50	.69	14.00	1.0467	274.66	.50	.43	.57	1.0986	288.20
75	17.07	.32	1.0478	274.94	.75	.83	.90	1.0997	288.50
					24.00	36.23	30.23	1.1008	288.80

CLOSER CALCULATION.

For calculations required to be more exact the power for transmission of heat by the walls and windows, as well as the difference of temperature within and without, must be taken into consideration.

For calculations of this kind the same rules apply which have been given under the head of cold storage, pages 153, etc.

The number of units of refrigeration found to be required must be divided by 284,000 to express tons of refrigeration.

COOLING BRINE AND SWEET WATER.

The amount of refrigeration required to cool brine or sweet water to supply the attemperators in the fermenting tubs is included in the estimate for the refrigeration required to neutralize the heat of fermentation.

TOTAL REFRIGERATION.

Therefore the total amount of refrigeration required is composed of the first three items mentioned in the second paragraph of this chapter, and by adding them we find the actual capacity of the machine or machines required in a given case. It may be verified in accordance with the considerations mentioned in the paragraph on "Increased Efficiency for Wort Cooling."

DISTRIBUTION OF REFRIGERATION.

The practical distribution of the refrigeration in the brewery is carried out on different principles, and should follow the figures obtained in the above calculations.

Formerly the cooling of rooms in breweries was frequently effected by the circulation of air, which was furnished direct by compressed air refrigerating machines. Later on the air to be used for this purpose was refrigerated in separate chambers with the aid of ammonia compression machines. At present, however, the chief means for cooling brewery premises are coils of pipe into which the ammonia is allowed to expand directly as it leaves the liquid receiver. These coils are generally placed overhead, in which position they assist greatly in keeping the air dry.

DIMENSIONS OF WORT COOLER.

The amount of refrigeration destined to do the cooling of the wort takes care of itself, provided the cooler, which, as already described, is generally constructed after the Baudelot pattern, is large enough to do the cooling in the proper time. The proportions frequently employed for the ammonia portion of the wort cooler are about ten lengths of 2-inch pipe, each length sixteen feet long, for fifty barrels of wort to be cooled from about 70° to 40° F. within three to four hours.

For 100 barrels of wort to be cooled the ammonia portion of the cooler consists of fourteen lengths of pipe sixteen feet long; for 180 barrels, of fifteen lengths twenty feet long; and for 360 barrels, twenty lengths twenty feet long, all pipes to be 2-inch. These are practical figures, and given with a view to afford ample cooling surface.

The amount of refrigeration which must circulate through the wort cooler within that time has been determined by the above calculation.

In the case of brine circulation, salt brine being used in the wort cooler, the surface of pipe should be made 20 per cent more than given above; in other words, a cooler of the above dimensions will answer for forty barrels of wort, instead of fifty, in case brine circulation is used.

DIRECT EXPANSION WORT COOLER.

In case of brine circulation, to which the foregoing dimensions apply, the pipes of the wort cooler may be of copper, but in case of direct expansion being used, the inside of the pipes cannot be copper, but must be iron or steel, and, therefore, copper plated steel pipe or polished steel pipe is used in this case, the latter being given the preference by most manufacturers on account of cheapness and relative efficiency.

The ammonia portion of the wort cooler should be made in two or more sections, having separate and direct connections for inlet of liquid ammonia and outlet of expanded vapor.

PIPING OF ROOMS.

The balance of refrigeration, that is, the whole amount, less that used for wort cooling, must be distributed over the store and fermenting rooms in due proportion. In doing so the time within which the refrigeration is to be dispensed must be considered foremost. The subsequent figures are based on the assumption that during every day the machine or brine pump is active for twenty-four hours to circulate refrigeration; if less time is to be used for that purpose more distributing pipe must be used in proportion.

As a general thing too much piping cannot be employed, for the nearer the temperature of the room to be cooled is to that within the pipe, the more economical will be the working of the ice machine.

In case of direct expansion it is frequently assumed that in order to properly distribute one ton of refrigeration about storage and fermenting rooms, it will require a pipe surface of 80 square feet, which is equivalent to 130 feet of 2-inch pipe, and to about 190 feet of 1¼-inch pipe. Smaller pipe than that it is not advisable to use. If radiating disks are employed less pipe may be used.

For brine circulation much more piping, even as much as 200 square feet of surface, are allowed per ton of refrigeration to be distributed.

In very close calculations allowance should be made for the difference in temperature in the different vaults, which for fermenting rooms is about 42° F., for storage rooms about 33° F., and for final storage or chip cask about 37° F.

HEAT OF FERMENTATION AGAIN.

In addition to the piping allowing for the transmission of heat through the walls, the balance of piping, *i. e.*, that which is to convey the refrigeration required to neutralize the heat during fermentation, must be apportioned according to the amount of heat which is developed in the different rooms. This can also be calculated very closely after the above rules, if the method of fermentation to be carried on is known.

But as a rule this is not the case, and to supply this deficiency it may be assumed that from the heat generated during fermentation about four-fifths is generated in the fermenting room, and about one-fifth in the ruh and chip cask cellar together. In this proportion the additional piping in these rooms may be arranged after due allowance has been made for the refrigeration conveyed by the attemperators.

EMPIRICAL RULE FOR PIPING ROOMS.

More frequently than the foregoing method empirical rules are followed in piping rooms in breweries, it being assumed that nearly all of the heat generated in the fermenting room proper (during primary fermentation) is carried off by the attemperators. On this basis it is frequently assumed that one square foot of pipe surface will cool about 40 cubic feet of space in fermenting room, and about 60 to 80 cubic feet of space in ruh and chip cask cellar (direct expansion).

These figures then apply to direct expansion; for brine circulation, about one-half of the above named spaces will be supplied by one square foot of refrigerating surface.

This figure appears to contemplate a range of about 9° F. difference between the temperature of rooms and that of refrigerating medium within pipe. Much more and much less pipe is frequently used for the same purpose, which is to be accounted for by reasons given on pages 135 and 136.

Here we allow more space per square foot of refrigerating pipe surface than is done in the rule at the bottom of page 135 for storage rooms in general to keep the same temperature. This is partially explained by the fact that brewery vaults are less frequently entered from without, and that their contents are less frequently changed than is the case with general storage vaults. Furthermore it is evident that the size of vaults is also a matter for consideration in this respect.

ATTEMPERATORS.

The attemperators are coils of iron pipe, one to two inches thick, the coil having a diameter of about two-thirds of the diameter of the fermenting tub, in which it is suspended, and a sufficient number of turns to allow about twelve square feet pipe surface per 100 barrels of wort, corresponding to about nineteen feet of 2-inch pipe. The refrigeration is produced by means of cooled water or brine circulating through the attemperators. The attemperators are suspended with swivel joints so that they can be readily removed from the fermenting tub.

There is a great variety in the form of attemperators, box or pocket coolers being also frequently used. On the whole the pipe attemperator as described seems to be the simplest and most popular.

It has also been proposed (Galland) to cool the fermenting wort by the injection of air, purified by filtration through cotton and refrigerated artificially. This plan, however, does not seem to be followed practically to any great extent.

REFRIGERATION FOR ALE BREWERIES.

While the general calculations relating to heat of fermentation, cooling of the wort and cooling of rooms are the same for ale as for lager beer, the specific data relating to piping, etc., in above paragraph, are given with special reference to lager beer, and must be modified when applied to ale.

This is due to the fact that the ale wort is cooled to a temperature of about 55° F. only, and that the storage rooms are to be kept at a temperature of 50° F., or thereabouts.

Accordingly, for ale wort cooling one ton of refrigeration will be required for every seventy-five barrels. For keeping the rooms at the temperature of 50° about twenty B. T. units and less of refrigeration for every cubic foot in twenty-four hours will be sufficient.

The refrigeration necessary to remove the heat of fermentation is calculated in the same manner as above.

The piping of store rooms in ale breweries is frequently done at the rate of one running foot of 2-inch pipe per sixty cubic feet of space.

The tables on refrigeration and piping discussed in the chapter on cold storage may also be consulted in this connection.

SWEET WATER FOR ATTEMPERATORS.

The circulation of refrigerated brine in the attemperators is not considered a safe practice by brewers in general, as a possible leak of brine would be liable to cause great damage to the beer. For this reason cooled or ice water (it is also termed sweet water to distinguish it from salt water or brine) is circulated in the attemperators, generally by means of an automatic pump which regulates the proper supply of sweet water to the attemperators, no matter how many or how few of them are in operation at the time. The ice or sweet water is cooled in a suitable cistern or tank which contains a cooling pipe in which ammonia is allowed to expand directly, or through which refrigerated brine is allowed to circulate. In some breweries the wort is also cooled by refrigerated sweet water made in the above way. This method absolutely precludes the possibility of contamination of ammonia or brine, but at the same time it is very wasteful in regard to the very indirect mode of applying the refrigeration; and for this reason brine in circulation is now mostly used for this purpose, experience having shown that the danger of contamination is practically excluded.

CHILLING OF BEER.

Recently it has been found desirable to subject the ready beer to a sort of chilling process immediately before racking it off into shipping packages. This process, however, is of no practical utility if the beer is not filtered after it has been chilled and before it goes into the barrels. In this case much objectionable albuminous matter, still contained in the ready beer, is precipitated by chilling and separated from the beer by filtration, while without filtration this matter would redissolve in the beer and cause subsequent turbidities, especially if the beer is used for bottled goods.

BEER CHILLING DEVICES.

The chilling was first effected by passing the beer through a copper worm placed in a wooden tub which was filled with ice. But by this the desired object was attained only partially. Therefore, the ice was mixed with salt to obtain a still lower temperature in the beer passing through the worm. Still more recently, and of course in all breweries where mechanical refrigeration is employed, the pipes through which the beer passes are cooled by brine or by direct expansion.

Special apparatus are also made for this purpose, and generally consist of a series of straight pipes provided with manifold inlet and outlet, and placed in a cylindrical drum, through which refrigerated brine or ammonia is allowed to pass in a direction opposite to the beer.

COOLING OF WORT.

Coolers of the same construction are now also frequently used for wort cooling instead of the Baudelot coolers. For both purposes, *i. e.*, the chilling of the ready beer and the cooling of the wort, the refrigerated brine appears to act as the best cooling medium, at least so with some makes of this kind of coolers as they are constructed and operated at present. If direct expansion is used it has been found impracticable (at least in the cases reported to the author) to effect a thorough chilling in the desired time. If used for wort cooling, direct expansion has also caused some trouble when used with some kinds of these new coolers, but it has been overcome in a measure by allowing the ammonia to enter the cooler almost one-half to one hour before the wort is passed through the same.

SAFEGUARDS TO BE EMPLOYED.

It has also been experienced that the expanded ammonia, especially if the expansion valve (one of which must be provided for each of these coolers) is not manipulated very carefully, enters the compressor in an oversaturated condition if allowed to pass directly to the same. Under such conditions the compressor will operate in an irregular manner, and even the cylinder head may be blown out in extreme cases. To guard against such calamities it is necessary to carry the expanded ammonia to the compressor in proper condition by allowing the same to mix with the expanded ammonia coming from the expansion pipes in other parts of the brewery, before reaching the compressor. To do this the expanded ammonia from the wort cooler and that from the cellar may enter a common conduit pipe at a sufficient distance from the compressor to insure a thorough mixture of the gases.

CAUSES OF TROUBLE.

The foregoing contains, we believe, the principal safeguards known at present to be of service to overcome the troubles with these coolers; troubles which, while they are not gainsaid by their makers, are nevertheless, we understand, declared by some of them so paradoxical in their action that they upset the entire theory of transmission of heat as given by the scientists at present. On the other hand, and to partly offset a statement so derogatory to the engineering profession, it may be permissible to suggest that the chief of the apparatus makers, while being expert practical coppersmiths, are perhaps not sufficiently versed in the intricate details offered by problems of heat transmission to give the construction of apparatus of a novel tendency the proper consideration.

It is not unlikely that the relative sizes of direct expansion pipes and brine pipes in the refrigeration of rooms have been taken as cases parallel to these coolers, while in fact the transmission of heat proceeds at a rate entirely different in both cases.

DIRECT REFRIGERATION.

Instead of refrigerating the fermenting and storage rooms of the brewery it has also been proposed to refrigerate the contents of the tubs and casks separately and in a more direct manner, just as the surplus heat of fermenting tubs is now withdrawn, by means of attemperators or similar devices. At first sight there would seem to be a source of considerable saving in this proposition, but it would be at the expense of cleanliness, dryness and reliable supervision of the brewery. Therefore it must be considered a change of very doubtful expediency.

BREWERY SITE.

In former times it was generally considered that the best location for a brewery site was on a hill side, to enable the fermenting and storage rooms to be built into the hill into natural rock, in order to profit by the natural low underground temperature in the summer and the higher underground temperature in the winter time; in other words, by the even temperature all the year around. This position was certainly well taken when the beer was made exclusively by top fermentation, and the position still holds good in a measure for ale breweries. As the great majority of breweries, however, are operated for the production of lager beers which have to ferment, and are stored at temperatures much lower than those obtaining in natural vaults (at least, in the moderate zones), artificial refrigeration or ice has to be resorted to. In either case the natural vaults offer very little advantage to overground structures, well insulated, especially if the larger cost of construction of natural vaults, their inconvenience as to room, and generally also as to accessibility, is considered. For these reasons the site for a brewery nowadays is generally selected with sole reference to convenience as to shipment of produce, reception of material and quality and accessibility of water supply.

ICE MAKING AND BREWERY REFRIGERATION.

Very frequently it happens that a brewery is to be operated in connection with an ice plant, and, generally speaking, it is doubtless not only more convenient, but also good economy to have more than one refrigerating machine in such cases on account of different expansion or back pressures that we have to work with.

STORAGE OF HOPS.

To keep hops from degeneration their storage at 32°—34° F, in a dry, dark, insulated room has been found the only successful way. The hops should be well dried, sulphurized and well packed before being placed in cold storage. Artificial refrigeration, as well as ice, may be used, but special precaution has to be used to keep the room dry in the latter case.

REFRIGERATION IN MALT HOUSES

The cold air which is required in malting, especially in the so called pneumatic methods of malting, it has also been proposed to furnish by means of refrigerating machinery, but it does not appear that it can be done successfully from a financial point of view, except, perhaps, under very exceptional circumstances.

ACTUAL INSTALLATIONS.

The following figures are taken from actual measurements of an existing installation in a brewery having a daily capacity of 375 barrels lager beer, which has the following appointments:

One ammonia compression machine of fifty tons, chiefly for wort cooling, direct expansion, reduces temperature of whole output, 375 barrels, from 70° to 40° F. in four hours (the ammonia portion of Baudelot cooler consisting of twenty pieces of 2-inch pipe, each twenty feet long).

One ammonia compression machine, 50 tons capacity, for storage attemperators, etc. (direct expansion).

Fermenting room, 90×75 feet, fourteen feet high, is piped at the rate of one foot 2-inch pipe for every twenty-seven cubic feet space. Each one of the sixty-five fermenting tubs contains an attemperator coil of twenty-one feet 2-inch pipe.

Ruh cellar, 90×74 feet, and twenty feet high, is piped at the rate of one foot 2-inch pipe for every forty cubic feet of space.

Chip cask cellar, 90×73 feet, and sixteen feet high, is piped at the rate of one foot 2-inch pipe for every fifty-two cubic feet of space.

A fifty-barrel lager beer brewery was equipped with machinery to furnish refrigeration in accordance with the following estimates:

3,200,000 B. T. units for storage.
416,000 B. T. units for cooling wort.
300,000 B. T. units for attemperators.

Total, 3,916,000 B. T. units=13.8 tons, or in round figures equal to fifteen tons refrigerating capacity. The whole capacity is calculated to cool the wort in four hours.

CHAPTER VIII.—REFRIGERATION FOR PACKING HOUSES, ETC.

AMOUNT OF REFRIGERATION REQUIRED.

The application of refrigeration in slaughtering and packing houses is quite similar to its application to cold storage in general, and the amount of refrigeration required in a special case may be estimated on the same principles.

THEORETICAL CALCULATION OF SAME.

The refrigeration required to keep the rooms at the required temperature is found after the rules given on page 173, etc. The additional refrigeration to chill or freeze the meat can be calculated after the rules given on page 183, etc.

PRACTICAL RULES FOR SAME.

The temperature of the chilling rooms is below 32° F. and the fresh slaughtered meats are stored in them until they have acquired the storage temperature in storage rooms, to which they are then removed.

For practical estimates it is frequently assumed that a refrigeration equivalent to about 80 B. T. units is required for every cubic foot of chilling room capacity in twenty-four hours.

The refrigeration for meat storage rooms is the same as that required for ordinary storage, *i. e.*, from 20 to 50 units (40 units being calculated on an average) for every cubic foot of space in twenty-four hours.

For crude estimates calculations are frequently made on the basis of allowing 3,000 to 5,000 cubic feet space per ton of refrigeration in twenty-four hours in chilling rooms, and 5,000 to 8,000 cubic feet space per ton of refrigeration in twenty-four hours in storage rooms, according to insulation, size of rooms and other conditions.

FREEZING ROOMS.

The freezing of meat is performed in rooms kept at a temperature of 10° F. and below. Considerable additional refrigeration is required for freezing, not only on account of the latent heat of freezing, which has to be withdrawn, but also on account of the low temperature at which the rooms have to be kept. For rough estimates at least 200

B. T. units of refrigeration should be allowed for every cubic foot of freezing room capacity.

CALCULATION PER NUMBER OF ANIMALS.

If the average number and kind of animals to be disposed of daily in slaughtering house is known, calculations are also made on a basis similar to the following:

From 6,000 to 12,000 cubic feet of space are allowed per ton of refrigerating capacity to offset the loss of refrigeration by radiation through walls and otherwise, and in addition to that, the extra refrigeration to be allowed in the chilling room for the chilling proper is arrived at in accordance with the assumption that one ton of refrigeration will take care of the chilling of

15-24 hogs (average weight, 250 pounds).
5- 7 beeves (average weight, 700 pounds).
45-55 calves (average weight, 90 pounds).
55-70 sheep (average weight, 75 pounds).

In actual freezing one ton of refrigeration will take care of one ton of meat (in twenty-four hours).

PIPING OF ROOMS.

The piping of rooms in packing houses may be arranged after rules referred to already. Not infrequently, however, other empirical rules are followed, viz.:

For chilling rooms, for instance, one running foot of 2-inch pipe (or its equivalent) is allowed for thirteen to fourteen cubic feet of space; that is, in case of direct expansion, and for seven to eight cubic feet of space for brine circulation.

For storage rooms, one running foot of 2-inch pipe is allowed for forty-five to fifty cubic feet in case of direct expansion, and for fifteen to eighteen cubic feet in case of brine circulation.

For freezing rooms, one running foot of 2-inch pipe is allowed for six to ten cubic feet of space for direct expansion, and for three cubic feet of space in case of brine circulation.

Others proportion the piping by the number of animals slaughtered, allowing thirteen feet of 2-inch pipe per ox, and six feet 2-inch pipe per hog in case of direct expansion in chilling room.

In case of brine expansion thirteen feet 1¼-inch pipe are allowed per hog, and twenty-seven feet 1¼-inch are allowed per ox in chilling room. (Large installations.)

STORAGE TEMPERATURES FOR MEATS.

The temperatures considered best adapted for the storage of various kinds of meats are given in the following table:

ARTICLES.	°F.
Brined meats	35-40
Beef, fresh	37-39
Beef, dried	36-45
Hams, ribs, shoulders (not brined)	30-35
Hogs	30-33
Lard	34-45
Livers	30
Mutton	32-36
Ox tails	32
Sausage casings	30-35
Tenderloins, butts, ribs	30-35
Veal	32-36

OFFICIAL VIEWS ON MEAT STORAGE.

The report of an official commission created by the French government to investigate the cold storage of meats, etc , closes with the following conclusions:

First.—Whenever meat is to be preserved for a comparatively short time, for market purposes, the animals being slaughtered close to the cold storage or not having to be transported, after slaughtering, for a distance involving more than a few hours (as much as twelve), in transit, congelation is not required to insure the conservation. It should be avoided, as by such a practice, that is, the temperature being kept in the storage above the freezing point, the meats are sure to retain all their palatable and merchantable qualities.

Second.—In special circumstances, such as for a protracted conservation, in case of a transportation of the slaughtered animals from very long distances, involving days or weeks in transit, congelation appears to be preferable and safer It does not necessarily render the meats less merchantable, wholesome or palatable, if they are frozen and thawed out, very slowly, gradually and carefully; and only after they have been deprived partially of the excess of moisture of their tissues.

Third.—Cold, dry air should be the vehicle of cold; it should circulate freely around the meats.

FREEZING MEAT.

The same commission recommends that in case the meat must be frozen it should be done in such a way that the fiber is not altered; it should preserve its elasticity as long as possible, up to the very moment when the liquid elements of the meat begin to solidify, so that, at the point of congelation, the dilatation of the water, in changing state, should not cause the bursting of the organic cells, leaving a uniform mass of disagreeable appearance at the thawing out. The congelation must proceed very slowly from the start, progressing gradually and very regularly through the mass, as soon as the freezing point has been reached; the temperaturo should be carefully watched, very evenly lowered without any sudden depression. Once congealed, the temperature of the meats can be carried very low without detriment.

CIRCULATION OF AIR IN MEAT ROOMS.

The required circulation of air in the meat rooms is either produced by natural draft or (especially in Europe) by means of blowers or fans, which circulate air, cooled artificially. The cooling of air used for the latter purpose is generally done in a separate room in which the air is brought in contact with the surfaces of pipes which are refrigerated by direct ammonia expansion. The warmer air is continuously exhausted from the meat rooms by means of a blower, which forces it through the cooling apparatus and thence back to the meat rooms in a cold and dry condition.

See also what has been said on ventilation, etc., in the chapter on cold storage.

BONE STINK.

As already stated, the freezing of meat must be done very carefully, in order to avoid any injury to the meat. Moro particularly the chilling and freezing must be done very gradually, for when the meat is plunged at once in a chamber below the freezing point, the external parts are frozen more quickly than the internal parts, and the latter are cut off by this external frozen and poorly conducting zone from receiving the same intensity of cold. The external frozen zone contracting on the internal portion causes many of the cells to be ruptured and the contents to escape, and on cutting into meat so frozen a pulpy consistency of the meat is found near the bones. This is particularly the case when whole carcasses are treated, but also parts of the animal show similar defects when frozen carelessly. The so called "bone stink," which shows itself as decaying marrow in the interior of the bones of many frozen meats, is also generally due to the too hasty freezing. However, the condition of animal at the time of killing (exhaustion by a long journey, injudicious feeding, excitement, delay in skinning, etc.) appears to favor the liability to bone stink.

Hanging the animals too closely together after they are slaughtered and dressed is said to be a fruitful source of bone taint, for when they are throwing off the animal heat and gases contained in the bodies, if hung too closely together they will steam one another and prevent this animal heat and gas from getting away. The absence of proper ventilation and an insufficient circulation of fresh air is also a likely cause, bearing in mind that what has to be aimed at is the driving away of this animal heat and gas as it passes out of the carcass. While the temperature of the cooling chamber should be kept moderately low, it should not be too low; a free circulation being of far more importance than lowness of temperature during this early cooling or chilling process.

Bone taint can be detected without actually cutting up a carcass, in the following way: A long wooden skewer is inserted at the point of the aitch bone; this passes the cup bone and enters the veins that divide the silver side from the top side, where, if any taint exists, it is sure to be found, the wooden skewer bringing out the taint upon it. For testing while in a frozen state a carpenter's brace and bit should be used. This must be inserted as above described.

FREEZING MEAT FROM WITHIN.

It has also been proposed to prevent the bone stink, etc., by freezing meat from the center by introducing into the same a pipe shaped like a hollow sword divided by a partition around which refrigerated brine or ammonia is permitted to circulate.

DEFROSTING OF MEAT.

The importance of doing the defrosting of meat with the same care as the freezing is well illustrated by a number of patents taken out for this operation. One of these processes subjects the meat to a continuous circulation of dry air formed by mixing cold air at a temperature of 19° and dry air heated to 70°, the combined current at about 26°, increased to about 60°, being forced through the thawing chamber by a fan. Time required for thawing, two to five days. This process is in use at Malta and Port Said.

Another patent provides for the circulation of air,

dried by arrangement of pipes containing cooling medium, and suitably heated by steam pipes, passing over the meat by natural means, and, by gradually increasing temperature, abstracting the frost without depositing moisture. Time required for defrosting: Beef, four days; sheep, two days. Process has been in continuous use in London for two and one-half years; it is also used in Paris and in Malta for meat supplied to troops.

MOLDY SPOTS ON MEAT.

The white mold spots which sometimes form on meat in cold storage are due to the growth of a fungus (*Oidium albicans*) the germs of which are quite common in the air. For this reason the formation of this mold may be prevented by providing a circulation of air which has passed over the cooling pipes (St. Clair's system, described under "Cold Storage"), by which the moisture and mold germs are withdrawn from the air.

KEEPING OF MEAT.

Meat, if kept constantly at 31° in a properly ventilated room from the time it has been slaughtered can be kept fresh at least six months, but if the temperature goes up at times as high as even only 33° the meat might not keep over a month; however, if the ventilation and humidity are properly regulated it should keep about two months in good condition in the latter case.

Beef should be placed in cold storage within ten hours after killing.

SHIPPING MEAT.

Meat properly prepared may be kept at a temperature between 32° and 35° F. for any length of time, but to insure against a break down of the refrigerating machinery aboard the vessel, the meat is generally frozen before it is loaded, thus providing for a deposit of cold (100 tons of frozen meat being equivalent for refrigerating purposes to seventy tons of ice) that can be drawn on in case the machinery fails temporarily.

REFRIGERATION FOR OTHER PURPOSES.

From the data, rules and examples given under the heads of cold storage, packing house and brewery refrigeration, and on refrigeration in general, it will be practicable to make the required approximate estimates for most of the other numerous applications of refrigerating machinery.

REFRIGERATION IN OIL WORKS.

In oil refineries artificial refrigeration has become indispensable for the purpose of separating the paraffine wax and refining the oil. Stearline, India rubber works, etc., can no longer be without artificial refrigeration.

DAIRY REFRIGERATION.

In the dairy practice, the cooling and freezing of milk, in butter making, etc., there is a great future for artificial refrigeration.

Refrigeration has also been patented for the special purpose of freezing the water out of milk in order to concentrate the same without heat.

REFRIGERATION IN GLUE WORKS.

Some glue manufacturers have found it to their interest to improve their product by drying their gelatine in rooms artificially refrigerated, thus permitting them to use glue solutions less concentrated.

VARIOUS USES OF REFRIGERATION.

Manufacturers of oleomargarine, of butterine, soap, chocolate, etc., derive great benefit from artificial refrigeration. For seasoning lumber it is also employed to some extent already.

Skating rinks, ice railways, etc., are kept in working order all the year now by artificial refrigeration.

Young trees are kept in cold storage to hold back unseasonable and premature growth.

The preservation of the eggs of the silkworm, so as to make the eclosion of the eggs coincide with the maturity of leaves of the mulberry tree has also become a subject of artificial refrigeration.

Many transatlantic vessels are equipped with gigantic refrigerating apparatus to enable them to transport perishable goods, chiefly meat, but also fruits, beer, etc.

In dynamite factories for maintaining the dynamite at a low temperature during the process of nitrating.

In manufactories of photographic accessories, for cooling gelatine dry plates.

In the establishments of wine growers and merchants for reducing the temperature of the must or unfermented wine, and for the obtainment of an equable temperature in the cellars, etc.

Wool and woolen garments, as likewise furs and peltry, are preserved from the attacks of moths by artificial refrigeration.

Beds in summer time may be cooled by pans filled with ice in the same way as they are warmed by warming pans in winter. This cooling of beds is said to produce immediate sleep and rest, and is especially recommended in cases of insomnia and other afflictions.

Decorative effects, quite novel and artistic, to adorn the dining table, etc., may be produced by freezing flowers, fishes, etc., tastefully grouped in clear crystal ice blocks of convenient shapes.

For refrigeration of dwellings, hospitals, hotels, public institutions, etc.:

This subject has been much written about, but in the practice of refrigerating dwellings and hotels during the hot season little progress has been made so far, many being of the opinion that it would be too expensive for general use. While this may be so, there is doubtless a great field open in this direction for the application of refrigeration in those cases in which expense is a secondary consideration.

The value of ice in therapeutics is generally recognized. From among the more recent applications in this direction may be mentioned the following: Ice is used for the induction of failing respiration by rubbing slowly the mucous membrane of the lips and mouth with a piece of ice to the rhythm of normal respiration.

Ice is said to moderate inflammation of the brain or its membranes, and also the severe headache of the early stages of acute fevers, also to relieve the pain and vomiting in cases of ulcer or cancer of the stomach. It is also excellent for the sore throat of fevers, and in cases of diphtheria. Sucked in small pieces, it checks secretions of the throat. Ice also arrests hemorrhage in a measure.

Artificial refrigeration is also very extensively used in the shipping of all sorts of produce, especially meat, eggs, etc., and the refrigerating installations in vessels crossing the ocean, and in railroad cars crossing the plains, are subjects of special study and detail which it would be beyond the scope of this book to enter into here. We may add, though, that the refrigeration during transit is not confined to railroad cars and steamboats, but that small delivery wagons for meat, eggs, etc., are now constructed with special reference to the keeping of their refrigerated contents until delivered to the consumer or retailer.

In distilleries for keeping the spirits in the store tanks cool during hot weather, and thereby obviating the very serious loss that is otherwise experienced through evaporation.

In chocolate and cocoa manufactories to enable the cooling room to be maintained at a low temperature in summer, and the process to be worked continuously all the year around. A great saving is likewise effected by the rapid solidification which is rendered possible, and the waste thus avoided; and furthermore, as the chocolate leaves the molds readily and intact, a considerably fewer number of the latter are required to do the same amount of work.

In sugar factories and refineries for the concentration of saccharine juices and solutions by freezing or congealing the water particles, which are then removed, leaving the residuum of a greater strength.

In India rubber works for the curing and hardening of India rubber blocks, thereby facilitating the cutting of same into sheets for manufacture of various elastic articles. The material in that state admitting of its being worked up in a much superior manner, and, moreover, at a far lower cost

REFRIGERATION IN CHEMICAL WORKS.

Some of the chemical industries in which artificial refrigeration is extensively used have been mentioned already, and to these may be added ash works, asphalt and tar distilleries, nitroglycerine works, etc. In fact, all chemical operations which depend largely on differences in temperature, notably all those involving crystallization processes, can in most cases be greatly assisted by the use of artificial refrigeration. This is particularly true of substances which it is difficult to obtain in a pure state, and which do not pass into the solid state, except at very low temperature. To successfully purify such substances—and there are a great many of them—artificial refrigeration is the most valuable auxiliary, and very remarkable results have been obtained already in this direction. The most successful purification of glycerine is an instance of this kind. Chloroform is another still more remarkable example. This substance, although considered pure, was nevertheless of a very unstable character. Time, action of light, heat and other unavoidable conditions, caused its degeneration, until it was shown by Pictet that an absolutely pure article of chloroform could be obtained by crystallizing the same at a temperature of about −90°. This is a very low temperature, considering practical possibilities of the present day, but it accomplishes the object, and there are many more equally useful applications not yet thought of, or beyond the reach of practical refrigeration at present.

CONCENTRATION OF SULPHURIC ACID.

The concentration of sulphuric acid, which is accomplished in expensive platinum vessels, can be accomplished, according to Stahl, in leaden vessels, if artificial refrigeration is used to crystallize the strong acid, which can then be separated from the weak mother acid. Another interesting chemical change brought about by artificial refrigeration is the decomposition of the acid sulphate of soda into neutral salt and free sulphuric acid.

DECOMPOSITION OF SALT CAKE.

Another interesting application of refrigeration in chemical manufacturing is the decomposition of the so called salt cake (acid sulphate of soda) into sulphuric acid and neutral sulphate of soda, which takes place when a watery solution of the said salt is subjected to a low temperature.

PIPE LINE REFRIGERATION.

In many cities refrigeration is furnished to hotels, butchers, restaurants, private houses, etc., by a pipe line which carries liquid ammonia; another pipe line returning the expanded ammonia to the central factory, at which a large supply of liquid ammonia is kept in store to regulate inequalities in the demand for refrigeration.

REFRIGERATION AND ENGINEERING.

When making excavations in loose soil, it has been found expedient to freeze the ground by artificial refrigeration, and this artifice is now extensively applied in mining operations, in the sinking of bridge piers, in tunneling through loose or wet soil, etc.

One of the greatest pieces of engineering with the aid of refrigerating machinery was accomplished about two years ago in the opening of a coal mine in Anzin, France. The coal was over 1,500 feet below the surface, and below strata strongly saturated with water, and impassable without artificial solidification.

CHAPTER IX.—THE ABSORPTION SYSTEM.

THE CYCLE OF OPERATIONS.

As in the compression system of ammonia refrigeration, the operations performed in the absorption system constitute what has been termed a cycle of operations, the working medium, ammonia liquor, returning periodically to its initial condition, at least theoretically so.

A COMPOUND CYCLE.

It is, however, not a reversible cycle, but rather two cycles merged into one, or a compound cycle. The anhydrous ammonia after leaving the still at the top, passes through the analyzer, condenser, receiver and refrigerator to the absorber, where it meets the weak liquor coming through the heater and exchanger from the still, and then after having been absorbed by the latter, passes as rich liquor from the absorber through the ammonia pump to the exchanger, and through the heater to the still, entering the latter by first passing through the analyzer, generally located at the top of the still.

APPLICATION OF FIRST LAW TO CYCLE.

Owing to the complexity of the operations of the double or compound cycle, its theoretical working conditions cannot be expressed by so simple a formula as in the case of a reversible cycle. Nevertheless, the tenets of the first law of thermodynamics apply in this case also, and therefore the heat and work which is imparted to the working substance while performing the operations of one period of the cycle must be equal or equivalent to the heat and work which are withdrawn during the same period—all quantities to be expressed by the same kind of units.

EQUATION OF ABSORPTION CYCLE.

Hence, if W_1 is the heat imparted to the liquid in the still, and W_2 the heat imparted to the anhydrous ammonia in the refrigerator, and W_3 the heat equivalent of the work of ammonia pump, we find—

$$W_1 + W_2 + W_3 = H_1 + H_2$$

H_1 being the heat withdrawn from the anhydrous ammonia in the condenser, and H_2 being the heat withdrawn from the working substance in the absorber.

As all the quantities in the above equation (besides W_1) can be readily determined, it enables us to find, if not a simple at least an artless expression for W_1 (*i. e.*, the heat which must be imparted to the liquid in the still).

WORKING CONDITIONS OF SYSTEM.

For the purpose of determining the theoretical values of the quantities which determine the efficiency of an absorption machine, we make the following stipulations which, we hold, are such as to be within the theoretical possibility of realization, although practically they have not as yet been fully realized, viz.:

That the apparatus is provided with efficient analyzer and rectifier, so that the ammonia when entering the condenser is practically in an anhydrous condition.

That the poor liquor when entering the absorber is only 5° warmer than the rich liquor when leaving the absorber.

That all the heat of the poor liquor, except that brought into the absorber, is imparted to the rich liquor on its way to the still in the exchanger.

That the uncompensated heat transfers from the atmosphere to the colder portions of the plant, and from the warmer portions of the plant to the atmosphere, are so well guarded against that they may be neglected in this connection.

HEAT ADDED IN REFRIGERATION.

The above premises being granted, the different items of the above equation are readily expressed. The heat, W_2, added to the working fluid in the expansion or refrigerating coils, is theoretically equal to the amount of refrigeration which is produced by its evaporation.

The refrigeration, r, in B. T. units which may be produced by the vaporization of one pound of anhydrous ammonia in an absorption machine is the same as in a compression machine, and is therefore expressible by the same formula:

$$r = h_1 - (t - t_1)s \text{ units,}$$

h_1 being the heat of volatilization of one pound of ammonia at the temperature t_1, of the refrigerator; t is the temperature of the liquid anhydrous ammonia, *i. e.*, the temperature of the condenser, and s the specific heat of ammonia.

For the purpose of this calculation the temperature of the outgoing condenser water may be taken for t, but in order to find the maximum theoretical refrigerating effect, the temperature of the incoming condenser water, or rather, about 5° added to that, should be taken for t, as the liquid anhydrous ammonia can be cooled to that degree by the condenser water. This also applies to the same calculation for compression system.

HEAT INTRODUCED BY PUMP.

The heat, W_3, imparted to the working medium by the operation of the ammonia pump is equivalent to the work required to lift the rich liquor from the pressure of the absorber to that of the still. It is not a very important quantity in this connection, and may be neglected in approximate calculations. However, it may be determined by the formula:

$$W_3 = \frac{P_2 \times (z - z_1)}{S \times 772} \text{ units.}$$

for each pound of anhydrous ammonia which is volatilized in the expander. In this formula P_2 stands for the number of pounds of rich liquor which must be moved for every pound of ammonia volatilized in the expander; and z and z_1 being in feet the heights of columns of water corresponding to the pressure in the still and pressure in absorber, respectively. S represents the specific gravity of the rich liquor, and 772 the equivalent of the heat unit in foot-pounds. In exact calculations the heat due to friction of pumps should be added.

RICH LIQUOR TO BE CIRCULATED.

The number of pounds of rich liquor, P_2, which must pass the ammonia pumps in order that one pound of liquid anhydrous ammonia may be disposable in the expander or refrigerator coils, depends on the concentration or strength of the poor and rich ammonia liquor, and if the percentage strength of the former be a, and that of the latter be c, we find—

$$P_2 = \frac{100}{c - \frac{(100-c)\,a}{(100-a)}} = \frac{(100-a)\ 100}{(100-a)\,c - (100-c)\,a} \text{ lbs.}$$

STRENGTH OF AMMONIA LIQUOR.

The percentage strength of the rich liquor depends largely on the construction of the absorber. Theoretically it is determined by the temperature at which it leaves the absorber and the pressure in the latter as shown in the tables on solutions of ammonia given by Starr, pages 96 and 97.

The lowest possible percentage strength of the poor liquor depends in a similar manner on the temperature and pressure in the still, but is also greatly affected by the constructive detail and operation of this appliance.

HEAT REMOVED IN CONDENSER.

The amount of heat, H_1, which is taken away from the working substance in the condenser, while one pound of vapor is condensed into liquid ammonia, is equal to the latent heat of volatilization of that amount of ammonia at the temperature of the condenser (temperature of outgoing condenser water), and may be readily obtained from the table on saturated ammonia, page 92.

HEAT REMOVED IN ABSORBER.

The amount of H_2 which must be withdrawn from the working liquid in the absorber is composed of different parts, viz.:

The heat developed by the absorption of one pound of ammonia in the poor liquor, H_n.

The heat brought into the absorber by a corresponding quantity of poor liquor, H_g.

The negative heat brought into the absorber by one pound of the refrigerated ammonia vapor, H_v.

Hence we find—

$$H_2 = H_n + H_g - H_v \text{ units.}$$

HEAT OF ABSORPTION.

The heat developed by the absorption of ammonia vapor in the poor liquor may be obtained after the formula given, pages 99 and 100, viz.:

$$Q_3 = 925\,b - \frac{142\,(2b + b^2)}{n} \text{ units.}$$

In this formula n stands for the number of pounds of water contained in the poor liquor for each pound of ammonia, and $1 + b$ stands for the number of pounds of ammonia contained in the rich liquor for every n pound of ammonia. Under these suppositions Q_3 stands for the number of heat units developed by the absorption of b pounds ammonia vapor, or the heat developed by one pound is—

$$H_n = \frac{Q_3}{b} \text{ units.}$$

The last two formulæ may be united, to give a simpler expression for the amount of heat developed when one pound of ammonia is dissolved in a sufficient quantity of poor liquor, containing one pound of ammonia to n pounds of water, in order to obtain a rich liquor which will contain $b + 1$ pound of ammonia for each n pound of water. The formula then reads—

$$H_n = 925 - \frac{284 + 142b}{n} \text{ units.}$$

The amount of heat developed by the absorption of one pound of ammonia in some cases of different strength of poor and rich liquor, calculated after the foregoing formula, is given in the subjoined table, together with the number of pounds of rich liquor that must be moved for each pound of ammonia evaporated in the refrigerator.

Ammonia in poor liquor, per cent.	Ammonia in rich liquor, per cent.	Heat of absorption by one pound of ammonia in units.	Pounds of rich liquor for each pound of active ammonia.
a	c	H_n	P_2
10	25	812	6.0
10	36	828	3.45
12	35.5	828	3.74
14	25	854	7.8
15	35	811	4.25
17	28.75	840	7.0
20	25	840	16.0
20	33	819	6.1
20	40	795	4.0

HEAT INTRODUCED BY POOR LIQUOR.

The number of pounds of poor liquor which enters the absorber for each pound of active ammonia vapor is equal to the rich liquor less one, this being the amount or weight of ammonia withdrawn, and therefore the heat, H_g, which enters the absorber with that amount of poor liquor, when its temperature is 5° above that of rich liquor leaving the absorber, is—

$$H_g = (P_2 - 1)\, 5 \times S \text{ units,}$$

S being the specific heat of the poor liquor, which may be taken at 1.

NEGATIVE HEAT INTRODUCED BY VAPOR.

The negative heat, H_v, brought into the absorber with every pound of ammonia vapor is—

$$H_v = (t - t_1)\, 0.5 \text{ units},$$

t being the temperature of the strong liquor leaving the absorber, and t_1 being the temperature in refrigerator coils.

HEAT REQUIRED IN GENERATOR.

From the above it is evident that the strength of strong and weak liquor, the pressure in still and absorber, and all other quantities, depend in a perfectly constructed plant in the last end on the temperature of cooling water and brine. Accordingly, it would be possible to express the heat required in the still or generator as a function of these temperatures, but the formula required to do this would be so complicated as to be without any practical value, nor would it possess any theoretical significance.

As all the quantities (excepting W_1) of the equation of the absorption cycle can be determined numerically in the manner shown, the quantity, W_1, or the heat required in the generator, can be readily determined after the formula—

$$W_1 = H_1 + H_2 - W_2 - W_3$$

WORK DONE BY AMMONIA PUMP.

The power, F (in foot-pounds), required to run the ammonia pump is theoretically expressed by the formula:

$$F = \frac{P_2\,(z - z_1)}{S} \text{ foot-pounds},$$

for every pound of active ammonia, *i. e.*, anhydrous ammonia evaporating in refrigerator. (See page 224.)

ANHYDROUS AMMONIA REQUIRED.

The number of pounds, P_1, of anhydrous ammonia required to circulate to produce a certain refrigerating effect, say m tons in twenty-four hours, is—

$$P_1 = \frac{m \times 284000}{r} \text{ pounds.}$$

HORSE POWER OF AMMONIA PUMP.

The power, F_1, to run the ammonia pump while producing a refrigerating effect of m tons in twenty-four hours, is, therefore—

$$F_1 = \frac{P_2 \times m(z - z_1)\,284000}{r \times S} \text{ foot-pounds,}$$

and expressed in horse power F_2, S being taken equal to 1:

$$F_2 = \frac{P_2 \times m \times 284000 \times (z - z_1)}{r \times 33000 \times 24 \times 60} \text{ horse power,}$$

33,000 being the equivalent of a horse power in foot-pounds per minute.

The formula for F_2 may be simplified to—

$$F_2 = \frac{P_2 \times m(z - z_1)\,0.006}{r} \text{ horse power.}$$

This is the horse power required theoretically, to which must be added the friction, clearance and other losses of the pump, as well as of the engine which operates the pump, to find the actual power and the equivalent amount of steam required for this purpose.

AMOUNT OF CONDENSING WATER.

The water required in the condenser expressed in gallons, G, for a refrigerating capacity of m tons in twenty-four hours is—

$$G = \frac{h_1 \times m \times 284000}{8.33 \times r(t - t_1)} \text{ gallons}$$

or approximately per minute in gallons, G_1 —

$$G_1 - \frac{h_1 \times m \times 24}{r(t - t_1)} \text{ gallons}$$

in which formula h_1 is the latent heat of volatilization of ammonia at the temperature of the outgoing condenser water, t, and t_1 the temperature of the incoming condenser water; r is the refrigerating effect of one pound of ammonia.

WATER REQUIRED IN ABSORBER.

The amount of heat to be removed in absorber for each pound of ammonia vaporized in refrigerator being H_2, as found in the foregoing, the amount of water required in absorber for a refrigerating capacity of m tons in twenty-four hours, expressed in gallons, G_2, is—

$$G_2 = \frac{H_2 \times m \times 284000}{8.33 \times r(t - t_1)} \text{ gallons}$$

or expressed per minute in gallons, G_3—

$$G_3 = \frac{H_2 \times m \times 24}{r(t - t_1)}$$

ECONOMIZING WATER.

When water is scarce or expensive, the same water after it has been used in condenser is used in the absorber, which, of course, raises the temperature of the ingoing and outgoing absorber water correspondingly. The water may also be economized by using open air condensers or by re-cooling the same by gradation, etc.

ECONOMIZING STEAM.

As the poor liquor is less in volume and weight than the rich liquor, it cannot possibly heat the latter to the temperature of still, other reasons notwithstanding. For this reason the waste steam of the ammonia pump may be used to still further heat the rich liquor on its way to the generator after it has left the exchanger. This is done in the heater, and the heat so imparted to the working fluid should be deducted from the heat to be furnished to the generator direct in theoretical estimates. The condensed steam from generator may be returned to boiler if it is not used for ice making.

AMOUNT OF STEAM REQUIRED.

The theoretical amount of steam required in generator expressed in pounds P_8 per hour for a refrigerating capacity of m tons in twenty-four hours is approximately found after the formula

$$P_8 = \frac{W_1 \times m \times 284000}{24 \times r \times h_8}$$

h_8 being the latent heat of steam at the pressure of the boiler, or, closer still, at the temperature of the generator.

As stated in the beginning, these calculations are based on ideal conditions, which are never met with in practical working, and therefore the quantities found must be modified accordingly, and the theoretical amount of steam as found must be increased by from 20 to 40 per cent, and even more, to arrive at the facts in most practical cases.

The amount of steam used by the ammonia pump must be added to the above. It is generally about $\frac{1}{8}$ to $\frac{1}{4}$ of the steam used in the generator.

ACTUAL AND THEORETICAL CAPACITY.

In order to compare the actual refrigerating capacity of an absorption plant with the theoretical capacity, the amount of steam used in the still, as well as the amount of rich liquor circulated by the ammonia pump, may be taken as a basis. The first case is practically disposed of in the foregoing. In the latter case the amount of liquid moved by the ammonia pump is equal to its capacity per minute, which is found by calculation, as in the case of a compressor, and reduced to pounds per minute. If this quantity is called C, and if P_2 is the number of pounds of rich liquor which must be circulated for each pound of active anhydrous ammonia, as found from the strength of the poor and rich liquor (see foregoing table), the refrigerating capacity of the machine, R, should be—

$$R = \frac{C \times r}{P_2} \text{ units per minute.}$$

The theoretical and actual heat balances can also be compared by determining the heat removed in the condenser and absorber, as well as the heat brought into the refrigerator and to the generator by actual measurement.

SIMPLER EXPRESSION FOR W_1.

If we neglect the work of the liquor pump and assume that the poor liquor arrives at the absorber at the absorber temperature, we can express the amount of heat W_1, theoretically required in the generator for each pound of anhydrous ammonia circulated by the formula—

$$W_1 = H_n - (h_2 - h) \text{ units,}$$

h_2 being the latent heat of volatilization of ammonia at the temperature of the absorber, and h, the latent heat of volatilization of ammonia at the temperature of the condenser.

It is frequently argued that an equivalent of the whole heat of absorption must be furnished to the generator, but this is only the case (theoretically speaking) when the temperature of the absorber is equal to that of the condenser.

EXPRESSION FOR EFFICIENCY.

The maximum theoretical efficiency E, of an absorption machine may be expressed in accordance with the above.

$$E = \frac{r}{W_1} = \frac{h_1 - (t - t_1)\,s}{H_n - (h_2 - h)}$$

and if we include the work of the ammonia pumps, etc., we have also—

$$E = \frac{r}{W_1} = \frac{h_1 - (t - t_1)\,s}{H_1 + H_2 - W_2 - W_3}$$

COMPARABLE EFFICIENCY OF COMPRESSOR.

In order to compare the maximum theoretical efficiency of an absorption plant with that of a compression plant the foregoing formula:

$$E = \frac{r}{W_1}$$

may be used, when in the case of compression W_1 stands for the amount of heat theoretically necessary to produce the work required from the engine for the circulation of one pound of ammonia.

If the absolute temperature of steam entering the engine is T, and that of the steam leaving the engine is T_1, and if the work of the engine which operates the compressor is expressed by Q_1 (in heat units), we find for W_1 the expression—

$$W_1 = \frac{Q_1\,T}{T - T_1}$$

If we omit friction of compressor and engine and insert for Q_1 the theoretical work of the compressor (page 111) we find—

$$Q_1 = \frac{(\tau - \tau_1)\,h_1}{\tau_1}$$

τ and τ_1 being the absolute temperatures of condenser and refrigeration respectively. It is then—

$$W_1 = \frac{h_1\,(\tau - \tau_1)\,T}{\tau_1\,(T - T_1)}$$

and for the maximum theoretical efficiency of the compression machine, leaving out friction, etc., we find—

$$E = \frac{[h_1 - (t - t_1)]\,s \times \tau_1 (T - T_1)}{h_1\,(\tau - \tau_1)\,T}$$

CONSTRUCTION OF MACHINE.

The construction details of the absorption plants vary so much that in this place we can only give the general outlines touching the appliances and contrivances which by a concert of action make up the refrigerating effect. The dimensions of parts vary also very greatly, and those given in the following paragraphs and tables are based on data reported from machines in actual operation where not otherwise stated.

THE GENERATOR.

The generator, retort or still is generally an upright cylinder heated with a steam coil in which the concentrated or rich liquor is heated. The rich liquor passes in at the top and leaves at the bottom. The retort and dome is made of steel plate, sometimes of cast iron; and this vessel, the same as other parts containing ammonia gas, should be capable of withstanding a liquid pressure of 400 pounds per square inch.

SIZE OF GENERATOR.

The size of the still or generator depends on the size of the machine, and for a 10-ton machine (actual ice making capacity) is about two to two and one-half feet wide and fifteen to eighteen feet high, and a little over half of this height is generally occupied by the steam coil. An English author gives the following table of dimensions for generators or stills of absorption machines, but they appear rather small compared with American structures for the same object:

Ice Made in 24 Hours.	Gallons of .880 Ammonia.	Size of Generator.	
		Diameter.	Length.
1	27	13.5 inches.	5 feet 6 inches.
2	54	17.0 "	6 " 0 "
3	80	21.5 "	6 " 0 "
4	108	22.5 "	6 " 6 "
6	162	22.5 "	10 " 6 "
8	216	25.0 "	12 " 0 "
10	252	26.0 "	12 " 0 "
12	270	28.0 "	13 " 0 "
15	405	29.5 "	14 " 0 "
24	540	35.0 "	14 " 0 "

BATTERY GENERATOR.

Generators have also been constructed on the battery plan, three or more cylinders being connected to form one generator, the rich liquor passing gradually from the first cylinder to the last, which it leaves as poor liquor. In this manner it is possible to attain a wider difference between the strength of the rich and poor liquor, it is claimed.

COILS IN RETORT.

The heating coils in retort or still are placed in the lower part of the retort, and consist of one or more spiral coils of pipe placed concentrically. According to Coppet, their connections should be at both the bottom entrance and exit, and should be made right and left handed, the object being to prevent the steam (when rushing down in the coils) from imparting a gyrating motion to the liquor, thus shaking the retort. The coils should be made of purest charcoal iron, free from defects or spots, as the hot ammonia liquor is very apt to penetrate such bad places and cause leaks. The space in still occupied by steam coil should always contain ammonia liquor, so that the coil is never exposed to the vapors. For this reason a gauge is provided, which shows the height of the liquor in the generator. As a further precaution there is placed above the steam coils an inverted cone, with a large central opening, placed so that the liquor will be deflected to the center of still, and not fall upon the coils, if ever the liquor should stand below them. A valve is provided at the bottom of the retort to empty same, if necessary, and also one at the poor liquor pipe leading to exchanger. The heating surface of the coil in retort varies considerably, and for a 10-ton machine it covers from eighty to 100 feet.

THE ANALYZER.

In the upper part of the still the so called analyzer is located. In it the rich liquor is made to pass over numerous shelves or disks into corresponding basins, over which it runs in a trickling shower from one disk through the next basin over the following disk, and so on, until it reaches the top of the boiling liquid in retort. While the rich liquor runs downward over these devices, the vapor from the retort passes them in its upward course and constantly meeting the rich liquid over an extended area, is enriched in ammonia, and deprived of water. Thus the ammonia vapor is rendered almost free of water when it reaches the top of the analyzer. At the same time the temperature of the rich ammonia liquor is increased from about 150° to 170°, at which it reaches the analyzer, to about 200°, more or less, when it reaches the body of liquor in the retort.

The passages in the analyzer must be amply large for the passage of water and ammonia vapor in opposite directions in order to avoid foaming, overloading, etc. The best iron or steel plate must be used in the construction of the analyzer. As also stated elsewhere, galvanized iron pipes and zinc surfaces in general must be avoided wherever they come in contact with ammonia. The surface in the analyzer runs from fifty to seventy square feet in a 10-ton machine.

THE RECTIFIER.

Frequently the vapor on its way from analyzer to condenser passes the so called rectifier, which is a small coil partly surrounded by cooling water, the lower end of which is connected with the condenser coil, but has also a liquid outlet to a separate liquor receiver which receives all watery condensation which may have formed in the rectifier. In this manner the vapors, when they enter the condenser proper, are as nearly anhydrous as they can practically be made. About twenty-five square feet of cooling surface is allowed in the rectifier for a machine of ten tons ice making capacity. The liquid separated from the vapor in the rectifier, after passing through a separate cooler, is returned to the ammonia pump, whence it passes back to the generator or still.

The following table, giving the heating surfaces of generator coils and surface in analyzer and rectifier for machines of different sizes, is also given on English authority, and these figures also fall short of the sizes employed in the United States:

Size in Tons of Ice Made in 24 Hours.	Surface in Generator Coils.	Surface in Analyzer Disks.	Surface in Rectifier Coil.
Tons	Square Feet.	Square Feet.	Square Feet.
2	16	14	4
6	43	34	11
12	81	68	20
15	160	133	40
30	214	169	50
50	304	262	74

THE CONDENSER.

The vapor after leaving the still or rectifier enters the condenser which is constructed on the same principles as the condenser in a compression machine. Besides the submerged condenser and the open air or atmospheric condenser (the latter, on account of accessibility, simplicity and cleansability, now most generally adopted) it has also been proposed to use condensers exposed to the atmosphere alone, thus to save the cooling water. Such condenser requires a considerable surface, at least over eight times that of the submerged condenser, and over five times that of the atmospheric condenser. The material for condenser coils, as well as for all other coils in the absorption machine, should be the very best iron.

Still another form of condenser consists of one pipe within another, in which the water surrounds the outside pipe and also runs through the internal pipe, while the gas passes through the annular space between the two pipes. This is a very effective form of condenser, but the difficulty of keeping it clean is very great, and it is almost impossible when the water is liable to leave a deposit. For sizes of condenser coils the same subject under compression machines should be referred to, also the subsequent table on general dimensions.

LIQUID RECEIVER, ETC.

The vapors after having passed the condenser, reach the receiver in a liquid form and thence pass through the expansion valve to the coils in freezing or brine tank. These parts of the plant, their construction and the mode of operating them are quite the same as in case of the compression plant. The liquid receiver for an absorption machine should be at least large enough for the storage of sufficient liquid ammonia to bring the poor liquor at the bottom of the retort to between 18° and 20° Reaumur when the machine is in operation.

THE ABSORBER.

In the absorber the vapor of ammonia, after having done its duty in the freezing tank or expansion coils, meets the poor liquor coming from the generator, and is reabsorbed by the latter. The absorber should be constructed in such a manner as to allow the ammonia solution as it gets stronger to meet the cooling water flowing in an opposite direction, so that the warmer water cools the weaker solution and the colder water cools the stronger solution. In compliance with this condition the vapors of ammonia should be in constant contact with the liquor, and the surface of contact ought to be of reasonable area.

This may be accomplished by passing the ammonia and weak liquor over traps or disks, similar to those in the analyzer, or through a series of pipes or coils, where they are in constant contact with each other, the pipes being efficiently cooled from the outside by water (spent water from condenser generally), in order to remove the heat of solution of the ammonia as fast as it is formed. Generally the ammonia gas and the poor liquor are mixed together into a manifold at the lower end of the coils. The surface of these pipes exposed to the cooling water in a tank in which they are submerged (atmospheric cooling, as in the case of atmospheric condensers, may also be used), is variously estimated at 300 to 500 square feet for a machine of ten tons ice making capacity.

THE EXCHANGER.

In the exchanger the heat which the poor liquor carries away from the still should be imparted to the rich liquor on its way to the still. As a matter of course the two liquids should flow in opposite directions, so that the hottest rich liquid meets the poor liquid when it is hottest, and the cold poor liquid meets the rich liquid when it is coldest.

The exchanger is also to be made of the best sheet steel, and the coils within should be extra heavy, and the whole apparatus must be able to sustain the same pressure as the retort. It should stand upright, and the liquor pump should force the rich liquor through these coils to the top of the retort or to the heater, and the poor liquor should pass in the opposite direction. In causing the liquors to take this course the pressure in the body of the exchanger can be regulated by the valve on the poor liquor pipe coming from the retort.

The amount of surface between the poor and rich liquor in exchanger varies according to its construction, all the way from twenty-five to fifty square feet for a 10-ton plant (ice making capacity). This statement covers those plants of which we have knowledge. According to Starr, who assumes the heat transfer to amount to 40 B. T. units per square foot surface per hour, for each degree Fahrenheit difference in temperature, about 120 square feet of exchanging surface would be required for an ice making plant of ten tons daily capacity.

THE HEATER.

The heater is another contrivance frequently used to further the objects of the exchanger. It consists of a coil of pipe through which the rich liquor passes from the exchanger before it reaches the retort. This pipe is located in a drum in which steam (generally spent steam from liquor pump) is circulated. It is constructed on the same principles as the other receptacles and coils. The surface of the heater coil is about thirty to fifty square feet in a 10-ton ice making plant.

THE COOLER.

The cooler is an arrangement frequently used to do for the poor liquor what the heater does for the rich liquor, *i. e.*, to promote the objects of the exchanger by withdrawing all the heat possible from the poor liquor before it reaches the absorber. This contrivance is built on the same principles as a condenser, and consists of a coil or series of coils, submerged in a tank through which cooling water circulates, or placed over a vat to allow the cooling water to trickle over them, similar to an atmospheric condenser. The surface of the cooler may be from sixty to eighty feet for a 10-ton ice making machine, and larger or smaller for different capacities, as the case may be.

THE AMMONIA PUMP.

The ammonia pump, which takes up the rich liquor from absorber to force it through the exchanger and heater to the generator, is generally a steam pump, the engine and pump cylinder being mounted on a common base. A pump driven by belt may also be used. The size and number of strokes of pump depend on the size of plant, but also largely on the strength of poor and rich liquor. (See table, page 139.)

For a 10-ton plant (ice making capacity) the pump has generally a diameter of three inches, the stroke being from six to ten inches and the number of strokes from twenty-five to fifty per minute. The ammonia pump is generally single-acting, in order to relieve the pressure on stuffing box, which latter fixture requires particular care in order to secure proper working of the pump.

MISCELLANEOUS ATTACHMENTS.

Like the condenser, the refrigerator, expansion coils, as also the brine tank (and brine pump) or the freezing tank, are constructed on the same lines in an absorption as in a compression plant, and therefore need no further mention here. The same may be said of the expansion valve, and of other valves required when desirable to shut off certain portions of the machine, of the required pressure gauges, thermometers and other attachments. In the use of the absorption plant for various purposes the same rules apply as in the use of a compression machine. As the spent steam from the generator is used for distilled water, and as the same cannot be contaminated with lubricating oil, the steam filter or oil separator is superfluous if the boiler feed water is of ordinary purity.

OVERHAULING PLANT.

In order to keep an absorption plant in the best possible order for the longest possible time it is necessary that the different parts be opened and overhauled from time to time (according to the water used and as other conditions may indicate) every alternate season or so in order to thoroughly clean and inspect the interior part, and to repair them in order to anticipate any possible breakdowns, etc. In all cases, before starting up to open a new season, the coils and traps should be tested.

COMPRESSION VERSUS ABSORPTION.

The question is frequently asked as to which kind of refrigerating plant—a compression or absorption plant—is the most profitable and the most economical; and many different answers are given to these questions. Different as the two kinds of machines look at first sight, the theoretical principles as well as defects are the same, as has been already explained, although the natural facilities, as relative price of coal and cooling water, etc., may be more favorable in certain localities for one class of machines than for another. Taking this into due consideration, the principal difference between the two machines in a given case must be sought in the more or less greater care and perfection with which they are built and operated, more particularly also in the quality, quantity and proper distribution of material, the workmanship and the life of the plant, considering also the kind of water and ammonia to be used.

When it is considered how difficult it is to give due regard to all these circumstances in the valuation or planning of an individual plant, the apparently conflicting results of different kinds of plants working in different localities and conditions, and the different opinions on them are explained in a great measure.

TABULATED DIMENSIONS, ETC.

The great variations in the dimensions of the various parts of absorption machines of different makes find expression in the following table, which purports to give the dimensions, capacity, etc., of different machines. For the correctness of these figures we are unable to vouch, as the manner in which we obtained them does not exclude clerical errors, hence we must submit them for what they are worth:

TABLE SHOWING DIMENSIONS, ETC., OF ABSORPTION MACHINES.

Actual ice making capacity in tons of ice..................	3	8	12	15	25	10
Number and size of steam boiler horse power or dimensions..........	15	30	40"x20'	50	2 42" x21⅓'	2 42" x10'
Pounds of coal used per hour.......... . .	65	140	135	220	504	168-180
Number and size of generators..........	30"x10'	30"x16'	24"x18'	44"x14'	2 30" x17¼'	28"x15'
Size of coil in generator in square feet	24	48	91	96	400	80
Surface of disks, etc., in analyzer in square feet..... ...	10	20	64	34	125	24
Cooling surface in exchanger in square feet.........	34	51	22½	68	65	25
Cooling surface of traps in absorber in square feet.........	130	260	191	470	1900	673
Cooling surface in condenser in square feet..................	345	690	220	1380	1220	544
Surface in expander or refrigerator in square feet.........	410	1200	726	2100	4000	1600
Cooling surface in rectifier in square feet..................	....	....	25	....	....	....
Cooling surface in heater..............	....	....	41	....	....	....
Temperature of water in degrees F.	70	70	80	70	76	80-94
Temperature of brine in degrees F.	10-20	10-20	10-12	10-20	7	10-14

From the foregoing table it appears that in absorption machine one pound of coal will make from four to seven pounds of ice. On the continent it is assumed that one pound of coal will make about ten pounds of ice in an absorption machine; the evaporative power of the coal being taken at eight pounds of water per pound of coal.

CHAPTER X.—THE CARBONIC ACID MACHINE.

GENERAL CONSIDERATIONS.

Among the refrigerating machines which use other refrigerating media than ammonia, those compression machines using carbonic acid have found favor for many specific purposes, especially so for the refrigeration of storage rooms in hotels and restaurants, where the impeccability of the gas to victuals is prominently valued. The non-corroding action of carbonic acid on any of the metals, and the fact that it cannot be decomposed during compression, etc., speak principally in favor of its use. The fact that a leak of carbonic acid is not demonstrated by its smell might be overcome by the addition of some odoriferous substance. The capacity of the compressor may be very small as compared with other refrigerating plants (see page 89), but the parts of the machine must also be made correspondingly stronger on account of the high pressure of the gas.

The cheapness of liquefied carbonic acid is also quoted in its favor as a refrigerating agent, as also its lesser danger to respiration in case of leaks. It is claimed that air containing 8 per cent of carbonic acid gas can be inhaled without danger, while an atmosphere containing only ½ per cent of ammonia is said to be decidedly dangerous. On the other hand, the presence of the least amount of ammonia in the air demonstrates itself by the smell, while this is not the case with carbonic acid.

Not only the neutrality of carbonic acid toward metals and packings, but also toward water, meat, beer and other products subjected to cold storage, should be mentioned in this connection.

The use of carbonic acid in refrigerating machines of the compression type has been somewhat stimulated by the cheap manufacture of liquid carbonic acid as a by-product of the brewing industry, especially in Germany, where over 400 such machines (1894) are said to be working satisfactorily.

PROPERTIES OF CARBONIC ACID.

The carbonic acid, which is a gas of 1.529 specific gravity (air = 1) at the atmospheric pressure, becomes liquid at a temperature of —124° F. at that pressure. At 32° F. it is liquid under a pressure of 36 atmospheres, and then has a specific weight of 0.93 (water = 1). The specific weight of the liquid at different temperatures, according to Mitchel, is at 32° F. = 0.93, at 42° F. = 0.8825, at 47.3° F., = 0.853, at 65.3° F. = 0.7385, and at 86° F. = 0.60.

The specific heat of carbonic acid gas by weight = 0.2167 (air = 0.2375). Of the liquid it is 1.

The author's attention has been called to the apparent inconsistency existing between the specific gravity of liquid carbonic acid, as given in the foregoing paragraph (0.6 at 86° F.), and the amount of carbonic acid contained in the cylinders in which the same is shipped. The cylinders have a capacity of 805 cubic inches (29.11 pounds of water) and are made to contain 20 pounds of liquid carbonic acid, and some manufacturers are said to crowd in 21 and 22 pounds, although this is doubtless a very risky proceeding. But even at 20 pounds the cylinders contain over 2½ pounds more (at 86° F.) than what is consistent with the above specific gravity. The fact that the drums do not burst with such a charge tends to show that the foregoing specific gravity is not correct (too low) or that different densities exist for different pressures at or near the temperatures characterizing the critical condition of carbonic acid (88° F.).

PROPERTIES OF SATURATED CARBONIC ACID GAS.

Transformed to English units from a metric table computed by Prof. Schroter, by Denton and Jacobus.

Temperature of ebullition in deg. F.	Absolute pressure in lbs. per sq. in.	Total heat reck'n'd from 32° Fahr.	Heat of liquid reck'n'd from 32° Fahr.	Latent heat of evaporation.	Heat equivalent of external work.	Incr'se of volume during evaporation.	Dens'y of vapor or weight of one cu. ft.
t	$P \div 144$	y	q	r	APu	u	
—22	210	98.35	—37.80	136.15	16.20	.4138	2.321
—13	249	99.14	—32.51	131.65	16.04	.3459	2.759
— 4	292	99.88	—26.91	126.79	15.80	.2901	3.265
5	342	100.58	—20.92	121.60	15.50	2438	3.853
14	396	101.21	—14.49	115.70	15.08	.2042	4.535
23	457	101.81	— 7.56	109.37	14.58	.1711	5.331
32	525	102.35	0.00	102.35	13.93	.1426	6.265
41	590	102.84	8.32	94.52	13.14	.1177	7.374
50	680	103.24	17.60	85.64	12.15	.0960	8.708
59	768	103.59	28.22	75.37	10.91	.0763	10.356
68	864	103.84	40.86	62.98	9.29	.0577	12.480
77	968	103.95	57.06	46.89	7.06	.0391	15.475
86	1,080	103.72	84.44	19.28	2.95	.0147	21.519

A, in the column heading, stands for the reciprocal of the mechanical equivalent of heat.

The preceding table, showing the properties of saturated carbonic acid, may be used in connection with the formulæ given in the chapter on the ammonia compression system. However, the results obtained in this manner are only approximations, since the carbonic acid is in a superheated condition during several stages of the cycle constituting the refrigerating process, as a reference to the practical data, given hereafter, will amply show.

CONSTRUCTION OF PLANT.

The refrigerating plants operated with carbonic acid are built on the same general plan as the ammonia compression plants, compressor, condenser and refrigerator being the identical important parts, specified as follows by a leading manufacturer:

THE COMPRESSOR.

The compressor is either of the horizontal or the vertical type (for smaller machines generally the latter). It should be made of the best material, steel or semi-steel, and it is provided with a jacket through which the return gas passes, which arrangement gives additional strength to the cylinder and tends to keep it cool. The piston rods, connecting rods, crank pins and valves should be made of forged steel, and so as to be interchangeable at any time.

STUFFING BOX.

The stuffing box is made gas tight by means of cupped leathers on the compressor rod. Glycerine is forced into the spaces between these leathers at a pressure superior to the suction pressure in the compressor, so that whatever leakage takes place at the stuffing box is a leakage of glycerine either into the compressor or out into the atmosphere, and not a leakage of gas.

What little leakage of glycerine takes place into the compressor is advantageous, inasmuch as it in the first place lubricates the compressor, and in the second place fills up all clearances, thereby increasing the efficiency of the compressor.

In order to replace the glycerine which leaks out of the stuffing box of the horizontal machine, there is a belt driven pump which operates continuously. The smaller machines are fitted with a hand pump, a few strokes of which are required to be made every four or five hours.

GLYCERINE TRAP.

Any glycerine which passes into the compressor beyond what is necessary to fill the clearance spaces is discharged with the gas through the delivery valves. In order to prevent this going into the system, all the liquid passes through a trap in which the glycerine drains to the bottom, whence it is drawn off from time to time.

It may be remarked here that the glycerine has no affinity for carbonic anhydride, hence it undergoes no change in the machine, and therefore there is no chance of the condenser coils becoming clogged.

CONDENSER.

The condenser consists of coils of wrought iron extra heavy pipes, which are either placed in a tank and surrounded by water, or are so arranged that water trickles over them, forming the well known atmospheric condenser. The coils are welded together into such length as to avoid any joints inside the tank, where they would be inaccessible.

In connection with the condensers, where sea water only is available for condensing purposes, one very important advantage of carbonic anhydride machines is claimed: As carbonic anhydride has no chemical action on copper, this metal is used in the construction of the coils, giving same longer life.

EVAPORATOR.

The evaporator consists of coils of wrought iron extra heavy pipe, welded into long lengths, inside which the carbonic anhydride evaporates. The heat required for evaporation is usually obtained either from brine surrounding the pipes, as in cases where brine is used as the cooling medium, or else from air surrounding the pipes, as in cases where air is required to be cooled direct.

Between the condenser and evaporator there is a regulating or so called expansion valve for adjusting the quantity of the liquid carbonic anhydride passing from the condenser.

SAFETY VALVE.

In order to enable the compressor to be opened up for examination of valves and piston without loss of carbonic anhydride, it is necessary to fit a stop valve on the suction and delivery sides so as to confine the carbonic anhydride to the condenser and evaporator. It is, of course, possible for a careless attendant to start the machine again without opening the delivery valve, and in such cases an excessive pressure would be created in the delivery pipe, from which there would be no outlet. To provide against this danger a safety device is adopted, consisting of a housing, at the base of which is a thin disk, which is designed to blow off at a pressure considerably below that to which the machines are tested.

JOINTS.

All joints should be made with special flange unions and brass bushings. They should be made absolutely tight with packing rings of vulcanized fiber, which withstand the heat and still have the necessary elasticity to insure the joint being perfectly tight when either hot or cold.

STRENGTH AND SAFETY.

The working pressure varies from about fifty to seventy atmospheres. Owing to the very small diameter of all parts, even in large machines, there is no difficulty in securing a very ample margin of strength. All parts of the machine subject to the pressure of the carbonic anhydride should be tested at three times the working pressure.

APPLICATION OF MACHINE.

Both the direct expansion and the brine system are used in connection with a carbonic acid refrigerating machine, but for most purposes the former is deemed preferable, as is also the case with ammonia compression. For ice making the can or plate system may be used, and also for other refrigerating purposes the application of the carbonic acid refrigerating plant is quite similar to that of any other compression or absorption plant. A plant quite similar, or rather identical in its main feature with a carbonic acid refrigerating plant is also used for the manufacture of liquefied carbonic acid, as it may be obtained from breweries, distilleries, calcination of lime and other sources.

EFFICIENCY OF SYSTEM.

The efficiency of the carbonic acid machine is somewhat lessened by the high specific heat of the liquid, and therefore decreases with greater divergence of temperature. It has been proposed to reduce this loss in efficiency by introducing a motor between the condenser and refrigerator, which would perfect the cycle of operations. After another method, the loss of efficiency due to the specific heat of liquid is reduced by allowing the liquid during its flow to expand from the condenser pressure to an intermediate pressure, and to return the vapors so produced after having cooled the remaining liquid to the condenser by an auxiliary compressor.

It has frequently been argued that carbonic acid compression machines could not be operated successfully when the temperature of the condenser water exceeds 88° F., the critical temperature of carbonic acid. According to the present conception of the critical condition, above the said temperature carbonic acid can only exist in the gaseous form, and cannot be converted into a liquid by means of the withdrawal of the latent heat of volatilization. This being the case, the refrigerating effect of a carbonic acid machine working with condenser water above 88° F. would only be that of a compressed gas while expanding against resistance, which would be comparatively small when compared with refrigerating effect produced by the volatilization of the liquefied medium. These considerations and arguments are, however, in direct conflict with the statements of Windhausen, according to which carbonic acid machines operated with condensing water of 90° to 94° F. and in tropical countries produce refrigerating effects ten times larger than what they would be if the carbonic acid acted simply as a compressed gas at such temperatures.

Experiments cited by Linde show that a carbonic acid machine working with a temperature of 92° F. at the expansion valve gives a refrigerating effect about 50 per cent less than when the temperature at the expansion valve was 53° F.

CAUSE OF APPARENT INCONSISTENCIES.

The foregoing and other apparent inconsistencies between the theory and practice of the working of the carbonic acid refrigerating plant have recently been fully explained on the basis that the carbonic acid is in the state of a superheated gas in the compression stage; in fact, it must be so if the condensing gas reaches a temperature over 80°, in order to produce refrigerating effects at all. The loss due to the absence of an expansion cylinder (completing a perfect reversible cycle) to reduce the temperature of the liquefied carbonic anhydride from the temperature of the condenser to that of the refrigerator, which constitutes the chief difference in the economy between ammonia and carbonic acid refrigerating machines, has also been somewhat overestimated in derogation of the carbonic acid machine as shown by Mollier.

COMPARISONS OF EFFICIENCY.

The calculation on the former basis (specific heat times weight of carbonic acid circulated is unit of time) gave this loss as about 0.80 per cent of the whole theoretical refrigerating effect for every degree difference between the temperature of the condenser and that of the refrigerator, as compared with 0.18 per cent loss in the case of ammonia. The accompanying table was calculated and published by Ewing several months ago, showing the relation between the ammonia and carbonic acid refrigerating plant with reference to the loss due to cooling of the liquid. In this table the upper limit of temperature in the condenser, or rather immediately before the expansion valve, is taken at 68° F., while the temperature in the refrigerator varies from 50° to —4° F.

THEORETICAL CO-EFFICIENT OF PERFORMANCE IN VAPOR COMPRESSION MACHINES, UNDER WET COMPRESSION, UPPER LIMIT OF TEMPERATURE BEING 68° F.

Lower Limit of Temperature, Deg. F.	Theoretical Co-efficient of Performance.		Co-efficient of Performance in Carnot Cycle.
	Ammonia.	Carbonic Acid.	
50	27.8	25.7	28.3
40	18.1	20.	18.5
32	13.2	11.4	13.6
23	10.2	8.5	10.7
14	8.3	6.8	8.8
—4	5.9	4.5	6.3

It will be noticed that with ammonia the theoretical performance—namely, that of a compression machine without an expansion cylinder—is only a little less than the ideal performance which would be obtained by following Carnot's cycle. Hence with this substance almost nothing would be gained by adding an expansion cylinder to the machine—nothing, certainly, that would in any way compensate for the increase of complexity and friction and cost which an expansion cylinder would involve.

With carbonic acid there is considerably more falling away from the ideal of Carnot, for the reason that the specific neat of the liquid bears a greater proportion to the latent heat of the vapor. But even then the saving in work which an expansion cylinder would bring about is not great, and in practice the expansion cylinder, even in carbonic acid machines, is never used so far.

PRACTICAL COMPARATIVE TESTS.

Quite a number of practical tests published by Linde several years ago led him to the compilation of the following table, which shows the excess of efficiency in per cents of ammonia refrigerating machine over and above that of a carbonic acid machine, both working at different temperatures before the expansion valve, the temperature in the brine surrounding expansion coil being the same (about 23° F.) in all cases.

Temperature before expansion valve ° F.............	54°	63°	72°	81°	90°
Excess of efficiency of ammonia plant...............	17%	23%	31%	47%	101%

The tests referred to by Linde, on which the foregoing table is based, were made in the Experimental Refrigerating Station in Munich, Germany, by Schroeter, and in the following little table are compiled some of the actual results of these experiments obtained in the case of an ammonia and of a carbonic acid refrigerating machine:

	AMMONIA MACHINE.				CARBONIC ACID MACHINE.			
NO. OF TEST.	1	2	3	4	5	6	7	8
Temp. in brine tank, degrees Celsius...	—6.1	—6.4	—6.4	—4.8	—4.	—4.8	—4.8	—6.7
Temp. in condenser, degrees Celsius ...	21.4	21.4	21.4	34.9	20.9	21.2	22.2	30
Temp. before expansion valve, degrees Celsius............	—6.5	11.6	18.4	28.3	—7.9	10	16.8	28.8
Refrigeration per hour per horse power of steam engine in calories ...	3,897	3,636	3,508	2,237	3,832	3,178	2,867	1,477

The correctness of these figures has never been doubted, and in view of these facts the efficiency of a carbonic acid machine now in the market, which is given at 4,300 and 3,700 calories for temperatures of 10° and 20° Celsius before the expansion valve per indicated horse power, must be considered as something phenomenal indeed. This machine has no expansion cylinder, and therefore its efficiency is comparable to the efficiencies given under tests 6 and 7 in the above table, which are nearly 25 per cent less.

CHAPTER XI.—OTHER COMPRESSION SYSTEMS.

AVAILABLE REFRIGERATING FLUIDS.

Besides ammonia other liquids are used, and still others have been proposed as working fluids in refrigerating machines. Most of these liquids are used on the same plan as ammonia in the compression system, and the machines, barring certain details, are constructed on the same principles as the ammonia compression machine, and the same rules and calculations apply to all of them. The following table shows the pressure and boiling point of some liquids available for use in refrigerating machines as given by Ledoux. (Denton and Jacobus' edition.)

Temperature of Ebullition.	Tension of Vapor, in pounds per square inch, above Zero.					
Deg. Fahr.	Sulphuric ether.	Sulphur dioxide.	Ammonia	Methylic ether.	Carbonic acid.	Pictet fluid.
(1)	(2)	(3)	(4)	(5)	(6)	(7)
—40	....	....	10.22	....	....	....
—31	....	....	13.23	....	....	....
—22	....	5.56	16.95	11.15	....	....
—13	....	7.23	21.51	13.85	251.6	....
— 4	1.30	9.27	27.04	17.06	292.9	13.5
5	1.70	11.76	33.67	20.84	340.1	16.2
14	2.19	14.75	41.58	25.27	393.4	19.3
23	2.79	18.31	50.91	30.41	453.4	22.9
32	3,55	22.53	61.85	36.34	520.4	26.9
41	4.45	27.48	74.55	43.13	594.8	31.2
50	5.54	33.26	89.21	50.84	676.9	36.2
59	6.84	39.93	105.99	59.56	766.9	41.7
68	8.38	47,62	125.08	69.35	864.9	48.1
77	10.19	56.39	146.64	80.28	971.1	55.6
86	12.31	66.37	170.83	92.41	1,085.6	64.1
95	14.76	77.64	197.83	....	1,207.9	73.2
104	17.59	90.32	227.76	....	1,338.2	82.9

MACHINES IN ACTUAL OPERATION.

Of those compression machines which are in actual use besides the ammonia and carbonic acid machine, which have been described already, those operated with sulphur dioxide, Pictet liquid, ethylic ether (sulphuric ether), ethyl chloride and methyl chloride may be mentioned especially. The latter machine is comparatively new, and not so far in practical use to any extent, and therefore no special account can be given of the same in the following short remarks.

Recently we have found some accounts given of a machine operated with chloride of methyl in an ice factory at Algiers. We are informed that the size of the engine is 30 horse power, that about eighty pounds of the chemical at about fifty cents per pound were needed to operate the plant during 5,000 hours without the least disturbance, and we are informed of a number of other details, but as to the actual amount of ice produced we are left in the dark entirely. The temperature of the brine is —4° F. The pressure in the expander appears to be very low

THE ETHYL CHLORIDE MACHINE.

A refrigerating machine using ethyl chloride as a refrigerant has been in use to some extent lately. The ethyl chloride evaporates at a quite high temperature; the machine works under a vacuum, and condensing pressures are very low, about fifteen pounds (gauge pressure) as a maximum. The refrigerating coils are made of sheet copper, flat, several inches broad, and about an inch thick in an experimental plant in operation in Chicago. The machine appears to be designed for small work only, fruit rooms, creameries, small butcher shops, etc., and is operated by any sort of a small motor.

REFRIGERATION BY SULPHUR DIOXIDE.

The sulphurous acid refrigerating machines are also in practical operation to some extent. They require, however, a much greater compressor capacity than the ammonia compressors (nearly three times as much), and give a low efficiency at very low refrigerator temperatures.

PROPERTIES OF SULPHURIC DIOXIDE.

The specific heat of liquid sulphurous acid is 0.41; the critical pressure 79 atmospheres, and the critical temperature 312° F. The specific gravity of the gaseous acid is 2.211 (air = 1), and the specific gravity of the liquid at – 4° F = 1.491.

The relation of the specific gravity, *s*, of the liquid to the temperature, *t*, is expressed by the following formula given by Andreef:

$$s = 1.4333 - 0.00277\,t - 0.000000\,271\,t^2$$

The specific heat of liquid sulphurous acid is 0.41 (water — 1).

LEDOUX'S TABLE FOR SATURATED SULPHUR DIOXIDE GAS

Temperature of Ebullition in deg. F. t	Absolute Pressure in lbs. per sq. in. $P \div 144$	Total Heat Reckoned from 32° F. λ	Heat of Liquid Reckoned from 32° F. q	Latent Heat of Evaporation. r	Heat Equivalent of External Work APu	Increase of Volume during Evaporation. u	Density of Vapor or Weight of One Cub. Ft.
Deg. Fah.	Lbs.	B.T.U.	B.T.U.	B.T.U.	B.T.U.	Cub. Ft.	Lbs.
—22	5.56	157.43	—19.56	176.99	13.59	13.17	.076
—13	7.23	158.64	—16.30	174.95	13.83	10.27	.097
— 4	9.27	159.84	—13.05	172.89	14.05	8.12	.123
5	11.76	161.03	— 9.79	170.82	14.26	6.50	.153
14	14.74	162.20	— 6.53	168.73	14.46	5.25	.190
23	18.31	163.36	— 3.27	166.63	14.66	4.29	.232
32	22.53	164.51	0.00	164.51	14.84	3.54	.282
41	27.48	165.65	3.27	162.38	15.01	2.93	.340
50	33.25	166.78	6.55	160.23	15.17	2.45	.407
59	39.93	167.90	9.83	158.07	15.32	2.07	.483
68	47.61	168.99	13.11	155.89	15.46	1.75	.570
77	56.39	170.09	16.39	153.70	15.59	1.49	.669
86	66.36	171.17	19.69	151.49	15.71	1.27	.780
95	77.64	172.24	22.98	149.26	15.82	1.09	.906
104	90.31	173.30	26.28	147.02	15.91	.91	1.046

USEFUL EFFICIENCY.

Exceptional care has to be taken to maintain tight joints in a sulphur dioxide machine, as any leakage might produce sulphuric acid, which would become destructive to the metal of the plant.

	No. of Test.	Temp. in degrees Fahr. corresponding to pressure of vapor.		Ice melting capacity per pound of coal, assuming three pounds per hour per horse-power.		
		Condenser.	Suction.	Theoretical friction* included.	Actual.	Per cent loss due to cylinder superheating.
Sulphuric Dioxide. Schroeter	11	77.3	28.5	41.3	33.1	19.9
	12	76.2	14.4	31.2	24.1	22.8
	13	75.2	—2.5	23.0	17.5	23.9
	14	80.6	—15.9	16.6	10.1	39.2
Ammonia. Schroeter	1	72.8	26.6	50.4	40.6	19.4
	2	70.5	14.3	37.6	30.0	20.2
	3	69.2	0.5	29.4	22.0	25.2
	4	68.5	—11.8	22.8	16.1	29.4
Ammonia. Denton	24	84.2	15.0	27.4	24.2	11.7
	26	82.7	—3.2	21.6	17.5	19.0
	25	84.6	—10.8	18.8	14.5	22.9

*Friction taken at figures observed in the tests which range from 14 to 20 per cent of the work of the steam cylinder.

For a comparison of the sulphur dioxide and the ammonia compression plants the foregoing table, abstracted from Schroeter and Denton's tests, may be consulted.

ETHER MACHINES.

Compression machines, with sulphuric ether as the working fluid, were in great favor in former days, but have been abandoned to a great extent, owing, probably, to the enormous size of compressor required, it being required to be about seventeen times as large as an ammonia compressor of the same capacity. The great inflammability of the ether is another objection. The formula and rules given for the ammonia compressor apply also for ether, with the exception that the specific heat of the saturated vapor of ether (unlike that of ammonia, steam, carbonic acid and sulphur dioxide), is positive, and therefore superheats during expansion and condenses during compression. An ether machine, therefore, needs no protection against superheating, and is always operated with dry vapor. Specific heat of liquid, 0.51.

TABLE SHOWING PROPERTIES OF SATURATED VAPOR OF ETHER.

Temperature Degrees F.	Pressure in Pounds Per Square Inch.	Heat of the Liquid.	Total Heat.	Heat of Vaporization.	Heat Equivalent of Internal Work.	Heat Equivalent of External Work.	Specific Volume.	Weight in Pounds of One Cubic Foot.
		B. T. Units.	B. T. Units.	B. T. Units.	B. T. Units.	B. T. Units.		
32	3.54	0.00	376 00	376.00	345.80	30.20	1.278	.048
50	5.51	21.28	393.76	372.48	341.48	31.00	0.844	.073
68	8.31	42.80	411.12	368.32	336.52	31.80	0.574	.107
86	12.20	64.56	428.00	363.44	330.88	32.56	0.401	.154
104	17.46	86.42	444.44	357.92	324.60	33.32	0.287	.232
122	24.32	88.76	460.44	351.68	317 64	34.04	0.210	.294
140	33 17	131.20	476.00	344.80	310.12	34.68	0.158	.392
158	44.32	153.92	491.12	337.20	301.96	35.24	0.120	.515
176	58.13	176 84	505.76	328.92	293.28	35.64	0.093	.705
194	74.96	200.00	520.00	320.00	284 12	35.68	0.073	.848
212	95.25	223 44	532.76	310,32	274.48	35.84	0.057	1.074
230	119.51	247,08	547.12	300.04	264.52	35.32	0.005	1.350
248	148.44	270.96	560 00	289.04	254.28	34.76	0.036	1.703

EFFICIENCY OF ETHER MACHINES.

The following data relating to the working of an ether machine are not the result of a careful test, but represent practical working, it is claimed:

For a production of fifteen tons of ice in twenty-four hours 245,000 B. T. units were abstracted per hour, and the indicated horse power of the engine was eighty-three, of which forty-six indicated horse power was used for the ether compressor and the balance for friction in compressor, pumping water, working cranes, etc. The temperature of the cooling water entering the condenser was 52° F. in this case.

REFRIGERATION BY PICTET'S LIQUID.

This liquid, which is also used in compression machines, is a mixture of carbonic acid and sulphurous acid, which, according to Pictet, who introduced the same, corresponds to the formula CO_4 S. According to Pictet, the pressure of this mixture or compound at higher temperature is less than the law of pressure relating to ordinary mixtures would indicate. The following table shows the relations of pressures and temperatures of this substance:

Temperature, Degrees F.	Pressure (Absolute) in Atmospheres.	Temperature, Degrees F.	Pressure (Absolute) in Atmospheres.
—22	0.77	50	2.55
—13	0.89	59	2.98
— 4	0.98	68	3.40
— 2.2	1.00	77	3.92
5	1.18	86	4.45
14	1.34	95	5.05
23	1.60	104	5.72
32	1.83	113	6.30
41	2.20	122	6.86

If the Pictet liquid were an ordinary mixture its pressure would gradually rise from 0.77 to 13.98 atmospheres from the temperature—22 to +112 degrees Fahrenheit. Instead of that the pressure increases from 0.77 to 6.86 atmospheres only, and at 77° F. is less than that of the sulphurous "acid" or sulphur dioxide alone.

ANOMALOUS BEHAVIOR OF PICTET'S LIQUID.

It is claimed that a compression plant, if operated with Pictet's liquid, will produce a greater effect than what is compatible with the familiar thermodynamic formula given on page 71 of this compend. This anomalous behavior is sought to be explained by the physical or chemical work done by the liquids while combining into one substance in the condenser, which work it is argued replaces part of the work which would have to be done if a simple working fluid were used. If this explanation were correct we would have to assume that while a certain amount of work (*i. e.* heat) is given off in the condenser, an equivalent amount of heat must be absorbed in the refrigerator, thus increasing the efficiency of the machine in two directions, a most happy coincidence, but one which is in no wise corroborated by the second law of thermodynamics.

OTHER EXPLANATIONS FOR THE ANOMALY.

In accordance with thermo-chemical tenets, the combination of carbonic and sulphuric dioxide should absorb heat while being formed in the condenser, and should generate heat while being decomposed in the refrigerator. Such a behavior would bring the working of a machine with Pictet's liquid within the scope of the second law, but it would hardly account for the alleged anomalous efficiency of such a machine.

Generally it is supposed that the influence of heat on chemical combinations is such that they become less permanent with increase of temperature, and that at a very high temperature they are dissolved in their elements. This is quite correct for such combinations which are formed by the development of heat, and which absorb heat while being decomposed. But the contrary takes place in the case of combinations which are formed under absorption of heat. These latter combinations become more permanent with the increase of temperature.

BLUEMCKE ON PICTET'S LIQUID.

According to experiments made by Bluemcke the pressure of Pictet's liquid is always higher than that of sulphurous acid at all temperatures. Furthermore he claims that the commercial "Pictet's liquid" is not compounded after the formula $CO_4\,S$, but that it contains only 3 per cent of CO_2 by volume. The mixture $CO_{16}\,S_7$, for which Pictet has established—76° as the boiling point has a tension of four atmospheres at a temperature of—17° C. Such conflicting statements as these are hardly calculated to remove the doubts connected with the use of Pictet's liquid, and more authentic experiments by disinterested parties and with liquids of well known composition will be required to definitely settle this matter.

MOTAY AND ROSSI'S SYSTEM.

Previous to Pictet's invention Motay and Rossi had operated a refrigerating machine on a similar plan with a compound of two liquids, one of which liquefies at a comparatively low pressure and then takes the other in solution by absorption. Their mixture consisted of ordinary ether and sulphur dioxide and has been termed ethylo-sulphurous dioxide. It is stated that the liquid ether absorbs 300 times its volume of sulphur dioxide at ordinary temperature and at 60° F. the tension of the vapor of the mixture is below that of the atmosphere. The compressing pump has less capacity than would be required for ether alone, but more than for pure sulphur dioxide.

Before exact formulæ can be given for the dimensions and efficiency of machines working with compound liquids their chemical and physical, and especially their thermo-chemical behavior, must be more definitely settled by experiments.

CRYOGENE—REFRIGERATING AGENTS.

Cryogene is another name for refrigerating medium, and literally translated means ice generator. Certain hydrocarbons, naphtha, gasoline, rhigoline or chimogene have also been recommended and used to some extent as refrigerating media. These liquids are used in much the same way as ether, in common with which they have a great inflammability; but they are much cheaper to start with. Van der Weyde's refrigerating machine consists of an air pump and a force pump, a condenser and two refrigerator coils, one of which also serves as a reservoir for the condensed liquid. The water to be frozen is placed in molds which are surrounded by a glycerine bath. The glycerine bath in turn is surrounded on the outside by the refrigerating medium,' naphtha, gasoline, chimogene, etc., which is evaporated by means of the air pump, thereby abstracting sufficient heat to cause the formation of ice.

ACETYLENE.

Acetylene, which has lately been so prominently mentioned as the illuminating agent of the future, has also been talked of as a refrigerating agent. It is a combination of hydrogen and oxygen after the formula $C_2 H_2$. It is highly inflammable and said to require a pressure of 48 atmospheres to be liquefied at freezing point of water.

CHAPTER XII.—AIR AND VACUUM MACHINES.

COMPRESSED AIR MACHINE.

Air is used in various ways as a working fluid in refrigerating plants, but on the whole to a limited extent only.

The compressed air machine is based on the utilization of the reduction of temperature which takes place when compressed air expands while doing work in an air engine. The air is compressed by a compressor and the heat which is generated by compression is withdrawn by cooling water. The cold air leaving the expansion engine is used for cooling purposes.

CYCLE OF OPERATIONS.

This may be done in such a way that the air having served for refrigerating purposes is periodically returned to the compressor in the same condition. In this case the operations of the refrigerating system constitute what is termed a perfect cycle, and the thermodynamic laws applicable to such a cycle obtain also in the case of the compressed air machine.

Practically it is far more convenient to reject the working fluid (air) along with the refrigeration, but for the purposes of the following calculations, which are rendered after Ledoux, we will assume that the operations of a cycle are fully performed.

WORK OF COMPRESSION.

For the work, W_r, of compression of the air, which is supposed to be done adiabatically (without losing or gaining heat), Ledoux gives the following formula:

$$W_r = \frac{k}{k-1}(P_1 V_1 - P_0 V_0) \text{ foot-pounds;}$$

and also—

$$W_r = \frac{m\,k\,c}{A}(T_1 - T_0) \text{ foot-pounds.}$$

In these equations P_0 and T_0 are the initial pressure and temperature of the air, counted from absolute zero.

V_0 is the volume described by the piston of the compressor cylinder.

V_1 is the volume described by the same piston during the outflow of the compressed air.

P_1 and T_1 are the temperature and pressure of the compressed air when leaving the compressor.

A is the reciprocal of the mechanical equivalent of heat $= \frac{1}{772}$.

k is the ratio of specific heat of constant pressure to the specific heat of constant volume.

$$k = \frac{0.23751}{0.16844} = 1.41$$

In the following equations:

m stands for the weight of air (in pounds) whose volume passes from V_0 to V_1.

c stands for the specific heat of air of constant volume.

P_2 and T_2 are the pressure and temperature of the air after expansion.

V_2 is the volume of the expansion cylinder.

TEMPERATURE AFTER COMPRESSION.

The temperature, T_1, of the air after adiabatic compression may be found after the following formulæ:

$$T_1 = T_0 \left(\frac{P_1}{P_0}\right)^{\frac{k-1}{k}}$$

and

$$T_1 = T_0 \left(\frac{V_0}{V_1}\right)^{k-1}$$

COOLING OF THE AIR.

The air after having been compressed is cooled down from the temperature T_1 to the temperature T_3, and volume V_3, and the quantity of heat, Q_1, which must be withdrawn from the air to accomplish this is—

$$Q_1 = m\,k\,c\,(T_1 - T_3) \text{ units.}$$

AMOUNT OF WATER REQUIRED.

The amount of cooling water, P_1, required is—

$$P_1 = \frac{Q_1}{8.3\,(t - t_1)} \text{ gallons.}$$

t and t_1 being the respective temperatures of incoming and outgoing condenser water.

WORK DONE BY EXPANSION.

The work, W_m, which may be obtained theoretically by allowing the air, after being cooled, to expand against a piston adiabatically until the temperature T_2 is reached is:

$$W_m = \frac{k}{k-1}(P_1 V_3 - P_2 V_2) \text{ foot-pounds.}$$

or

$$W_m = \frac{m k c}{A}(T_3 - T_2) \text{ foot-pounds.}$$

TEMPERATURE AFTER EXPANSION.

The temperature, T_2, of the air after expansion is found after the formula:

$$T_2 = T_3 \left(\frac{P_2}{P_1}\right)^{\frac{k-1}{k}}$$

T_3 and P_1 being the temperature and pressure of the air when entering the expansion cylinder.

REFRIGERATION PRODUCED.

The refrigeration, H, which is produced by the air during adiabatic expansion is expressed by—

$$H = m k c (T_0 - T_2) \text{ units,}$$

T_0 being the temperature of the air after it leaves the refrigerator.

WORK FOR LIFTING HEAT.

The net work, *W*, therefore which is theoretically required to lift the amount of refrigeration, *H*, is expressed by the formula—

$$W = W_r - W_m \text{ foot-pounds, or also-}$$

$$W = \frac{m k c}{A}\left[(T_1 - T_0) - (T_3 - T_2)\right] \text{ foot-pounds.}$$

EQUATION OF CYCLE.

If the quantities, Q_1, H and W_r and W_m are expressed in the same (thermal) units, the equation of the cycle of operations may be expressed by—

$$Q_1 = W_r - W_m + H$$

if W_r and W_m are expressed in foot-pounds.

$$Q_1 = A(W_r - W_m) + H.$$

EFFICIENCY OF CYCLE.

The theoretical efficiency, E, of this refrigerating cycle may be expressed by the formula:

$$E = \frac{H}{W} = A \frac{T_0 - T_2}{(T_1 - T_3) - (T_0 - T_2)}$$

and $\frac{T_2}{T_3}$ being equal $\frac{T_0}{T_1}$, we also find—

$$E = A \frac{T_0}{T_1 - T_0} = A \frac{T_2}{T_3 - T_2}$$

This expression is the same as that found for the maximum theoretical efficiency of a reversible refrigerating machine, page 71.

The above formulæ apply also in case any other permanent gas is employed in place of air.

SIZE OF CYLINDERS.

From the above equations the relative sizes V and V_2 of compression and expansion cylinders, for a given amount of refrigeration in a given time, can be readily ascertained for theoretical conditions.

The ratio which should exist between the volumes of the two cylinders in order that the air is expelled at atmospheric pressure is expressed by the following equations:

$$\frac{V_2}{V_3} = \frac{V_0}{V_1}$$

$$\frac{V_2}{V_0} = \frac{T_3}{T_1}$$

V_3 standing for the volume of air after compression and after subsequent cooling, when it has the temperature T_3.

ACTUAL EFFICIENCY.

Owing to the bulkiness of air, the compression and expansion cylinders have to be very large, a fact which tends to increase the friction considerably. Besides this there is considerable clearance, and the moisture contained in the air also decreases the efficiency, all of which circumstances, combined with others of minor importance, reduce the actual performance of the air machine much below the theoretical efficiency.

RESULTS OF EXPERIMENTS.

The foregoing remarks are forcibly illustrated by the following tests of compression machines, which were published by Linde some time ago. The figures in this table show that in the most favorable experiment (Lightfoot) the actual efficiency is scarcely 33 per cent of the theoretical efficiency. (After Ledoux the friction alone reduces the theoretical refrigerating for about 25 per cent.)

ACTUAL PERFORMANCE OF COLD AIR MACHINES.

SYSTEM	Bell-Colem'n.	Lightfoot.	Haslam.
TEST No.	1	2	3
Diameter of compression cylinder	28″	27″ s'gle act'g	25¼″ 2-cylinder
Diameter of expansion cylinder	21″	22″	19½″ 2-cylinder
Diameter of steam cylinder	21″	....	20″ H. P. 31″ L. P.
Stroke of all cylinders	24″	18″	36″
Revolutions per minute	63.2	62	72
Air pressure in receiver, pounds (absolute)	61	65	64
Temperature of air entering the compression cylinder	65½° F.	52° F.	....
Temperature of air after expansion	−52.6° F.	−82° F.	−85° F.
I. H. P. in compression cylinder	124.5	43.1	346.4
I. H. P. in expansion cylinder	58.5	28.0	176.2
I. H. P. in steam cylinder	84.4	24.6	332.7
B. T. U. abstracted per hour and I. H. P. of steam cylinder at 20° F.	668	1,554	954

The figures for test No. 1 have been observed and published by Professor Schroeter (*Untersuchungen an Kæltemaschinen verschiedener Systeme*, Munich (1887); those for No. 2 are published in minutes *Proc. Inst. Mech. Eng.*, London, 1881. The data for trial No. 3 are taken from a paper read last year before the Manchester Society of Engineers.

WORK REQUIRED FOR ISOTHERMAL COMPRESSION.

If the compression of air takes place isothermically, in which case the air is kept at constant temperature during compression by injection of cold water and a cold water jacket, the work of compression is lessened. The work W_2 in foot-pounds required in theory to compress isothermically V cubic feet of air under a pressure of

P pounds (per square foot) to the volume of V_1 cubic feet is—

$$W = P\,V \times 2.3026 \log \frac{V}{V_1} \text{ foot-pounds.}$$

WORK DONE IN ISOTHERMAL EXPANSION.

The work, W_1, in foot-pounds which can be done theoretically by the isothermal expansion of V_1 cubic feet of air to the volume of V cubic feet, and the pressure P is—

$$W_1 = P\,V \times 2.3026 \log \frac{V}{V_1}$$

OTHER USES OF COMPRESSED AIR.

The isothermal expansion of air is employed in cases where compressed air is used, not for refrigeration, but for the production of power, as in tunneling, drilling in mines, transmission of power by compressed air, etc. These are purposes for which the compressed air has been extensively used.

TABLE SHOWING LOSS OF PRESSURE BY FRICTION OF COMPRESSED AIR IN PIPES.

(F. A. Halsey.)

Diameter of Pipe.	Cubic Feet of Free Air compressed to a Gauge Pressure of 60 lbs. per Square Inch, and passing through the Pipe per Minute.									
	50	75	100	125	150	200	250	300	400	600
	Loss of Pressure in Pounds per Square Inch for each 1,000 Feet of Straight Pipe.									
Ins.	Lbs.	Lbs.	Lbs.	Lbs.	Lbs.	Lbs.	Lbs.	Lbs.	Lbs.	Lbs.
1	10.40	......	......	......	......	......	......	......	......	......
1¼	2.03	5.90	......	......	......	......	......	......	......	......
1½	1.22	2.75	4.89	7.65	11.00	......	......	......	......	......
2	.85	.79	1 41	2.20	3 17	5.64	8.78	......	......	......
2½	.14	.32	.57	.90	1.29	2.30	3.58	5.18	9.20	......
3	......	.11	.20	.31	.44	.78	1.23	1.77	3.14	7.05
3½	......	......	......	.15	.21	.38	.59	.85	1.51	3.40
4	......	......	......	......	......	.20	.31	.45	.80	1.81
5	......	......	......	......	......	......	.10	.15	.26	.59
6	......	......	......	......	......	......	......	......	......	.23

CALCULATED EFFICIENCY.

The best working pressure for a compression air machine appears to be at 4½ atmospheres, and the calculations for this pressure give, according to Denton, a theoretical efficiency of 17.5 pounds ice melting capacity per pound of coal (assuming three pounds of coal per horse power). Allowing for friction, one pound of coal should give a refrigerating effect equivalent to eleven pounds ice melting capacity with a consumption of nine gallons of water. ($T_0 = 59°$ F. and $T_3 = 64.4°$ F. weight of one cubic foot of air with 0.0357 pounds of moisture = 0.07524 pounds.)

LIMITED USEFULNESS.

In consequence of the low practical efficiency the air compression system is impracticable for indirect refrigeration, and can best be used where cold, dry air is the ultimate object, and even in this case its economical adaptability seems to depend on circumstances.

One of the chief difficulties in cold air machines, says Gale, is the presence of moisture held in suspension by the atmosphere. Moisture in the air occasions loss of efficiency in two ways. If the air enters the expansion cylinder in a saturated condition, when the air is cooled by expansion while performing work, a certain amount of vapor is condensed and thrown down—the point of saturation being dependent on the temperature. The vapor in changing to a liquid state gives its latent heat of vaporization to the air; and as the expansion of the air continues and the temperature is still further diminished, the liquid freezes and accumulates in the form of snow or ice in the valves and passages, giving up its heat of liquefaction to the air. Thus does not only the presence of moisture in the air produce mechanical difficulties, choking the air passages and impeding the action of the valves, but, for the same expenditure of energy, the cold air leaves the machine at a higher temperature than would have been the case if there had not been a superabundance of moisture in the air during expansion.

VACUUM MACHINES.

With the name of vacuum machines is designated a class of refrigerating apparatus in which water is used as a refrigerating agent. In their most simple form they work on the same principle as a compression machine. The vaporization of the water at a temperature low enough to cause the freezing of the water must take place under vacuum. The vacuum is formed by a vacuum pump which acts exactly like a compressor, withdrawing the vapors or vapor from the refrigerator where the pressure is about 0.1 pound per square inch, and compressing the same into a condenser against a pressure of about 1.5 pounds per square inch.

REFRIGERATION PRODUCED.

The refrigeration produced by the evaporation of a part of the water in the refrigerator causes a corresponding portion of the water to turn into ice. As the latent heat of ice is about 142 and that of the watery vapor about 940, theoretically the evaporation of one pound of water would be able to produce from six to seven pounds of ice.

EFFICIENCY AND SIZE.

The efficiency, dimensions, etc., of a vacuum machine if worked on the plan of reversible cycle may be calculated by the same rules given for the ammonia compressor. As the latent heat of watery vapor is very great in comparison to the specific heat of the liquid (see page 87) the theoretical efficiency of a vacuum machine will be found considerably greater than that of the other compression machines.

This seeming advantage, however, is more than counterbalanced by the enormous size of the compressor required on account of the low tension of the water vapor at the temperature of the refrigerator. It is found that the compressor or vacuum pump of a vacuum machine of a certain capacity will have to be about 200 times as large as that of an ammonia compression machine of the same capacity.

If the temperatures to be produced by a vacuum machine are to be lower than that of freezing water, a solution of salt has to be placed in the refrigerator instead of pure water, to prevent the freezing of the refrigerating agent.

COMPOUND VACUUM MACHINE.

In order to avoid compressors of such an enormous size the foregoing form of a vacuum machine has been complicated by the addition of an absorbent, preferably concentrated sulphuric acid, which, by means of its absorbent power for watery vapor, releases the work of the compressor or air pumps. A machine of this construction works on nearly the same principle as an absorption machine, and its efficiency, etc., may be discussed on the same basis.

In the machines constructed on the latter principle, which vary considerably in detail, the fuel used to reconcentrate the sulphuric acid (which has become diluted from 60° to 52° Beaume) represents one of the principal expenses. The vacuum pump is small, but in continuous operations there must also be a pump for the exchange of the diluted and concentrated acid.

This exchange is performed in such a way that the cold, weak acid leaving the absorber withdraws the heat from the strong acid coming from the evaporator.

EXPENSE OF OPERATING.

The larger part of the heat withdrawn from the water or salt brine in the refrigerator appears again in the absorber as heat of combination between the sulphuric acid and the vapor. It is removed by cooling water.

It is stated that for the production of 100 pounds of ice it will take about eight pounds of coal in the evaporator and about twelve gallons of cooling water. Besides this we must allow for the power required to operate the vacuum and acid pumps.

OBJECTIONS TO SULPHURIC ACID.

The vessels and pipes containing or carrying the sulphuric acid must be of lead or lead lined, and on the whole the handling of this liquid is considerable of an inconvenience. For this and other reasons the use of the vacuum machine will probably be confined to special cases. The making of ice in connection with some other industry requiring the production of diluted sulphuric acid on a large scale, and at a great distance from the sulphuric acid factory, would be such a case.

SOUTHBY'S VACUUM MACHINE.

The apparent simplicity and directness of action of a vacuum machine for the direct production of ice has produced several inventions in this direction. In a machine designed by *Southby & Blyth*, the freezing can, or cans, containing the water to be frozen is placed in a box, which can be closed air tight, and from this box the air, and eventually the watery vapor, is exhausted by means of two pumps of peculiar construction. One is an air pump which is designed to draw all the air from the interior of the machine, and the vacuum so formed fills itself with watery vapor from the water in the freezing can. A second larger pump then compresses the vapor and forces the same into the condenser. But in order to do this effectually the condensation of vapor in the compressor has to be prevented, as otherwise the tension of the compressed vapor to be ejected would be so small in quantity that it would not be forced through the exit valve. To accomplish this the cylinder of the large pump is heated to a temperature above that at which the vapor will condense, and in this way the compressed vapor is almost entirely forced into the condenser. The water forming in the condenser, together with the air drawn over from the water, etc., is ejected by the small air pump.

The small air pump, in connection with the large compressor, and the heating of the latter, are the two principal new features which are claimed to insure the success of this machine. Owing to the low pressures (from 0.15 to 2 inches, average pressure on piston 1-6 pound per square inch) the fricture of the compressor can be made very small.

OPERATING SOUTHBY'S MACHINE.

When starting the machine air at a comparatively high pressure has to be dealt with, occasioning an adverse pressure on the piston of say seven pounds, or over thirty times that of the working pressure; and the air being non-condensible will not disappear on compression, as is the case with watery vapor. For this reason, provision has been made that both ends of the vapor pump cylinder can be kept open for any necessary length of time during the first portion of the delivery stroke, so as to permit the air to return to the under side of the piston and thereby lessen and regulate the expenditure of power to be expended in obtaining a vacuum. This is accomplished by means of a by-pass and valve, which can be opened at starting, and kept open for about nine-tenths of the piston stroke, being closed gradually as soon as the vacuum becomes more perfect, and altogether as soon as all the air has been got rid of. According to British writers the manufacturers of this machine intend the same to be used in confined places, on board ship, or where the escape of injurious gases would be dangerous, also for making ice by hand power. The quantity of cooling water for the condenser is said to be very small indeed.

CHAPTER XIII.—LIQUEFACTION OF GASES.

HISTORICAL POINTS.

The liquefaction of the formerly so called permanent gases has always attracted considerable attention on the part of physicists and chemists as a means of studying matter in different states of aggregation, and also as means of producing extremely low temperatures. The names of Faraday, Thilorier, Natterer, De La Tour, and, more recently Pictet, Cailletet, Wroblewski, Olszewski, Dewar and others have become famous in connection with this subject.

The former methods used for liquefaction of gases on a larger scale than the bent tubes used by Faraday, etc., and which until recently were also employed by Dewar in the production of larger quantities of liquid oxygen, etc., were practically identical with those originated by Pictet and Cailletet, who in addition to pressure used a succession of various cooling agents (liquefied gases), one cooling the next, and so on, until at last a temperature was reached low enough to liquefy the gas in hand.

Although large quantities of liquid gases could be prepared in this manner, and could be experimented upon, still their production was extremely expensive, and therefore the whole subject was confined to scientific studies and experiments.

These costly methods, however, have been replaced within recent years by more practical operations, which render the liquefaction of the most permanent gases an easy and comparatively inexpensive task, and have made the subject one of general and perhaps practical interest.

SELF-INTENSIFYING REERIGERATION.

This surprising result was consummated by Prof. Linde, the originator of the ammonia compression system of refrigeration, who inaugurated and perfected a self-intensifying refrigerating method, by which the liquefaction of gases, notably air, oxygen, nitrogen, can be carried out on a large scale and at moderate cost. The first large apparatus working on Linde's new plan was exhibited before a body of physicists, chemists and engineers in Munich in the month of May, 1895, and then and there large quantities of liquid air were produced at the rate of several quarts per hour.

The principles upon which Linde's apparatus works are very ingeniously conceived, and the ingenuity displayed in this direction are only equaled by the simplicity in the construction of the apparatus itself.

Linde dispenses entirely with the use of auxiliary refrigerants, but makes the gases themselves supply the refrigeration required for their liquefaction, by means exclusively mechanical; *i. e.*, by the use of an ordinary compressor, exchanger, water cooler, expansion valve and liquid receiver.

LINDE'S SIMPLE METHOD.

The gas to be liquefied, atmospheric air for example, is taken in by a compressor, and after compression is forced through an ordinary water cooler to dispose of the heat of compression; thence it is forced through a coil several hundred feet long, the end of which is provided with an expansion valve, which dips into a liquid receiver or collection vessel; from this vessel issues another pipe, which forms a coil surrounding (forming an annular concentric space) the coil previously mentioned, and which returns the air (after having expanded into the liquid receiver) to the compressor. The compressed air while expanding into the liquid receiver, against pressure, as it were, does a certain amount of (interior) work, and generates a corresponding amount of refrigeration; *i. e.*, it lowers its own temperature correspondingly. In this condition the air flows back to the compressor, and on the way, while passing around the coil through which the compressed air passes, cools the latter before it enters the liquid receiver. The air when it again reaches and passes the compressor and water cooler leaves the same with a higher pressure, and again enters the liquid receiver at lower temperature than it did before, and in this manner pressure and refrigeration gradually increase in the liquid receiver by what may be termed an accumulative effect, produced by constant repetitions of the cycle of operations just described, until finally the critical temperature is reached, at which the air liquefies and collects at the bottom of the liquid receiver, whence it may be withdrawn by means of a faucet.

As fast as the air becomes more compressed and is finally withdrawn from the cycle in its liquid form, other air must be supplied to the compressor, and as the efficiency of the cycle is at its best at very high pressure, the original air is already supplied to the same in a compressed state by an auxiliary compressor. The system of concentric coils forming the exchanger and the liquid receiver must be inclosed in a chamber especially well insulated in order to render the apparatus operative.

THE RATIONALE OF LINDE'S DEVICE.

Several schemes of "regenerative," accumulative or self-intensifying systems of refrigeration and liquefaction have been proposed before, but none succeeded in producing liquid air before Linde, who also was the first who clearly understood and pointed out the physical principles underlying the operation, and who gave numerical data regarding the efficiency of the cycle of operation involved therein.

Accordingly, the performance of interior work by the very gas to be compressed is the source of the refrigeration, which causes its temperature to fall below the critical point, at which it is readily liquefied by pressure.

It was known long ago, and it had been experimentally elaborated some thirty years by Joule and Thompson that the law of Gay Lussac did not strictly apply to air and some other gases, and that a certain amount of interior work (to overcome the mutual attraction of their molecules) was done on expanding; still, this amount of interior work (and corresponding refrigeration) was deemed so insignificant that expansion, while doing actual mechanical work (moving a piston in an expansion cylinder), was considered indispensable in an air refrigerating machine. Linde, however, pointed out that this refrigeration, due to the free expansion of a gas from a higher to a lower pressure, although small at low pressure, would increase very rapidly with the pressure in an apparatus working on the accumulative principle.

The increase of the heat elimination with the pressure, and the economic principle of Linde's method, become readily apparent when we analyze the formula which expresses the relation between the lowering of temperature d and the pressure p before and the pressure p_1 after expansion. In this formula—

$$d=0.476\,(p-p_1)\left(\frac{493}{T}\right)^2$$

T is the temperature at which the compressed gas expands in degrees absolute Fahrenheit; the pressures are expressed in atmospheres.

The fall of temperature of a gas during free expansion from a higher to a lower pressure is frequently referred to as the "Joule effect," or as the "Joule-Thompson effect."

VARIABLE EFFICIENCY.

This formula readily shows that the refrigeration of the gases increases with the increase of the difference $p-p_1$, that is, the difference of pressure on both sides of the expansion valve; and also with the decrease of T, that is, with the expanding temperature. As the latter is constantly lowered in accordance with the accumulative principle on which the apparatus works, the efficiency of the system evidently increases the nearer the temperature of the gas reaches its critical point.

While the degree of refrigeration depends on the difference, $p-p_1$, the amount of work or power required to operate the apparatus or to force the air round and round the circuit depends on the quotient $\frac{p}{p_1}$ or the ratio of pressure in front and back of the compressor piston. By making $\frac{p}{p_1}$ small and $p-p_1$ great, which can be done by working at very high pressures, the efficiency of the system may be brought near a maximum figure.

To accomplish this, in a measure, the air or gas to be liquefied is already brought to a pressure of some fifty atmospheres by an auxiliary compression before it is furnished to the compressor, which operates the liquefying circuit proper.

HAMPSON'S DEVICE.

In keeping with the foregoing consideration, Hampson has constructed a similar apparatus, which may be operated with compressed air or gases contained in cylinders alone, and without a compressor and water cooler. In this case, only that portion of the gas or air which is actually liquefied remains in the system; the other portion is exhausted or wasted, so to speak. This apparatus is specially adapted for lecture purposes, and is only a modification of Linde's, well foreshadowed in the latter's original observations on the subject.

OTHER METHODS.

Regarding the history of Linde's method of liquefaction, it may be mentioned that Siemens, as early as 1857, applied for a patent in Germany on a self-intensifying or regenerative process of refrigeration, in accordance with which the air is first compressed with an ordinary compressor, and then expanded in a motor cylinder, whereby the temperature is reduced; the air is then passed through an exchanger, in which it is cooled by the compressed air which enters the exchanger from the opposite side. Siemens did not attempt to carry out his invention, it appears, but in 1885 Solvay patented a similar device and put the same in operation, but did not succeed in obtaining temperatures lower than —140° F., and did not succeed in liquefying air.

In 1893 Tripler obtained an English patent for a gas liquefying apparatus, and for several years has been producing liquid air and experimenting with the same. On this fact it appears that some people try to establish the priority of Tripler for the production of liquid air by the self-intensifying process over Linde.

TRIPLER'S INVENTION.

In this connection, however, it must not be overlooked that Tripler, no more than Solvay or Siemens, made no mention in his specification of the effect due to the air expanding against pressure through a narrow orifice or expansion valve, nor is there any evidence on record that Tripler made any liquid air until a considerable time after Linde and even Mr. Hampson had made the same in large quantities. The latter, in writing to the *Engineer* (London, England), makes the following and apparently not unjust reference to Tripler's discoveries:

"So far as is known to the public, Mr. Tripler can only be credited with three attainments of any magnitude. In 1893 he patented in this country an invention for liquefying gases by cold, which involved an obvious fallacy so gross and so important to the invention that, instead of producing cold, it would actually produce heat. That is attainment No. 1. In 1897, having imitated on a larger scale my invention for a self-intensive liquefier, which had been made and illustrated in detail nearly two years before, he showed it as an original invention; and having performed, with but slight variations except their larger scale, experiments with which the scientific world on this side of the Atlantic had long been familiar, he omitted all reference to that fact. Thirdly, in 1899, in connection with the working of a liquid air engine, he overlooked the vital point in the liquefaction of air that the latent heat given out in liquefaction must be removed by some other substance than the liquefied portion."

USES OF LIQUID AIR.

Much has been written about the utilization of liquid air in various ways, especially as a motive power. It is entirely superfluous here to assert the impracticability of the use of liquid air as a vehicle for motive power under ordinary circumstances. A medium in which the motive power has to be stored up at such a low temperature, entailing the loss of considerable mechanical energy, could not be considered economical for the transfer of power, for this reason alone.

As a means for the storage of power, liquid air has also been prominently mentioned by the lay press, but the very fact that it is impracticable to store or maintain it for any length of time under ordinary conditions with any degree of safety or without losing the larger portion of the liquid precludes this idea altogether.

Another reason, moreover, for the unavailability of liquid air as motive power is to be sought in the fact that not only mechanical power, but also considerable refrigerative capacity, is stored up in this medium, for which no adequate return would be obtained if it were used as a motive power for ordinary purposes.

The circumstance may not exclude the possibility of the use of liquid air for motive power in cases where expense is of little consideration, and in which certain conveniences are aimed at, as for instance for the throwing of projectiles, for the preparation of high explosives, for the propelling of torpedoes, for aerial navigation and in other cases of emergency.

With regard to the use of liquid air as a refrigerating medium, similar considerations do obtain. The expense of its production is too high to render it available for ordinary refrigeration; but where very low temperature is required for specific purposes, as for the preparation and purification of certain chemicals, for medical uses, for physical experiments, etc., liquid air and doubtless other liquefied gases have certainly many advantages, and therefore this subject cannot be ignored by the progressive engineer.

SPECIFIC USES OF LIQUID AIR.

From among the specific uses of liquid air, which already have taken a more practical form, we may mention the production of liquid oxygen for which Linde also constructed a special apparatus which is based on the observation that when liquid air is allowed to evaporate under certain precautions, the nitrogen evaporates first, leaving a liquid containing 50 per cent and more of oxygen.

The apparatus used by Linde for this purpose is quite similar to his liquefaction apparatus, the principal novel feature of it being an arrangement whereby the nitrogen as well as the oxygen is enabled to leave the machine at ordinary temperature. Thus the whole refrigeration bestowed on the gases during liquefaction is returned to or retained in the system.

This liquid, consisting chiefly of oxygen, has already been put to practical uses in the production of very high temperatures. Inasmuch as in combustions with ordinary air the nitrogen, which has to be heated also, carries away much of the heat of combustion, the "Linde air" will work a great change in this direction.

Not only in ordinary combustion, but also in other chemical oxidizing processes in which the presence of nitrogen lessens the affinity, the Linde product will be of great service, and is already utilized in the manufacture of chloride after the "Deacon" process.

For illuminating purposes the "Linde liquid" (liquid air containing over 50 per cent oxygen) will doubtless also be made available, and it is possible that the electric furnace may soon have a rival in a furnace operated with "Linde air," for it has been reported already that calcium carbide has been prepared by such a furnace without the use of electricity.

Another interesting use of liquid air is the rapid production of high vacuum. For this purpose the vessel to be exhausted is filled with a gas more easily condensable than air, say with carbonic acid gas. The vessel is provided with an extension which can be sealed off very readily. The open end of the extension is then immersed into liquid air, when the carbonic acid is withdrawn from the vessel and deposited in the extension, which is then sealed off, leaving a high vacuum in the vessel.

TABULATED PROPERTIES.

The accompanying table shows the physical constants of a number of gases, which have also been studied in the liquid states, as compiled by Peckham.

TABLE OF PHYSICAL CONSTANTS OF GASES.

	Critical Temperature Centigrade.	Critical Pressure. Atmospheres.	Boiling Point at Ordinary Pressure. Centigrade.	Freezing Point. Centigrade.	Freezing Pressure Mm.	Density of Gas.	Density of Liquid at Boiling Point.	Color of Liquid.
Carbon dioxide CO_2	31°[1]	77.0	—78.2°[3]	—79°[2]	760[2]	22	0 83@0°[4]	Colorless
Ethylene, C_2H_4	95.0	44.0 58.0	—110[5]	....	....	14	....	....
Hydrogen, H_2	—234.5[6] (Theor.)	20.0	—243.5[6] (Theor.)	....	....	1	....	Colorless
Nitrogen, N_2	—146	35.0	—194.4	—203 to —214 Mean—208	60	14	0 885	Colorless
Carbonic oxide, CO	—139.5	35.5	—190.0	—207.0	100	14	....	Colorless
Argon, A	—121.0	50.6	—187.0	—189.6	....	19.9	About 1.5	Colorless
Air	—140.0	39.0	—191.0	—207[9]	....	....	0.933	Bluish
Oxygen, O_2	—118.8	50.8	—182.7	....	....	16	1.124	Bluish
Nitric oxide, NO	—93.5	71.2	—153.6	—167.0	138	15	....	Colorless
Marsh gas, CH_4	—81.8	54.9	—164.0	—185.8	80	8	0.415	Colorless
Helium, He	....	....	Below —264[7] (Theor.)	....	....	2.02[8]	....	....
Fluorine	....	....	—187	....	....	....	....	....

1. Andrews. Deschanel Nat. Phil., II, 352.
2. Villard & Jarry. Comptes Rendus, 1895, 120, 1413.
3. Regnault. Muspratt's Chemie, IV, 1626.
4. Thilorier. Muspratt's Chemie, IV, 1626.
5. Fownes. Elem. Chem., 12th ed., p. 534.
6. Olzewski. Phil. Mag., 1895 (5), 40; 202.
7. Olzewski. Ann. Phys. Chem., 1896 (2), 59, 184.
8. Clève. Compt. Rend., 1895, 120, 1212.
9. Dewar.

CHAPTER XIV.—MANAGEMENT OF COMPRESSION PLANT.

INSTALLATION OF PLANT.

The installation of a refrigerating plant comprises the proper mounting of all its parts, the proving of the pumps, piping, etc., and the charging of the plant with ammonia. A working test is also frequently made. For the mounting the same rules apply as in the case of other motive machinery.

PROVING OF THE MACHINE.

In order to prove a new plant, before it is charged with ammonia it should be filled with compressed air to a pressure of about 300 pounds. This is done by working the compressor, while the suction valves provided for this purpose are opened. Thick soap lather, which is spread over the pipes, etc., shows leaks by the formation of bubbles under the above pressure. The condenser and brine tanks, filled with water, show leaks by the bubbles of air escaping through the water. The air pressure thus obtained on the system may be used to blow out the pipes, valves, etc. After a pressure is pumped on the system, and after the temperature is equalized throughout the whole system, the pressure gauge ought to remain stationary if the plant is absolutely air tight.

PUMPING A VACUUM.

If the machinery is found to be perfectly air tight, all the air is discharged from the system by opening the proper valves and working the pumps. After a vacuum has been obtained all outlets are closed, and the constancy of the vacuum is observed on the vacuum gauge to see if the plant will withstand external pressure.

CHARGING THE PLANT.

After the vacuum is shown to be perfect, the drum with ammonia is connected to the charging valve. Before opening the valve on ammonia flask, the expansion valve between ammonia receiver and expander is closed. Now the liquid ammonia is exhausted into the system, while the compressor is kept running at a very slow speed with suction and discharge valves opened and water running on the condenser.

CHARGING THE PLANT BY DEGREES.

If the air is not completely exhausted from the plant, *i. e.*, if the vacuum is not perfect, it is advisable to charge the plant with ammonia by degrees. First about one-half of the total amount of ammonia is charged, and after this has thoroughly circulated in the system, most of the remaining air will have collected in the top of condenser, whence it can be blown off by a cock. After this has been done the balance of the ammonia is charged in a similar way in one or two additional installments.

OPERATION OF PLANT.

The proper working of a compression machine is chiefly regulated by the amount of ammonia passing through the same, which is done by the expansion valve, which must be manipulated very carefully.

The pipe conveying the compressed ammonia to the condenser should not get warm, and the temperature of the brine should be about 5° to 10° F. higher than the temperature corresponding to the indication of pressure gauge on refrigerator.

The temperature of the cooling water should be about 10° to 15° F. (sometimes as much as 20°) below the temperature corresponding to the pressure in condenser coils.

The sound of the liquid ammonia passing the regulating valve should be continuous and sonorous, this indicating the absence of a mixture of gas and liquid.

DETECTION OF LEAKS.

If any ammoniacal smell is discovered while charging the plant, it is probably due to leaks, and they should be instantly located and mended. It is of importance to discover the existence of a leak at the first inception. When in a machine in operation, the liquid in the tanks begins to smell, it shows either a very considerable leak or one of long standing, and in order to detect a leak readily under those circumstances it is best to test those liquors regularly from time to time with Nessler's solution, of which a few drops are added to some of the suspected liquid in a test tube or other small glass vessel, as described on page 103.

MENDING LEAKS.

It is a very efficient and simple method to close small leaks by soldering them up with tin solder, which is frequently employed and gives general satisfaction. The soldering fluid, in order to properly clean the iron, should contain some chloride of ammonia, and it is best and proper that its quantity should be such as to form a considerable proportion of a double chloride of zinc and ammonia. A soldering liquid of this kind can be made by dissolving in a given amount of muriatic acid as much zinc as it will dissolve, and to do this in such a manner as to be able to ascertain the weight of zinc that has been thus dissolved. An amount of chloride of ammonia or sal ammoniac approximately equal in weight to that of the zinc dissolved is then added to the solution of zinc in muriatic acid.

If the leaks are too large to be mended in this way, new coils or new lengths of pipe must be put in. In some cases, where conditions are favorable, electric welding may be resorted to. A cement made by mixing litharge with glycerine to a stiff paste is also recommended for closing leaks. In this case the cement must be fortified by the application of sheet rubber and sheet iron sleeves kept in position by iron clasps.

Generally the amount of ammonia is determined after a rule of thumb fashion, allowing one-third pound of ammonia for every running foot of 2-inch pipe (or its equivalent) in expansion coils. Thus a plant of twenty-five tons ice making capacity having about 5,000 feet of 2-inch pipe would require about $\frac{5000}{3} = 1,666$ pounds of ammonia, while a direct expansion plant of twenty-five tons refrigerating capacity having at the rate of 2,000 feet of 2-inch pipe would require about $\frac{2000}{3} = 700$ pounds of ammonia. A machine of the same capacity (twenty-five tons refrigeration) with brine circulation would require only about 275 pounds of ammonia.

Calculated for capacity, this would correspond to about forty-five pounds of ammonia per ton of ice making capacity, twenty-five pounds of ammonia per ton of direct expansion refrigerating capacity and twelve pounds of ammonia per ton of refrigerating capacity, brine circulation. These rules are arbitrary, some allowing much less ammonia, according to the location of pipes.

WASTE OF AMMONIA.

Another question of considerable interest to the practical operators of ice plants is in regard to the waste of ammonia that may be expected to be incurred. Theoretically speaking, no waste ought to take place, as the same quantity of ammonia is used over and over again, but in practice the anhydrous ammonia gives way in the course of time. This is due to leakage in a great measure, and partly also to decomposition of ammonia. The amount of wastage depends, of course, largely on the care with which the plant is operated, and in the absence of any actual leakage is altogether due to decomposition of ammonia, which can be obviated in a great measure by keeping down the temperature around the compressor as much as possible. The amount of ammonia wasted while a machine is running depends almost entirely on the care and watchfulness, and may run all the way up to 200 pounds per year on a plant of twenty-five tons capacity. In some cases it amounts to very little, but about fifty to 100 pounds is generally considered as an unavoidable waste for a 25-ton machine. Where there is a liquid receiver provided with a gauge glass, the attendant can readily tell when the ammonia is running low in the machine. Otherwise the insufficiency of ammonia is shown by a fluctuating pressure, variation in the temperature of the discharge pipe, and by the running of the valves in the compressor, which sometimes run smooth and easy, and at other times hard, showing that the supply of ammonia and the consequent resistance varies.

A rattling noise of the liquid while passing the expansion valve shows the passage of vapor along with the liquid ammonia, and proves that the ammonia in the system is deficient.

AMMONIA IN CASE OF FIRE.

It appears that the dangers of ammonia in case of fire have been greatly over-rated, and at least in the beginning of a fire it acts as an extinguisher rather than otherwise. For this reason it seems more advisable in case of fire to allow the ammonia to escape whenever it is deemed good policy to stand the loss of the ammonia rather than run the risk of fire. If the latter happened the ammonia would be lost anyhow, and that, too, most likely, at a temperature high enough to make it share in the conflagration, while when allowed to escape, as long as the fire is low it may help to stifle the same or extinguish it altogether.

Before resorting to such an expedient the *pros* and *cons* should, of course, be duly considered, and the attendant should properly protect himself by a mask or similar contrivance against the suffocating effect of the ammonia vapors to which he may be exposed while providing means for their escape in the free atmosphere. In order to further provide for such an emergency, the outlet valve at the lower end of the condenser should be conveniently located, as the liquid ammonia should be permitted to escape first. While countenancing such heroic measures, I will not dispute that under certain conditions decomposing ammonia may, through ignition, also become the cause of fire. When, for instance, the head of a compressor running very hot should be blown off, the escaping hot ammonia, especially if saturated with lubricating oil, may be in a condition prone to decompose, and in case these vapors should come in contact with the flame of a light, the fire under the boiler, or a lighted match, a flash of fire might take place, which amid the confusion generally attending an accident of this kind might give rise to a destructive conflagration. In view of this possibility, it has been recommended that the lamps in the engine room of a refrigerating plant should be protected by a fine wire screen, that the doors leading to the boiler door should be likewise made of fine wire cloth and be provided with a reliable self-closing contrivance. The lighting of matches, etc., should be avoided in the engine room for the same reason.

CONDENSER AND BACK PRESSURE.

The lower the pressure and temperature in condenser coil, and the higher the pressure and temperature in expanding coil (back pressure), the more economical will be the working of the plant. This is readily apparent from the formulæ given for the estimation of the compressor capacity; it is even more readily apparent from the subjoined tables, showing the actual result obtained by Schroeter in working an anhydrous ammonia compressor under different conditions. For these reasons the cooling water on the condenser should be used as cold as it can be had and in as ample profusion as possible. Likewise the expansion or back pressure should be held as high as possible.

In brewery refrigeration, cold storage and other establishments in which the temperature is to be kept at

32° F., or thereabouts, by direct expansion, a back pressure of about 33 pounds gauge pressure, corresponding to about 20° F., is generally maintained.

In case brine circulation is used for above purposes, the brine returns with a temperature of 24 to 26° F. and enters the room with a temperature of about 20°. The back pressure in ammonia coils in this case is 25 to 28 pounds, corresponding to a temperature of 10 to 15° F.

During the chilling stage in meat or other cold storage, the temperature in the room rises in the beginning to 50°, and a higher back pressure—about 60 pounds, corresponding to a temperature of about 40° in ammonia coil—is maintained. Gradually, as the temperature falls in the room, the back pressure also decreases until it reaches the point corresponding to the temperature of the room for cold storage, viz., about 30 pounds.

In freezing meat, for which purpose temperatures of 0° F. and below in rooms are required, the back pressure gets as low as 4 pounds, corresponding to a temperature of —20° F.

For ice making a temperature of 10° to 20° is maintained in the brine, and the back pressure in ammonia coils in this case is from 20 to 28 pounds, corresponding to a temperature of 5° to 15° F.

TABLE SHOWING EFFICIENCY OF PLANT UNDER DIFFERENT CONDITIONS.

No. of test		1	2	3	4
Temperature of refrigerated brine.	Inlet deg. F.	43.194	28.344	13.952	—0.279
	Outlet, deg. F	37.054	22.885	8.771	—5.879
Specific heat of brine (per unit of volume)		0.8608	0.8508	0.8427	0.8374
Quantity of brine circulated per hour, cu. ft.		1,039.38	908.84	633.89	414.98
Cold produced, B. T. U. per hour		342.909	263.950	172.776	121.474
Temperature of cooling water in condenser.	Inlet, deg. F...	48.832	49.476	48.931	49.098
	Outlet, deg. F...	66.724	68.013	67.282	67.267
Quantity of cooling water per hour in cu. ft.		338.76	260.83	187.506	139.99
Heat eliminated by condenser, B. T. U. per hour		378.358	301.404	214.796	158.926
I. H. P. in compressor cylinder.		13.82	14.29	13.53	11.98
I. H. P. in steam engine cylinder		15.80	16.47	15.28	14.24
Consumption of steam per hour in lbs		311.51	335.98	305.87	278.79
Cold produced per hour, B. T. U.	Per I. H. P. in comp. cyl....	24.813	18.471	12.770	10.140
	Per I. H. P. in steam cyl.....	21.703	16.026	11.307	8.530
	Per lb. of steam	1,100.8	785.6	564.9	435.82

PERMANENT GASES IN PLANT.

As long as their amount is small and as long as there is sufficient liquid in the condenser coil to act as a seal preventing the free circulation of the permanent gases in the system, their presence will only decrease the capacity of the condenser coil, as it were, requiring either a little more cooling water or increase the pressure in the condenser. If these gases are present in larger quantity, and especially when there is no excess of liquid ammonia in condenser coils, they will disseminate themselves through the whole plant and interfere both with the economical working of the plant and the correct indications of the gauges, etc. For these reasons the engineers ought to be watchful to prevent any accumulation of such gases. Sometimes they consist chiefly of atmospheric air, but sometimes also of hydrogen and nitrogen, due to the decomposition of ammonia. The best way to remove these gases from the system is by drawing them off at the top of the condenser coil. It is advisable when drawing off the permanent gases to make the condenser as cold as possible by using an excess of cooling water and by stopping the inflow of ammonia gas to the condenser for the time being. A small hose, or, better still, a permanent small pipe, may be attached to the top of the condenser or provided with a valve near the condenser, the other end dipping in cold water. If on opening the valve bubbles are seen to escape through the water the valve should be kept open as long as such bubbles appear in the water. If, however, the bubbles cease to appear in noticeable quantity, while a crackling noise in the water indicates that most of the gas escaping through the pipe is ammonia, which is absorbed by the water, then the valve should be closed, as all the permanent gases that can be removed at the time without undue loss of ammonia have been disposed of, at least for the time being.

FREEZING BACK.

The tendency of freezing back shown by certain machines and not by others, is explained by their mode of working. The former machines work by what is called the method of wet compression, and the others by the method of dry compression. The tendency to freeze back itself involves no loss, for a machine intended for wet compression may also be worked with dry gas, by opening the expansion valve very little, but in doing so the capacity of the machine is reduced and the power required to work the compressor is increased.

PRACTICE IN WET COMPRESSION.

In working with wet expansion the object is to deliver the gas from the compressor in a saturated condition, but if this were actually done we would never be sure that certain amounts of liquid were not mixed with the gas, which would constitute a severe loss. For this reason it is indicated to allow the temperature of the vapor leaving the compressor to be about 20° above that of the liquid leaving the condenser. Inattention to this point probably accounts for many differences of opinion in regard to dry and wet compression. Any liquid present under such conditions would fill the clearance space, and by expanding would destroy a corresponding percentage of compressor capacity ($\frac{1}{32}$-inch clearance filled with liquid ammonia would reduce the capacity over one-third).

ORIGIN OF PERMANENT GASES.

In the operation of a compression plant the undue heating of the gas during compression must be considered as the chief cause for the decomposition of ammonia and the origination of permanent gases. However, it also frequently happens that air is drawn into the system through leaks, in case a vacuum has been pumped, which some engineers are unnecessarily in the habit of doing whenever they stop the plant for a length of time.

CLEARANCE MARKS.

The clearance in the compressor is not a fixed quantity, but changes with the natural wear of cranks and cross-head. For this reason clearance marks should be provided for on the guides and cross-heads of compressors as well as engine. These will indicate if the clearance is equalized at the end of cylinders, and guide us in the matter of keying up the bearings. The clearance should not exceed $\frac{1}{32}$ part of an inch.

VALVE LIFT.

The lift of compressor valves must be carefully adjusted to the speed of piston (to get full discharge), supply of condenser water, etc.

If valves are not properly set and cushioned they pound, which may even cause the texture of the metal to change in such a way as to cause their breaking to pieces.

PACKING OF COMPRESSOR PISTON.

If the piston rod is of uniform diameter and well polished, the packing will last several months, otherwise it may have to be renewed every month.

If the compressor valves or pistons should leak, the refrigerator pressure will rise and the condenser pressure will fall.

When it becomes necessary to open any part of the plant the ammonia should be transferred to another part, or if this is impracticable it should be removed by absorption in water.

POUNDING PUMPS AND ENGINES.

Sounds that appear to proceed from first one place and then another about the engine and pumps can generally be located by the use of a piece of rubber tubing, one end of which is held to the ear while the other end is brought close to the suspected place. The opposite ear should be closed to shut out the sound.

An old yet very effective way to locate any noise inside of an engine or pump cylinder is to place one end of a wrench or other piece of metal between the teeth, and resting the other end on the cylinder head, close both ears. Every sound within the cylinder can thus be readily heard.

CLEANING CONDENSER.

If the condenser coils have a tendency to become incrusted by deposit from the water, they should be cleaned from time to time. On such occasions they may also be tested with a water pressure of some 400 pounds per square inch to discover corrosion, perforation and other bad places.

CLEANING COILS, ETC., FROM OIL.

If there is oil in parts of the system whence it cannot be removed by the oil traps, those parts may be blown out, and if consisting of pipe they can be blown out by sections, if practicable. Another way more strongly recommended, and more simple, to clean ammonia pipes from oil, consists in allowing high pressure ammonia gas to enter them; this warms and liquefies the oil sufficiently to permit of its being drawn (mixed with the ammonia) into the compressor, whence it passes to the oil traps, where it is separated from the ammonia. This method of cleaning the coils is said to be very effective if repeated from time to time, say once a week, or better still, every other day.

INSULATION.

The most important point in the economical running of a plant is insulation, and especially does this refer to the ammonia on its way from the refrigerator to the compressor, and from the condenser to the refrigerator through the liquid receiver, etc. For these reasons these conduits cannot be insulated too well. The same applies to brine tank, freezing tank, etc.

PAINTING BRINE TANKS, ETC.

Light colored surfaces radiate and absorb less heat than dark surfaces under the same conditions. Also smooth and bright surfaces will radiate and absorb less heat than rough and dead looking surfaces of the same color. That the differences in radiation brought about in this way are great enough to be quite observable about a refrigeration plant, for instance, on the efficiency of a brine tank or other vats, we make no doubt. For this reason light colors, possibly white, and smoothly varnished at that, are, doubtless, best adapted to all surfaces. Preferably a white earthy paint, like barytes, etc., but no white lead, should be used for this purpose.

LUBRICATION.

The oil used for lubricating the compressor differs from ordinary lubricating oil in that it must not congeal at low temperature, and must be free from vegetable or animal oils. For this reason only mineral oils can be used, and of these only such as will stand a low temperature without freezing, such as the best paraffine oil, will do. Regular cylinder oil, however, should be used for the steam cylinder, and a free flowing oil of sufficient body for all bearings and other wearing surfaces.

For heavy bearings on ice machines a heavy oil should be used, while small bearings, such as shafts of dynamos, should be lubricated by a very light oil, to avoid undue heating in either case. Graphite or black lead is also an efficient lubricant.

CHAPTER XV.—MANAGEMENT OF ABSORPTION PLANT.

MANAGEMENT OF ABSORPTION MACHINE.

The management of an ammonia absorption plant has many points in common with that of a compression plant. The detection and mending of leaks, lubrication, the management of ammonia, withdrawal of permanent gas, etc., are the same in both, and they have been enlarged upon in the foregoing. There are, however, many precautions and troubles peculiar to the absorption system, and the most important of them will be shortly mentioned hereafter, and some of these in turn will also apply to the operation of the compression plant.

INSTALLATION OF ABSORPTION PLANT.

The installation and testing of an ammonia absorption plant is generally attended to by the manufacturers. The plant before being put in operation should be tested to a pressure of about 300 pounds per square inch.

CHARGING ABSORPTION PLANT.

Before the ammonia is charged into the machine, it is necessary to expel from the entire apparatus the air which it naturally contains.

There are two methods of doing this, one of which consists in opening all the connecting valves in the machine; leave one open to the atmosphere, introduce direct steam in the retort until all the air is forced out, and then shut the outlet valve and let the apparatus cool off. When it becomes cold, there will be found to be a vacuum in the whole apparatus. It is then ready to receive the ammonia. This method, however, is not to be recommended, as the heat of the steam will soften the joints, especially if rubber is used.

The best way is to pump a vacuum by means of a good pump. The boiler feed pump or the ammonia pump may be used for this purpose, and when a vacuum of twenty-five inches is obtained, close all the valves. Then connect the charge pipe with the drum of aqua ammonia, taking care not to let any air enter the pipe after the drum is empty. Close the charge valve and repeat the operation with another drum, until the vacuum in the machine is gone, and then pump in the balance with the ammonia pump until nearly the requisite charge is put in; then heat the ammonia slowly by turning steam through the heater coils. When the pressure gauge indicates 100 pounds, more or less, open the purge cock and lead the discharge into a pail of cold water through a rubber tube until no air bubbles come out; then turn on the condensing water into the condenser cooler and absorber, and apply the steam until the liquefied gas shows in glass gauge. Then open distributing valve to freezing tank, and turn the poor liquor into absorber, and in a few minutes the ammonia pump may be started to pump the enriched liquor through the coils of exchanger and into the retort. Let the condensed steam into the deaerator and let cooling water run over the distilled water cooler coils. Let it run out until the water becomes clear and tasteless. Proceed in this way, carefully watching for ammonia leaks wherever there are joints. If none exist, keep on until all the pipes in the freezing tank become coated with frost, and the remaining air has consequently been driven out through the coils and out of the absorber purger. Then close down and proceed and make the brine solution, when the machine is ready to start again and the balance of the ammonia may be put into the machine and operated in the regular manner.

OVERCHARGE OF PLANT.

In charging an absorption machine with ammonia liquor, which is generally done when it is cold, it should be borne in mind that the liquid expands when heat is applied, and that if the machine is charged to its working point when cold, it will invariably be overcharged under working conditions. In such a case the liquor may go out of sight in the gauge and great variations of pressure take place, which are apt to damage the rectifying pans, and the proportionate strengths of poor and rich liquor are disturbed.

AMMONIA REQUIRED.

When the regular automatic operation of the absorption cycle has been inaugurated, a surplus of liquid ammonia should show itself in the liquid receiver. If there is a deficiency in this respect it can be supplied by the addition of anhydrous ammonia, or by the addition of strong ammonia liquor, and the withdrawal of a corresponding amount of weak liquor. The sound of the liquor passing the expansion valve should be continuous and sonorous, as in the case of the compression machine, indicating the absence of a mixture of gas and liquid.

RECHARGING ABSORPTION PLANT.

For the purpose of recharging an absorption plant *De Coppet* gives the following rational directions: When the gas has leaked out or the liquor has become impoverished, and knowing the original charge by weight and density, as for instance, say the original charge was 4,000 pounds at 26° B., there would be 1,040 pounds of ammonia in 2,960 of water; if the density through leakage or purging came down to say 23°, there would be a loss of 120 pounds in the original charge, which can be easily supplied by placing a drum of anhydrous ammonia on a scale, taking a long and small flexible pipe, say a half inch, connected between the drum and same part of the machine, say the feed pipe to freezing tank, weigh the drum accurately before opening the valve, let the liquid gas run in the machine until there are within a few pounds of the quantity missing; run out of the cylinder into the machine, say ten or fifteen pounds, then close the cylinder valve and try the machine by running it in the usual way for an hour or two. Then add the ten or fifteen pounds extra, and if all the air has been blown out of the tube, and if the ammonia is pure, his machine will work all right again. When the liquor is lacking it is best to recharge the machine with strong aqua at 26° to 28° until the original level is reached, which can easily be ascertained if a glass level or test cock has been placed on the generator or still. He has adopted this method for fifteen years, and finds it far preferable to that of concentrating the liquid and recharging it with rich ammonia afterward, securing the same amount of poor liquor, besides saving time and money.

When the question presents itself as to how much anhydrous ammonia, x, in pounds must be added to m pounds of ammonia liquor of the percentage strength a in order to convert it into ammonia liquor of the percentage strength b, it may be readily answered after the following formula:

$$x = \frac{m\,(b-a)}{100-b}$$

CHARGING WITH RICH LIQUOR.

When the absorption system is charged with strong aqua ammonia it happens sometimes that the pump will not readily take the strong liquor. This is due to the great tension of the ammonia in the strong solution, which fills the pump up with ammonia vapor in such a way that the liquid cannot be drawn in. The same thing frequently happens with boiler feed pumps, when the feed water becomes nearly boiling hot. Generally it is found that in such cases the pump stands too high; if it stands below the liquid to be pumped the latter will fill the pump in preference to the vapor, and the pump will generally work all right.

It should be noticed, however, that this artifice of elevating the receptacle containing the rich liquor above the pump will only be efficient if it is done in such a manner that the liquid will run into and fill the pump by its own gravity. If the liquid has to be syphoned over by the pump, it will make little difference whether the pump stands a little above or below the liquor, as in either case the vapor of the rich liquor will fill the syphon and pump in preference to the liquid if the pump is not in first-class working order. This tendency is increased when the pump is allowed to run dry and hot on starting, and for this reason the cooling of the pump with water frequently remedies the trouble. This, the cooling of the pump, so it will take the rich liquor, may be accomplished according to a practical operator by stopping the pump, while the machine otherwise is running as usual. In this way the absorber is cooled down in a short time; meanwhile the drum containing the rich liquor has also been connected with the pump which is now started first to pump cold liquor from the absorber for a few seconds when the absorber valve is closed and the pump started on the rich liquor, which will then be taken readily. If not the procedure may be repeated once or twice.

PERMANENT GASES IN ABSORPTION PLANT.

The permanent gases in the absorption plant may be due to decomposition of ammonia and also air which has found its way into the system. It appears, however, that the decomposition of water vapor in the presence of iron (and probably iron containing carbon in a greater quantity or in a more dissolvable form than other iron) is largely responsible for their presence. The carbon which is present in all iron may also combine with hydrogen, forming carburetted hydrogen. That the nature of the iron of still and condenser worms has some influence in this direction is proven by the fact that some plants are much more damaged by these corroding influences than others. This difference in behavior must be attributed to the iron rather than to the ammonia.

CORROSION OF COILS.

As may be inferred from the foregoing paragraph, it will not only be the permanent gases, thus found, which annoy the manufacturer, but also the corrosion and consequent destruction of the coils and tanks. This is, indeed, the case especially in the upper regions of ammonia still and in the condenser. As a precautionary measure it is well to have the coil in the still always covered with liquid.

ECONOMIZING CONDENSER COILS.

As has been stated, the iron of the coil or worm in condenser and in the ammonia still suffers much from pitting and corrosion, especially if the liquid does not always stand above the coil in the still. Coddington finds that the pitting takes place first at the top of the coils, and therefore he has found it a good practice to turn the condenser coil over after a certain period, say after it has been used about four years.

KINDS OF AQUA AMMONIA.

The difference between the different kinds of aqua ammonia in the market is only in strength and price, the latter differing like that of other commodities, according to the law of demand and supply. At present we find in the market (according to Beaumé hydrometer scale for liquids lighter than water, the latter showing 10°):

1. 16° aqua ammonia, often called by druggists F. F. F., containing a little more than 10 per cent of pure anhydrous ammonia.

2. 18° aqua ammonia, called by druggists F. F. F. F., containing nearly 14 per cent of anhydrous ammonia.

3. 26° aqua ammonia, called by druggists stronger aqua ammonia, and containing 29¼ per cent of pure anhydrous ammonia. This is the aqua ammonia generally used in absorption plants for the start. At last quoting the prices (in carboys) were about two and one-half cents per pound for the 16°, three and one-half cents per pound for the 18° and four and three-quarters cents per pound for the 26°, the latter not in carboys, but in iron drums.

It is also frequently supposed that a difference in the nature of ammonia is due to the different sources from which it is derived, viz., from gas liquor direct, or from intermediate sulphate of soda, but manufacturers claim, and with apparent reason, that this is not the case if both kinds are equally well purified.

LEAKS IN ABSORPTION PLANT.

If, while the pump and generator appear to work regularly, there is a great disproportion in the strength of the poor and the rich liquor, so that the strength of the former to the latter is 22 to 25, where it should be 17 to 28, or thereabouts, it is likely due to some leaks, more particularly in the exchanger or equalizer or in the rectifying pans.

LEAK IN EXCHANGER.

If there is a leak in the equalizer coil large enough to seriously affect the working of the machine, the pipe connecting the equalizer and the coil in the weak liquor tank will become cool when the pump is running fast, and the equalizer will be cool back to a short distance from the leak, where the cold ammonia from the absorber mingles with the weak liquor from the generator. And at times, when the pump is running very fast, the whole weak liquor line may cool back to within a few inches of the generator, showing that strong ammonia is being pumped into the bottom and top of generator, as well as into absorber. There will also be a ringing or hissing noise in the neighborhood of the leak. First locate the trouble in the equalizer by noticing the cooling of the pipes, and then find the place in the equalizer by feeling the different sections with the pump running slower, having also the assistance of an ear tube.

Another way to try an exchanger coil while the machine is running is as follows: Close poor liquor valve between the generator and exchanger; close absorber poor liquor feed, and run pump as slow as possible; open the poor liquor feed wide; if there is a leak, the pump will start faster. When the poor liquor feed is closed at the absorber and between retort and exchanger, the pump is working against the generator's pressure, while when the absorber feed is wide open the pump is working against a lower pressure (ten pounds per square inch) through the leaky coil of the exchanger, then to the absorber, thus forcing a by-pass circulation of rich or enriched poor liquor from the absorber through the exchanger, through the leak of the coil of the exchanger, back through the poor liquid cooler and to the absorber again. If the leak in the coil is of a large size, the machine will come to a standstill, and will stay that way until the leaky coil is not removed.

LEAK IN RECTIFYING PANS.

If under existing regularities in the relative strength of the poor and rich liquor the exchanger has not been found leaking, but perfect in its working, it is almost beyond doubt that the rectifying pans are out of order. In order to make sure on this point a certain small quantity of the liquefied ammonia may be withdrawn from the liquid receiver, and then be allowed to evaporate (the vessel containing the ammonia being placed in ice water). If under these conditions a remnant (water) amounting to 20 per cent and more is shown, then there is doubtless a leak in the rectifying pans, which should be repaired.

STRONG LIQUOR SYPHONED OVER.

When the ammonia is short in a machine the same may be absorbed so quickly in the absorber as to cause the contents of the still to be syphoned or drawn over in the absorber and (if not guarded against by check valve) into the refrigerator. Defective action of the ammonia pump may cause the same trouble. For this reason the gauge at still must be closely watched, so that the liquor always covers the steam coil, by which an undue decomposition of the ammonia and formation of permanent gases is also avoided.

This siphoning over of the ammonia from one part of the system, and absorption into another where it does not belong, is frequently called a "boil-over"; and besides the siphoning over of the liquid to the absorber, etc., it sometimes happens, also, that the liquid runs over from the generator into the condenser coils.

If the liquified or condensed ammonia collects promptly in the liquid receiver, which shows on the gauge glass of same, there is always pressure enough behind the expansion valve to hold the ammonia in the generator, and there will be no danger of a boil-over unless the ammonia pump receives the liquid from the absorber too fast. To avoid this the absorber is always supplied with a gauge glass, so the ammonia can be kept at a certain height by means of a valve commonly called the poor liquor valve. But if the engineer does not watch it very closely, the ammonia will get out of his sight, and sometimes even into the expansion coils. This is sometimes made worse by not having a governor on the ammonia pump, which is sure to vary with the variation in steam pressure, causing the pump to run faster or slower.

REMEDY FOR BOIL-OVER.

If, however, through carelessness on these points or otherwise, a boil-over into the expansion coils has taken place it may become necessary to nearly close the expansion valve long enough to pump a vacuum on the absorber, and then blow what gas is on hand through the coils. This generally cleans them and takes the ammonia back to the absorber. This is rather troublesome work, but the work will have to be done before the machine will work satisfactorily.

If the expansion coils are divided in sections supplied by manifolds, so that all the sections except one can be shut off, and all the ammonia gas be made to pass through one section at a time, each of the sections can be cleaned without pumping a vacuum on the absorber.

CORRECTION OF AMMONIA IN SYSTEM.

To avoid the boil-over or siphoning over, the generator gauge must be closely watched, as has already been mentioned, and if the liquid line is not visible in the generator the weak liquor should be cut off from the absorber, and the generator glass watched to see if the liquid rises; and if it does, and no part of the charge has gone over into condenser or brine tank coil, and the absorber has been pumped down below where it is usually carried, it is a plain case of shortage of aqua ammonia. If there is no frost on the pipes, and the receiver glass is full of liquid, the weak liquor valve should be left closed and the expansion valve opened wider; and if the absorber fills without much of the rumbling noise, it is filling with liquid from the brine tank coil. If the machine is found to contain enough ammonia, and there is no leak in the pans or the equalizer, and the head pressure is too low and the back pressure too high, the trouble is to be found in the pump. But if the high pressure is too low and the low pressure not too high, with everything else all right, the machine should have an addition of anhydrous ammonia.

CLEANING THE ABSORBER.

Most cooling waters used in the operation of absorbers in connection with absorption machines contain carbonates of lime, magnesia and iron in sufficient quantity to form a scale inside of the absorber. This scale consists of the carbonate of lime, etc., mentioned before, which becomes insoluble at the temperature of the absorber, owing to the volatilization of the free carbonic acid in the water which held them in solution. It is a matter of considerable trouble, but also of necessity, to remove this scale from time to time, which depends on the nature of the water.

This is generally done by taking the coils out and suspending them over a fire to be heated considerably above the boiling point of water (not red hot, however). While still hot, or better still, after cooling, the scale may be removed by hammering and rolling the coil about.

As a much simpler device Coddington recommends the use of crude hydrochloric acid (price two and a half cents per pound) diluted with six times its weight of water. With this mixture he fills up the coils and lets them stand until it ceases to digest the scale, which usually requires two hours. If one dose of acid does not clean the pipe thoroughly he repeats the same. In this case it is not required to remove the coils at all, but only the bottom and top of the absorber have to be disconnected. Some care, however, must doubtless be exercised, so as not to have the acid act for too long a time, as in that case the iron of the coil itself might be affected.

HIGH PRESSURE IN ABSORBER.

Too high pressure in the absorber, and, incidentally thereto, too high temperature in the refrigerator, may be due to too much liquid in the system, or to too little cooling water. Too high pressure in the absorber may also be due to air or permanent gases in the system. These must be withdrawn through the purge cock at the top of the absorber, through a pipe or hose leading into a bucket of water, as described under the head of compression plant.

OPERATING THE ABSORBER.

It is often claimed that the absorber runs too hot, which may be due to the presence of permanent gases, due to decomposition of ammonia or to the presence of air, or to incrustation of the pipes, all of which prevent the full utilization of the cooling surface of the condenser. It may also be that in such a case the exchanger does not do its full duty or that ammonia pump is not in good working order and that it does not displace a sufficient amount of liquid.

Another point of great importance in this respect is the proper regulation of the expansion valve, so as to prevent any excess of ammonia entering the refrigerator and the absorber. Any ammonia which enters the absorber in a non-volatilized or wet condition, means so much additional heat in the absorber, more cooling water and more waste all around. For this reason we are advised to so regulate our expansion valve that the pressure on absorber gauge is about three pounds, and not much over.

If, on the other hand, there is too little or no pressure on the absorber, the ammonia pump will not do its duty, and this will be prevented by the foregoing pressure on absorber also. In order to correct too low a pressure in the absorber the decrease of the water supply to the latter is generally the most convenient remedy.

PACKING AMMONIA PUMP.

The packing of the liquor or ammonia pump is done the same way as in case of any other pump, but owing to the pressure and the smell in case of leaks it ought to be attended to with special precaution. The packing used should be of the best kind, as it will wear least on the rods, and does not require to be pulled up so tight, which increases the work and the wear and tear. The pump rod should be turned true if unevenly worn, as it is next to impossible to pack a bad rod well.

Any good hemp packing is excellent for most pumps. It should be well packed into the stuffing box, but not too hard. If, after screwing down the nut in place, the box is not full, remove the nut again and put in more packing. Replace the nut and screw well down, not too tight. If properly done, thumb and finger will screw the nut tight enough. The piston rod should be kept properly oiled. The packing nuts should be tightened up from time to time, and the packing should be renewed occasionally without waiting till it is burned out. Some operators use pure gum rings that will slip into the stuffing box with light pressure. Square or rectangular gums will answer if the rings are not convenient to get. This packing must not be screwed down too tight, as the ammonia will swell the rubber, and in that case it may bind the rod so tightly that it will roll it out of the stuffing box. Use mineral oil for lubricating.

ECONOMIZING WATER.

The economizing of water is a question of even more importance with the absorption system than with the compression system, as it is used not only in the condenser and boiler, but also for the absorber. In this case also it can be recooled and re-used by gradation, and in localities where the water is warm, it may be good policy to cool it by gradation in the first place. The water after having passed the absorber is better for boiler feeding than the natural water, not only because it is heated to some extent already, but also because it has already deposited some or most of its mineral matter which would tend to form scale in the boiler. The cooling water after having left the absorber might be used to condense the moist steam from ammonia pump, in case this is also needed for ice making before it enters the boiler. Some absorption machines use the cooling water for the double purpose of cooling the absorber first, and then the condenser, or *vice versa.*

OPERATING BRINE TANK.

The principal information relating to brine and freezing tanks is given elsewhere. The following may be added relative to their operation: In order to be able to fully utilize the coils in brine tanks, they should be made in short runs, and kept free from ice. Sometimes when the brine is not strong enough, the formation of ice around the expansion coil may take place, and this greatly reduces the capacity of the freezing tank, and in some measure accounts for the great variation in pipe lengths required in different plants. No galvanized iron pipe should be used for direct expansion, and connections, etc., should be made with extra strong unions, flanged joints, etc. No right and left coupling, nor ordinary couplings should be used, and the element of uncertainty should be entirely avoided.

LEAKS IN BRINE TANKS.

Small leaks in brine tanks may sometimes be stopped by the application of bran or corn meal near the place where the leak is. The meal or bran should be carried (in small portions at the time) to the place where the leak is, by means of a short piece of open pipe.

In making repairs to coils while immersed in brine the workmen should besmear their arms and hands with cylinder oil, lard or tallow, as that will enable them to keep them in the cold brine without much inconvenience for some time.

TOP AND BOTTOM FEED BRINE COILS.

The expansion coils in brine tanks are fed from bottom or top according to the system of refrigeration, as mentioned elsewhere, but it is claimed that the disadvantages of both ways of feeding can be avoided by using what is called

TOP FEED AND BOTTOM EXPANSION.

This system is a combination of the best elements of the two systems above described. Each alternate coil in a tank is connected to a liquid manifold (provided with regulating valves) at the top of the tank, and the ammonia is evaporated downward through one-half of the coils in the tank. All of the coils in the tank are connected to a large bottom manifold (which might be called an equalizing expansion manifold), and the gas is returned up through the second half of the coils to a gas suction manifold at the top of the tank, located behind and a little above the liquid manifold. The suction manifold is provided with a tee for connecting the suction pipe leading to the compressors.

CLEANING BRINE COILS.

When the pipes in the brine tank are to be blown out by steam, the brine must be removed and the headers of the coils must be disconnected and each coil must be steamed out separately with dry steam, care being taken to let the steam blow through the coils long enough to heat them thoroughly, so that when the steam is shut off the coils are left hot enough to absorb all moisture inside.

DRIPPING CEILING.

Dripping ceiling is an awkward trouble liable to occur where rooms are to be refrigerated. There seems to be no universal cure for a dripping ceiling; even as to the causes of such occurrence the most experienced engineers seem to have only conjectures. In some cases it seems that in storage rooms located one above the other the ceiling of the lower drips on account of the cold floor above. In other cases it appears that the space between the ceiling and refrigerating coils is too small, allowing condensation to form on the ceiling which otherwise would have settled on the pipes again. It is asserted that porous ceilings, formed with brick arches laid in ordinary mortar, will prevent condensation overhead, while ceilings formed of sheet metal, wood painted, and varnish air tight and ditto cement ceilings are prone to condense moisture. The dripping from refrigerating coils should be caught in drip pans placed or hung below them, and, generally speaking, the drippings ought to be prevented from entering the fermenting tubs, dripping over meat, vegetables and cold storage goods in general.

REMOVING ICE FROM COILS.

The removal of ice from ammonia expansion coils can be best effected by allowing hot ammonia vapor to enter them, and a connection to permit this should be provided for. The ice can be thawed off in this way or loosened so that it can be knocked off. If the ice is removed soon after it has formed, say daily, it is sufficiently loose in itself, so that it can be cleaned off without any special artifices.

MANAGEMENT OF OTHER PLANTS.

The management of other refrigeration plants, notably of those which work on the compression plan, such as the sulphurous acid, the carbonic acid and "Pictet liquid" machines, is in most principal points like that of the ammonia compression machines. In the case of carbonic acid it is somewhat difficult to detect and locate leaks on account of its being free from odor. The best available means in this connection are soapsuds, smeared over the pipes, joints, etc., when leaks will demonstrate themselves by the formation of bubbles.

COST OF REFRIGERATION.

The principal expense in the production of artificial refrigeration and artificial ice is coal and labor. And as it takes much less labor in proportion to run a large plant than a small one, it is evident that larger plants, especially for ice making, are more profitable. Also less coal is required for larger than for smaller plants. While four men are required to operate ice plants of one to five tons capacity, it will take only five men to operate a 10-ton plant, and only eight men to operate a 35-ton plant.

CHAPTER XVI.—TESTING OF PLANT.

TESTING OF PLANT.

The testing of a plant is executed in different ways in accordance with what the test is intended to prove. When the question is simply as to what a plant can be made to do, independent of the use of coal, the use of condensing water and the wear and tear of machinery, the test is simply a matter of shoveling coal and pumping condenser water. However, the time of such tests has gone by, and the question nowadays is, as to what a machine will do under normal comparable conditions and as to how the refrigeration produced compares with the amount of work expended and the amount of coal consumed.

FITTING UP FOR TEST.

To make a test of this kind a number of preparations have to be made. The compressor as well as the steam engine has to be provided with indicators; the condensing water supply has to be connected with a meter, and the amount of brine circulated must be ascertained in a similar manner. The temperature of incoming and outgoing brine, of the incoming and outgoing condenser water, must be measured as exactly as possible, as also the actual temperature of the gas when entering and leaving the compressor, for which purpose mercury wells should be placed in the suction and discharge pipe near the compressor.

MERCURY WELLS.

A mercury well is simply a short piece of pipe, closed at one end and fitted tightly into a pipe or vessel, the temperature of which is to be ascertained. The pipe is filled with mercury, and an exact thermometer is placed in the latter.

THE INDICATOR DIAGRAM.

An indicator diagram shows the outline of a surface, limited on one side by a horizontal line, the length of which represents the length of the stroke of a piston (of a pump, engine, compressor, etc.), in reduced scale. A line connecting the two ends of the straight line overhead is formed by connecting the points, which by their vertical distance from the said horizontal line indicate the pressure working on the piston when passing their respective points on the horizontal line on a certain scale.

These diagrams are obtained by instruments called indicators, which are applied in accordance with instructions accompanying each instrument when bought.

The area of such a diagram limited by a straight line on one side and by a curve on the other sides, represents the work done by the compressor during one stroke in foot-pounds.

The area of the diagram may be found by calculation in dividing the same into convenient sections, measuring them and adding them up.

The area may also be measured by a machine constructed for this purpose, called a planimeter.

With proper precaution and an accurate scale, the area of these diagrams can also be ascertained by cutting them out carefully and weighing them. The weight so obtained can then be compared with that of a rectangular piece of paper of the same thickness and known surface.

In addition to the actual work done by or applied to a piston during each stroke, these diagrams show at a glance the conditions of pressure at the different positions of the piston, give also a ready idea of the regularity of its working, the working of the valves and the changes of temperature.

CALCULATION OF DIAGRAM.

Usually, and in the absence of a planimeter, the indicator diagram of the compressor is divided into ten vertical stripes, the median heights of which are added and divided by 10, whereby the median height of the diagram is found in inches or millimeters. As it is known for every indicator spring what pressure corresponds to one millimeter or to one inch or fraction of an inch, we can readily find the mean pressure of the compressor from the average height of the diagram. The average pressure in pounds per square inch multiplied by the area of the piston in square inches and by the number of feet traveled by the same per minute gives the work of the compressor in foot-pounds per minute, which may be divided by 33,000 to find the horse power of the compressor. In close calculations allowance must be made for the thickness of the piston rod in double-acting compressors, as the area of the piston is lessened on one side to that extent. It is also well to obtain a number of indicator diagrams at intervals of from ten to thirty minutes.

MEAN PRESSURE OF COMPRESSOR.

In the absence of an indicator diagram the mean pressure in the compressor, and indirectly the work of the compressor, may be found approximately in the accompanying table (De La Vergne's catalogue) from the refrigerator and condenser pressure and temperature.

Condenser Pressure.		103	115	127	139	153	168	184	200	218
Condenser Temperature.		65°	70°	75°	80°	85°	90°	95°	100°	105°
Refrigerator Pressure.	Refrigerator Temperature.									
4	—20°	41.46	43.91	46.34	48.77	51.23	53.68	56.11	58.54	60.99
6	—15°	42.72	45.38	47.90	50.74	53.40	56.08	58.86	61.40	64.08
9	—10°	44.40	47.38	50.33	53.29	56.25	59.20	62.16	65.14	68.09
13	— 5°	45.86	49.15	52.42	55.70	58.97	62.25	65.53	68.81	72.08
16	0°	46.94	50.56	54.16	57.78	61.40	65.00	68.62	72.22	75.84
20	5°	47.74	51.73	55.70	59.68	63.67	67.66	71.62	75.61	79.61
24	10°	48.04	52.40	56.77	61.13	65.51	69.86	74.24	78.59	82.97
28	15°	47.88	52.67	57.44	62.23	67.02	71.81	76.60	81.39	86.18
33	20°	47.08	52.30	57.53	62.75	67.98	73.23	78.46	83.68	88.91
39	25°	45.06	51.34	57.05	62.75	68.46	74.17	79.88	85.58	91.29
45	30°	43.16	49.71	55.92	62.14	68.35	74.56	80.77	86.98	93.19
51	35°	40.52	47.26	54.02	60.76	67.52	74.28	81.02	87.78	94.52

INTERPRETATION OF DIAGRAM.

In order to interpret the compressor diagram with regard to the working of the compressor, its valves, defects, etc., Lorenz gives the following outlines:

If all parts of the machine are in proper condition, the general appearance of the diagram will be that represented in Fig. 1. The suction line, *S*, is only a little below the suction pressure line, *v v*, and the pressure line, *d*, is only a little higher than the condenser pressure, *k k*, Fig. 1. The work required to open the compressor valves is indicated by small projections at the pressure and suction line, and the influence of clearance is shown by the curve *r*. This curve cuts the back pressure line after the piston has commenced to travel back, and therefore lessens the suction volume to that extent. The diagram also shows that the vapors are taken in by the compressor, not at the back pressure, but at what may be termed the suction pressure, which is somewhat lower. For this reason it is that the compression curve, c, does not intersect the back pressure line until after the piston has changed the direction of its movement. The theoretical volume of the compressor, as indicated by the line *v v*, is therefore lessened in practical working for vapors possessing a certain tension.

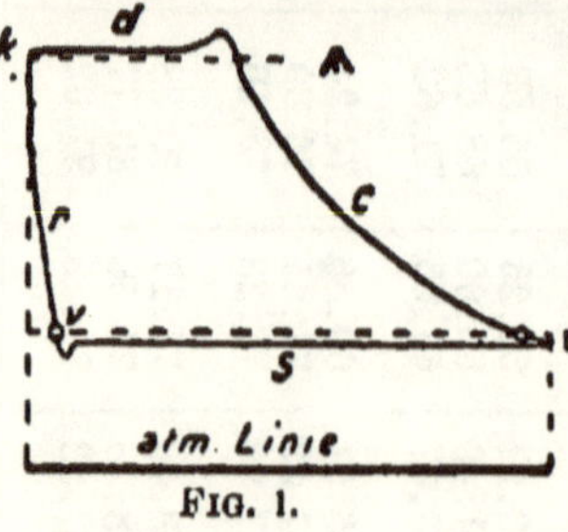

FIG. 1.

EXCESSIVE CLEARANCE.

The diagram of a compressor having an excessive amount of clearance is shown in Fig. 2. It is characterized by a flat course of the back expansion line *r*, Fig. 2, thus lessening the useful volume of the compressor.

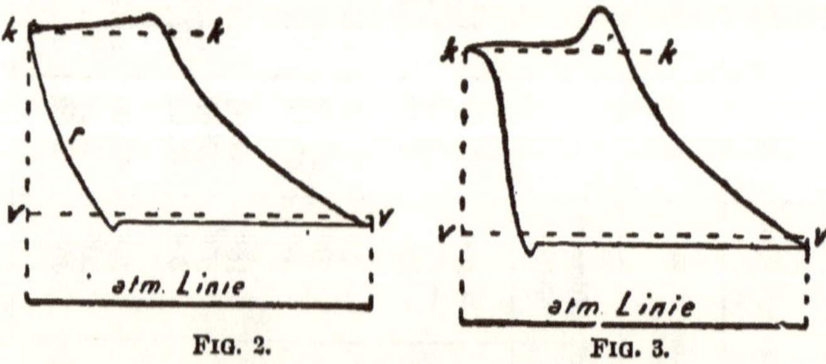

FIG. 2. FIG. 3.

In a similar manner the binding of the pressure valve

is shown by the diagram, Fig. 3, which may be caused by an inclined position of the guide rod of the valve. This same deficiency causes also frequently a delay in the opening of the pressure valves, which is indicated by a too great projection in the pressure line, as is also shown in Fig. 3. After the valve is once opened the pressure line pursues its normal course until the piston starts backward, when the defect is again shown in the back pressure line, as stated.

IRREGULAR PRESSURE AND STIFF VALVE.

Much work is also lost when the resistance in the pressure and suction pipes is too great, respectively, when the valves are weighted too much. In such cases the diagram has the appearance shown in Fig. 4, in which the

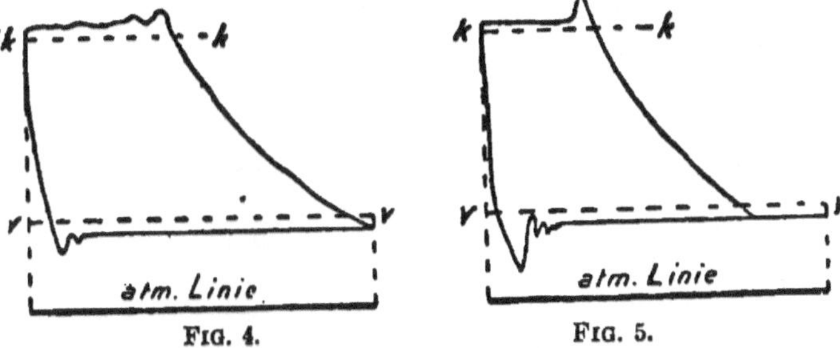

FIG. 4. FIG. 5.

pressure and suction line are at a comparatively great distance from the condenser pressure line and the back pressure line. If this happens the valve springs should be replaced by weaker ones; and if this does not have a noticeable effect, the pipe lines and shutting off valves must be thoroughly inspected and cleaned if necessary.

The binding of the suction valve causes a considerable decline in the pressure at the commencement of the suction, and is therefore shown by an increased projection in the beginning of the suction line, as shown in Fig. 5. At the commencement of compression this defect shows itself by a delay in the compression, which is also indicated in the diagram, Fig. 5.

LEAKY VALVE AND PISTON PACKING.

Leaking of the compressor valves is shown in the diagram illustrated in Fig. 6. The projections in the compression and suction line do not appear, but the compression line passes gradually into the pressure line, and the back expansion line passes gradually into the suction line. If the leak in the pressure valve predominates the compression curve is almost a straight line, and steep; and if the leak in the suction valve predominates the compression line runs in a rather flat course.

If the piston is not well packed, and leaks, the vapors are allowed to pass from one side of the piston to the

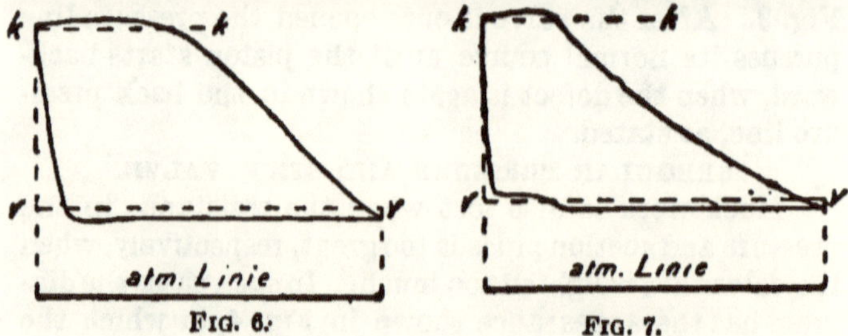

FIG. 6. FIG. 7.

other, thus causing a very gradual compression, and a consequent flat course of the compression line, as shown in Fig. 7. On the other hand, it will take a longer time before the suction line reaches its normal level on the backward stroke, as the suction valve is prevented from opening until the velocity of the piston is so great that the vapors passing the piston are no longer sufficient in amount to fill the suction space. Then the pressure decreases gradually, and the suction valve begins to play, which is also signified in the diagram, Fig. 7. Several of the defects mentioned may exist at the same time.

MAXIMUM AND ACTUAL CAPACITY.

The maximum theoretical capacity of a machine is the measure of what the same could possibly do under the existing conditions. (Temperature of brine, amount and temperature of cooling water.)

The actual capacity is expressed by the amount of refrigeration actually produced. It is all the way from 15 to 30 per cent less than the maximum theoretical capacity. This is a natural consequence of the impossibility of avoiding leakage, clearance, friction, transmission of heat to the refrigerating medium on its passage to, through, and from the compressor, etc.

COMMERCIAL CAPACITY.

Frequently the term commercial capacity is used, and is meant to indicate what a machine would do under what may be called average conditions as regards back pressure, condenser pressure, etc. It is readily understood, however, that as long as such average back pressure and condenser pressure is not generally agreed upon, the term commercial capacity is indefinite. A back pressure of about twenty-five pounds and a condenser pressure of about 140 pounds have been proposed by Richmond as such average conditions.

NOMINAL COMPRESSOR CAPACITIES.

The following table shows the approximate dimensions of a few compressors, together with what may be termed their nominal or commercial capacity. The theoretical capacity for various back pressures, etc., may be found by referring to the subjoined table, which gives the capacity in tons in 24 hours, for various compressor capacities per minute:

TABLE SHOWING NOMINAL COMPRESSOR CAPACITIES.

Number of compressor	1	2	3	4	5
Diameter of compressor in inches ($2r$)	5	6¼	9	10	18
Length of stroke in inches (b)	8	12	16	20	28
Volume of compressor in cubic feet $V=\frac{r^2\ b \times 3.14}{1728}$	0.09	0.23	0.58	0.89	4.12
Number of revolutions per minute (m)	90	90	70	68	60
Capacity of compressor (single acting) per minute in cubic feet (Vm)	8.1	20.7	41	60.5	247
Nominal or commercial capacity in tons of refrigeration in twenty-four hours about $\frac{Vm}{4}$	2	5	10	15	60

ACTUAL REFRIGERATING CAPACITY.

In case of brine circulation the actual refrigerating capacity, R, in twenty-four hours is found after the formula—

$$R=\frac{P \times s \times (t-t_1)}{284{,}000}\text{ tons.}$$

in which P is the number of pounds of brine circulated in twenty-four hours, and t the temperature of the returning, and t_1 the temperature of the outgoing brine; s is the specific heat of the brine.

FRICTION OF COMPRESSOR.

The amount of friction or lost work of compressor is equal to the difference of the work shown by the indicator diagram of the engine and that of the compressor. The total work used by the compressor is equal to that shown by the engine indicator diagram.

HEAT REMOVED BY CONDENSER.

The total heat, H, removed by the condenser is found by the formula—

$$H \quad P \times (t - t_1) \text{ units,}$$

in which P is the amount of condenser water circulated in a certain time (twenty-four hours), and t and t_1, the temperatures of the outgoing and incoming condenser water, respectively.

MAXIMUM THEORETICAL CAPACITY.

The maximum theoretical refrigerating capacity, R_1: of the compressor is found after the formula—

$$R_1 \quad \frac{C\,(h_1 - t + t_1)}{200\,v} \text{ tons,}$$

in which formula C stands for the compressor volume per minute, *i. e.*, for the space through which the compressor piston travels per minute; v is the volume of one pound of vapor at the temperature t_1, in cubic feet; t is the temperature in the condenser, and t_1 the temperature of the ammonia in suction pipe; h is the latent heat of volatilization of one pound of ammonia at the temperature of t_1.

The compressor volume per minute, C, is found after the formula—

$$C \quad d^2 \times l \times m \times 0.785,$$

in which m is the number of revolutions (single-acting), d the diameter, and l the length of stroke in feet.

If a compressor works with wet gas the volume, v, may be taken from the table for saturated ammonia on page 94; if it works with dry gas the volume, v, should be taken from the table on superheated ammonia vapor on page 311. In the latter case both the pressure and temperature of ammonia in suction pipe should be ascertained.

At the temperatures and pressures not filled out in table the ammonia exists as a liquid. Other approximate values may be found after the formula—

$$v = \frac{280 + 0.62\,t}{p} \text{ (see page 96).}$$

CORRECT BASIS FOR CALCULATION.

The foregoing method for the calculation of maximum theoretical capacity is based on the temperatures of the ammonia vapor in suction pipe and in the condenser. It is, however, argued (by Linde and others)—and with considerable force, we think—that the temperatures of brine leaving the brine tank, and of water leaving the condenser should be used instead. The latter method is followed in the calculation relating to the compression machine on page 115, etc. It is true the results obtained by the former method of calculation will come nearer to the practical results, but those obtained by the latter method will give more comparable results as regards the efficiency of different machines.

MORE ELABORATE TEST.

For more elaborate tests, the loss of refrigeration in engine rooms and a number of other details must be considered, and additional mercury wells will be necessary.

TABLE SHOWING DATA OF TEST.

The following table showing another series of tests made by Schroeter, at Munich, gives the different quantities which should be ascertained, and they also show the difference in efficiency of one and the same machine if worked under different conditions:

NUMBER OF EXPERIMENT.	1	2	3
STEAM ENGINE.			
Feed water per hour in gals.	48.3	49	64
Temperature of feed water, ° F.	71	70	84
Mean pressure (indicator), pounds per square inch	26.2	26.3	33.5
Revolutions per minute	44.91	45.10	44.97
Work done in horse powers	18.88	18.99	24.06
COMPRESSOR.			
Work done by compressor in horse powers	15.31	14.98	21.54
Pressure in condenser coils	135.2	131.2	199.2
Pressure in refrigerator coils	55.2	41.89	41.9
REFRIGERATOR.			
Temperature of incoming brine	42.8	28.37	28.35
Temperature of outgoing brine	37.2	22.97	22.99
Brine circulating per hour in lbs.	65,051	50,364	43,115
Specific gravity of brine	1.250	1.250	1.247
Specific heat of brine	.850	.846	.845
Heat absorbed in refrigerator in calories	310,335	230,657	195,920
CONDENSER.			
Temperature of condenser water ° F.	49.21	49.17	40.42
Temperature of outgoing condenser water	67.57	67.34	95.60
Amount of condenser water per hour, pounds	19.338	15,041	5,328
Heat absorbed in condenser in calories	355,950	273,891	248,680
Horse power produced by engine per 1,000 units refrigeration	.060	.082	.123
Horse power used in compressor per 1,000 units refrigeration	.049	.065	.109
Pounds of condensing water used per 1,000 units refrigeration	62	65	27

EFFICIENCY OF ENGINE AND BOILER.

To determine the efficiency of engine and boiler the amount of coal used per indicated horse power of engine must also be ascertained. Frequently also the amount of steam used by the engine is determined by means of calorimetric test. (See page 109.)

TEST OF ABSORPTION PLANT.

The testing of an absorption plant can be executed on similar lines, and the various movements of efficiency can be calculated from the elements of the test, reference being had to the formulæ given in the chapter on the absorption machine. For crude tests the amount of coal used within a certain time, to heat the ammonia still and to propel the ammonia pump, is directly compared with the amount of ice produced or with the refrigeration, as it can be measured by the work done in brine tank as shown in the foregoing.

MORE EXACT TESTS.

For more exact tests, the temperature and pressure in the different parts of the plant must be closely observed, the work done by the ammonia pump must be ascertained, the strength of weak and rich liquor and a number of other items must be recorded in order to obtain not only an idea of the actual capacity of the plant, but also to learn in what, if in any, respect the same is falling short, and in what direction a possible remedy may be looked for. To show more clearly what is wanted in this direction, we append the tabulated record of a test made of an absorption machine by Professor Denton some time; not that we think it represents an exemplary capacity, but simply to show how the items of the test may be arranged.

DISCUSSION OF TABLE.

The actual amount of coal used is not measured in the foregoing test. If we assume that one pound of coal makes about eight pounds of steam, the foregoing test shows that one pound of coal would give a refrigerating effect equivalent to the melting of somewhat less than fourteen pounds of ice, which would correspond to an actual ice making capacity of about seven pounds of ice per pound of coal. From a letter written from southern Louisiana, recently shown us, it appears that an absorption machine in regular operation in that locality fur-

nishes eight pounds of ice per pound of coal used as a minimum.

TABLE SHOWING RESULTS OF TEST.

Average pressures above atmosphere, generator lbs. per sq. in.	150.77
" " " " steam " "	47.70
" " " " cooler " "	23.69
" " " " absorber " "	23.4
" temperatures, deg. F., generator	272.
" " " condenser inlet	54¼.
" " " " outlet	80.
" " " " range	25½.
" " " brine inlet	21.20
" " " " outlet	16.14
" " " " range	5.06
" " " absorber inlet	80.
" " " " outlet	111.
" " " " range	31.
" " " heater upper outlet to generator	212.
" " " " lower " absorber	178.
" " " " inlet from "	132.
" " " inlet from generator	272.
" " " water returned to main boilers	260.
Steam per hour for boiler and ammonia pumps, lbs.	1,986.
Brine circulated per hour, cu. ft.	1,633.7
" " " pounds	119,260
" specific heat	0.800
" heat eliminated per lb., B. T. U	4.104
" cooling capacity per 24 hours, tons of melting ice	40.67
" " " " lb. of steam, B. T. U	243.
" ice melting capacity per 10 lbs. of steam, lbs.	17.1
Heat rejected at condenser per hour, B. T. U	918,000
" " " absorber "	1,116,000
" consumed by gen. per lb. of steam condensed, B. T. U.	932
Condensing water per hour, lbs.	36,000
" coil, approx. sq. ft. of surface	870
Absorber " " " " "	350
Steam " " " " "	200
Pump ammonia, dia. steam cyl., in.	9
" " " ammonia cyl., in.	3⅝
" " " stroke, in.	10
" " " revolutions per minute	.22
" brine steam cyl., diam., in.	9½
" " brine " " "	8
" stroke, in.	10
" revolutions per min.	70

Effective stroke of pumps 0.8 of full stroke.

ESTIMATES AND PROPOSALS.

By way of recapitulation it may be mentioned that in ordering refrigerating machines, or in asking for estimates or proposals, one cannot be too explicit in stating the conditions under which the plant is calculated to work and what it is expected to accomplish. Foremost should be stated:

First.—The temperature and quantity of the available water supply should be given under all circumstances, and also the average temperature during the different seasons, if possible.

Second.—If water power or a surplus of steam power is available, it should be specified; also the price and kind of coal, if possible.

Third.—The kind of machine that is required, whether absorption or compression, and whether ammonia or some other refrigerating agent is to be used.

In case the principal object of the plant is the production of ice, the following additional points should be clearly specified:

(*a*) If absolutely pure and clear ice is required, *i. e.*, ice made from distilled water, or whether opaque and relatively impure ice will answer.

(*b*) If the required buildings are to be erected in wood or masonry, or if already existing buildings are to be utilized, and in the latter case, dimensions and mode of construction.

(c) The amount of ice that is to be produced in twenty-four hours.

MISCELLANEOUS REFRIGERATION.

For the refrigeration of rooms in breweries, packing houses or cold storage establishments, etc., the following additional points should be specified, or as many of them as is practicable.

(*a*) If the rooms are to be refrigerated by direct expansion or by brine circulation.

(*b*) The size of rooms, the construction of the walls and the temperature at which they are to be held.

(c) The amount and kind of beer to be brewed, and the time it is proposed to be kept in storage in case of a brewery.

(*d*) In the case of a packing house, the number and kind of animals to be chilled daily, and the number and kind of carcasses to be frozen, and the length of time they are to be kept in storage.

(*e*) In the case of a cold storage establishment, the nature of the products to be stored, or the temperature at which they are to be held, and the amount of what is to be placed into cold storage daily.

CONTRACTS.

In case contracts are made for refrigerating machinery, the amount of coal and water to be used for a certain specified duty, should apply to a specified kind of coal, to the temperature of the actual water supply (not to fictitious conditions), and to a specified number of revolutions of compressor or pump for specified dimensions.

In order to ascertain the amount of refrigeration which may be expected from an existing compressor, the diameter, length of stroke and number of revolutions should be given. Also state whether single or double-acting; the temperature of the cooling water; the back pressure and pressure or temperature (both if practicable) in condenser.

UNIT OF REFRIGERATING CAPACITY.

In accordance with some British writers, the refrigerating capacity of one ton of melting ice is equivalent to 318,080 thermal units. In the United States 284,000 thermal units are allowed to be equivalent to one ton of refrigerating capacity, or to the refrigerating capacity of one ton of melting ice. This disagreement is due to the different amount of ice which is taken to make up a ton. In the former case 2,240 pounds are calculated per ton, and in the latter only 2,000 pounds are allowed per ton.

TEST OF OTHER MACHINES.

The testing of other refrigerating machines, such as are operated with sulphurous acid, carbonic acid, Pictet's liquid, etc., can be performed on the same lines as that of the ammonia compression machine. A similar course also applies in the case of air compression, vacuum machines and other devices, the principal question always being as to what amount of coal or power and of cooling water is required to produce a certain amount of refrigeration. In comparing the efficiency of machines in different localities due allowance must always be made for differences in the water supply, its temperature, its accessibility and available quantity.

APPENDIX I.—TABLES, ETC.

MENSURATION.

MENSURATION OF SURFACES.

Area of any parallelogram..... = base × perpendicular height.
Area of any triangle............ = base × ½ perpendicular height.
Area of any circle.............. = diameter² × .7854.
Area of sector of circle......... = arc × ½ radius.
Area of segment of circle...... = area of sector of equal radius, less area of triangle.
Area of parabola................ = base × ⅔ height.
Area of ellipse.................. = longest diameter × shortest diameter × .7854.
Area of cycloid.................. = area of generating circle × 3.
Area of any regular polygon... = sum of its sides × perpendicular from its center to one of its sides ÷ 2.
Surface of cylinder............. = area of both ends + length × circumference.
Surface of cone.................. = area of base + circumference of base × ½ slant height.
Surface of sphere............... = diameter² × 3.1415.
Surface of frustum............. = sum of girth at both ends × ½ slant height + area of both ends.
Surface of cylindrical ring..... = thickness of ring added to the inner diameter × by the thickness × 9.8698.
Surface of segment............. = height of segment × by whole circumference of sphere of which it is a part.

POLYGONS.

1. To find the area of any regular polygon: Square one of its sides, and multiply said square by the number in first column of the following table.

2. Having a side of a regular polygon, to find the radius of a circumscribing circle: Multiply the side by the corresponding number in the second column.

3. Having the radius of a circumscribing circle, to find the side of the inscribed regular polygon: Multiply the radius by the corresponding number in third column.

Number of Sides.	Name of Polygon.	1 Area = S^2 ×	2 Radius = S ×	3 Side = R ×	Angle contained between two sides.
3	Equilateral Triangle.	.433	.5774	1.732	60°
4	Square......	1.	.7071	1.4142	90°
5	Pentagon...	1.7205	.8507	1.1756	108°
6	Hexagon....	2.5891	1.	1.	120°
7	Heptagon...	3.6339	1.1524	.8678	128.57°
8	Octagon	4.8284	1.3066	.7654	135°
9	Nonagon....	6.1818	1.4619	.684	140°
10	Decagon	7.6942	1.618	.618	144°
11	Undecagon..	9.3656	1.7747	.5635	147.27°
12	Dodecagon .	11.1962	1.9319	.5176	150°

In the heads of the columns in above table, S = side, and R = radius.

PROPERTIES OF THE CIRCLE.

Diameter × 3.14159 = circumference.
Diameter × .8862 = side of an equal square.
Diameter × .7071 = side of an inscribed square.
Diameter^2 × .7854 = area of circle.
Radius × 6.28318 = circumference,
Circumference ÷ 3.14159 = diameter.

The circle contains a greater area than any plane figure bounded by an equal perimeter or outline.

The areas of circles are to each other as the squares of their diameters.

Any circle whose diameter is double that of another contains four times the area of the other.

Area of a circle is equal to the area of a triangle whose base equals the circumference, and perpendicular equals the radius.

MENSURATION OF SOLIDS.

Cylinder.................... = area of one end × length,
Sphere..................... = cube of diameter × .5236.
Segment of sphere......... = square root of the height added to three times the square of radius of base × height and .5236.
Cone or pyramid............ = area of base × ⅓ perpendicular height.
Frustum of a cone.......... = product of diameter of both ends + sum of their squares × perpendicular height × .2618.
Frustum of a pyramid...... = sum of the areas of the two ends + square root of their product, × ⅓ of the perpendicular height.
Solidity of a wedge.......... = area of base × ½ perpendic'r height.
Frustum of a wedge........ = ½ perpendicular height × sum of the areas of the two ends.
Solidity of a ring........... = thickness + inner diameter, × square of the thickness × 2.4674.

POLYHEDRONS.

No. of Sides	Names.	1 Radius of Circumscribed Circle. R = S X	2 Radius of Inscribed Circle. R = S X	3 Area of Surface. $A = S^2 X$	4 Cubic Contents. $C = S^3 X$
4	Tetrahedron...	.6124	.2041	1.7320	.1178
6	Hexahedron...	.866	.5	6.	1.
8	Octahedron....	.7071	.4082	3.4641	.4714
12	Dodecahedron	1.4012	1.1135	20.6458	7.6631
20	Icosahedron...	.951	.7558	86.602	2.1817

Side is length of linear edge of any side of the figure.

1. Radius of circumscribed circle = *side* multiplied by the number in first column corresponding to figure.

2. Radius of inscribed circle = *side* multiplied by the number in second column corresponding to figure.

3. Area of surface = square of *side* multiplied by the number in third column corresponding to figure.

4. Cubic contents = cube of *side* multiplied by number in fourth column corresponding to figure.

TABLE OF AMMONIA GAS (SUPER-HEATED VAPOR).

Temperature in Degrees F.

Number of Cu. Ft., v, Approximately Contained in 1 Lb. of Gas.

Press. p in Lbs. Abso.	0	5	10	15	20	25	30	35	40	45
15	18.81	19.05	19.20	19.48	19.68	19.87	20.06	20.25	20.544	20.74
16	17.50	17.85	18.09	18.24	18.43	18.52	18.81	18.90	19.20	19.44
17	16.60	16.70	16.96	17.08	17.28	17.48	17.66	17.85	18.09	18.31
18	15.54	15.84	15.93	16.12	16.32	16.51	16.70	16.89	17.08	17.32
19	14.78	14.97	15.16	15.26	15.45	15.64	15.84	15.93	16.12	16.36
20	14.01	14.25	14.40	14.49	14.68	14.88	14.97	15.16	15.36	15.58
21	13.34	13.53	13.63	13.82	14.01	14.11	14.30	14.40	14.59	14.80
22	12.76	12.86	13.05	13.15	13 34	13.44	13.63	13.72	13.92	14.12
23	12.19	12.28	12.48	12.57	12.76	12.86	13.05	13.15	13.34	13.54
24	11.71	11.80	11.90	12.09	12.19	12.38	12.48	12.57	12.76	12.96
25	11.23	11.34	11.42	11.61	11.71	11.80	11.90	12.09	12.19	12.38
26	10.75	10.84	11.04	11.13	11.23	11.32	11.52	11.61	11.71	11.85
27	10.36	10.46	10.56	10.75	10.84	10.94	11.04	11.23	11.32	11.45
28	9.98	10.08	10.17	10.36	10.46	10.56	10.65	10.75	10.84	10.94
29	9.60	9.69	9.79	9.98	10.08	10.17	10.27	10.36	10.46	10.57
30	9.2120	9.30	10.46	9.60	9.69	9.79	9.98	10.08	10.17	10.27
31	8.84	9.12	9.21	9.31	9.40	9.50	9.60	9.69	9.80	9.91
32	....	8.83	8.93	9.02	9.12	9.21	9.31	9.40	9.50	9.61
33	....	8.54	8.64	8.73	8.83	8.91	9.02	9.11	9.21	9.31
34	....	8.25	9.35	8.49	8.54	8.64	8.73	8.83	8.92	9.02
35	....	....	8.16	8.25	8.35	8.44	8.54	8.64	8.64	8.75
36	....	....	7.87	7.96	8.06	8.16	8.26	8.35	8.44	8.55
37	....	....	7.68	7 67	7.87	7.96	8.06	8.16	8.26	8.36
38	....	....	7.48	7.58	7.68	7.77	7.77	7.87	7.96	8.05
39	....	....	....	7.39	7.48	7.48	7.58	7.68	7.77	7.87
40	....	....	....	7.20	7.29	7.39	7.39	7.48	7.58	7.68
41	....	....	....	7.00	7.10	7.20	7.20	7.29	7.39	7.49
42	....	....	....	6.81	6.91	7.00	7.10	7.10	7.20	7.30
43	....	....	....	....	6.72	6.81	6.91	7.00	7.08	7.16
44	....	....	....	....	6.52	6.62	6.72	6.81	6.91	7.10
45	....	....	....	....	6.43	6.52	6.62	6.62	6.72	6.82

SQUARE ROOTS AND CUBE ROOTS OF NUMBERS.

FROM 1 TO 20.

No.	Sq.	Cube.	Sq. Rt.	C. Rt.	No.	Sq. Rt.	C. Rt.	No.	Sq. Rt.	C. Rt.
.1	.01	.001	.316	.464	.4	2.098	1.639	.5	3.240	2.189
.15	.023	.003	.387	.531	.5	2.121	1.651	.6	3.256	2.197
.2	.04	.008	.447	.585	.6	2.145	1.663	.7	3.271	2.204
.25	.063	.016	.500	.630	.7	2.168	1.675	.8	3.286	2.211
.3	.09	.027	.548	.669	.8	2.191	1.687	.9	3.302	2.217
.35	.123	.043	.592	.705	.9	2.214	1.699	11.0	3.317	2.224
.4	.16	.064	.633	.737	5.0	2.236	1.710	.1	3.332	2.231
.45	.203	.091	.671	.766	.1	2.258	1.721	.2	3.347	2.237
.5	.25	.125	.707	.794	.2	2.280	1.733	.3	3.362	2.244
.55	.303	.166	.742	.819	.3	2.302	1.744	.4	3.376	2.251
.6	.36	.216	.775	.843	.4	2.324	1.754	.5	3.391	2.257
.65	.423	.275	.806	.866	.5	2.345	1.765	.6	3.406	2.264
.7	.49	.343	.837	.888	.6	2.366	1.776	.7	3.421	2.270
.75	.563	.422	.866	.909	.7	2.388	1.786	.8	3.435	2.277
.8	.64	.512	.894	.928	.8	2.408	1.797	.9	3.450	2 283
.85	.723	.614	.922	.947	.9	2.429	1.807	12.0	3.464	2.289
.9	.81	.729	.949	.965	6.0	2.450	1.817	.1	3.479	2.296
.95	.903	.857	.975	.983	.1	2.470	1.827	.2	3.493	2.302
1.	1.000	1.000	1.000	1.000	.2	2.490	1.837	.3	3.507	2.308
.05	1.103	1.158	1.025	1.016	.3	2.510	1.847	.4	3.521	2.315
1.1	1.210	1.331	1.049	1.032	.4	2.530	1.857	.5	3.536	2.321
.15	1.323	1.521	1.072	1.048	.5	2.550	1.866	.6	3.550	2.327
1.2	1.440	1.728	1.095	1.063	.6	2.569	1.876	.7	3.564	2.333
.25	1.563	1.953	1.118	1.077	.7	2.588	1.885	.8	3.578	2.339
.3	1.690	2.197	1.140	1.091	.8	2.608	1.895	.9	3.592	2.345
1.35	1.823	2.460	1.162	1.105	.9	2.627	1.904	13.0	3.606	2.351
1.4	1.960	3.744	1.183	1.119	7.0	2.646	1.913	.2	3.633	2.363
.45	2.103	3.049	1.204	1.132	.1	2.665	1.922	.4	3.661	2.375
1.5	2.250	3.375	1.225	1.145	.2	2.683	1.931	.6	3.688	2.387
.55	2.403	3.724	1.245	1.157	.3	2.702	1.940	.8	3.715	2.399
1.6	2.560	4.096	1.265	1.170	.4	2.720	1.949	14.0	3.742	2.410
.65	2.723	4.492	1.285	1.182	.5	2.739	1.957	.2	3.768	2.422
1.7	2.890	4.913	1.304	1.194	.6	2.757	1.966	.4	3.795	2.433
.75	3.063	5.359	1.323	1.205	.7	2.775	1.975	.6	3,821	2.444
1.8	3.240	5.832	1.342	1.216	.8	2.793	1.983	.8	3.847	2.455
.85	3.423	6.332	1.360	1.228	.9	2.811	1.992	15.0	3.873	2.466
1.9	3.610	6.859	1.378	1.239	8.0	2.828	2.000	.2	3.899	2.477
.95	3.803	7.415	1.396	1.249	.1	2.846	2.008	.4	3.924	2.488
2.0	4.000	8 000	1.414	1.260	.2	2.864	2.017	.6	3.950	2.499
.1	4.410	9.261	1.449	1.281	.3	2.881	2.025	.8	3.975	2.509
.2	4.840	10.65	1.483	1.301	.4	2.898	2.033	16.0	4.000	2.520
.3	5.290	12.17	1.517	1.320	.5	2.916	2.041	.2	4.025	2.530
.4	5.760	13.82	1.549	1.339	.6	2.933	2.049	.4	4.050	2.541
.5	6.250	15.63	1.581	1.357	.7	2.950	2.057	.6	4.074	2.551
.6	6.760	17.58	1.613	1.375	.8	2.967	2.065	.8	4.099	2.561
.7	7.290	19.68	1.643	1.393	.9	2.983	2.072	17.0	4.123	2.571
.8	7.840	21.95	1.673	1.409	9.0	3.000	2.080	.2	4.147	2.581
.9	8.410	24.39	1.703	1.426	.1	3.017	2.088	.4	4.171	2.591
3.0	9.00	27.00	1.732	1.442	.2	3.033	2.095	.6	4.195	2.601
.1	9.61	29.79	1.761	1.458	.3	3.050	2.103	.8	4.219	2.611
.2	10.24	32.77	1.789	1.474	.4	3.066	2.111	18.0	4.243	2.621
.3	10.89	35.94	1.817	1.489	.5	3.082	2.118	.2	4.266	2.630
.4	11.56	39.30	1.844	1.504	.6	3.098	2.125	.4	4.290	2.640
.5	12.25	42.88	1.871	1.518	.7	3.115	2.133	.6	4.313	2.650
.6	12.96	46.66	1.897	1.533	.8	3.131	2.140	.8	4.336	2.659
.7	13.69	50.65	1.924	1.547	.9	3.146	2.147	19.0	4.359	2.668
.8	14.44	54.87	1.949	1.561	10.0	3.162	2.154	.2	4.382	2.678
.9	15.21	59.32	1.975	1.574	.1	3.178	2.162	.4	4.405	2.687
4.0	16.00	64.00	2.000	1.587	.2	3.194	2.169	.6	4.427	2.696
.1	16.81	68.92	2.025	1.601	.3	3.209	2.177	.8	4.450	2.705
.2	17.64	74.09	2.049	1.613	.4	3.225	2.183	20.0	4.472	2.714
.3	18.49	79.51	2.074	1.626						

TABLE OF SQUARES, CUBES, SQUARE ROOTS AND CUBE ROOTS OF NUMBERS FROM 1 TO 100.

No.	Square.	Cube.	Sq. Rt.	C. Rt.	No.	Square.	Cube.	Sq. Rt.	C. Rt.
1	1	1	1.0000	1.0000	51	2601	132651	7.1414	3.7084
2	4	8	1.4142	1.2599	52	2704	140608	7.2111	3.7325
3	9	27	1.7321	1.4422	53	2809	148877	7.2801	3.7563
4	16	64	2.0000	1.5874	54	2916	157464	7.3485	3.7798
5	25	125	2.2361	1.7100	55	3025	166375	7.4162	3.8030
6	36	216	2.4495	1.8171	56	3136	175616	7.4833	3.8259
7	49	343	2.6458	1.9129	57	3249	185193	7.5498	3.8485
8	64	512	2.8284	2.0000	58	3364	195112	7.6158	3.8709
9	81	729	3.0000	2.0801	59	3481	205379	7.6811	3.8930
10	100	1000	3.1623	2.1544	60	3600	216000	7.7460	3.9149
11	121	1331	3.3166	2.2240	61	3721	226981	7.8102	3.9365
12	144	1728	3.4641	2.2894	62	3844	238328	7.8740	3.9579
13	169	2197	3.6056	2.3513	63	3969	250047	7.9373	3.9791
14	196	2744	3.7417	2.4101	64	4096	262144	8.0000	4.0000
15	225	3375	3.8730	2.4662	65	4225	274625	8.0623	4.0207
16	256	4096	4.0000	2.5198	66	4356	287496	8.1240	4.0412
17	289	4913	4.1231	2.5713	67	4489	300764	8.1854	4.0615
18	324	5832	4.2426	2.6207	68	4624	314432	8.2462	4.0817
19	361	6859	4.3589	2.6684	69	4761	328509	8.3066	4.1016
20	400	8000	4.4721	2.7144	70	4900	343000	8.3666	4.1213
21	441	9261	4.5826	2.7589	71	5041	357911	8.4261	4.1408
22	484	10648	4.6904	2.8020	72	5184	373248	8.4853	4.1602
23	529	12167	4.7958	2.8439	73	5329	389017	8.5440	4.1793
24	576	13824	4.8990	2.8845	74	5476	405224	8.6023	4.1983
25	625	15625	5.0000	2.9240	75	5625	421875	8.6603	4.2172
26	676	17576	5.0990	2.9625	76	5766	438976	8.7178	4.2358
27	729	19683	5.1962	3.0000	77	5929	456533	8.7750	4.2543
28	784	21952	5.2915	3.0366	78	6084	474552	8.8318	4.2727
29	841	24389	5.3852	3.0723	79	6241	493039	8.8882	4.2908
30	900	27000	5.4772	3.1072	80	6400	512000	8.9443	4.3089
31	961	29791	5.5678	3.1414	81	6561	531441	9.0000	4.3267
32	1024	32768	5.6569	3.1748	82	6724	551368	9.0554	4.3445
33	1089	35937	5.7446	3.2075	83	6889	571787	9.1104	4.3621
34	1156	39304	5.8310	3.2396	84	7056	592704	9.1652	4.3795
35	1225	42875	5.9161	3.2711	85	7225	614125	9.2195	4.3968
36	1296	46656	6.0000	3.3019	86	7396	636056	9.2736	4.4140
37	1369	50653	6.0828	3.3322	87	7569	658503	9.3274	4.4310
38	1444	54872	6.1644	3.3620	88	7744	681472	9.3808	4.4480
39	1521	59319	6.2450	3.3912	89	7921	704969	9.4340	4.4647
40	1600	64000	6.3246	3.4200	90	8100	729000	6.4868	4.4814
41	1681	68921	6.4031	3.4482	91	8281	753571	9.5394	4.4979
42	1764	74088	6.4807	3.4760	92	8464	778688	9.5917	4.5144
43	1849	79507	6.5574	3.5034	93	8649	804357	9.6437	4.5307
44	1936	85184	6.6332	3.5303	94	8836	830584	9.6954	4.5468
45	2025	91125	6.7082	3.5569	95	9025	857375	9.7468	4.5629
46	2116	97336	6.7823	3.5830	96	9216	884736	9.7980	4.5789
47	2209	103823	6.8557	3.6088	97	9409	912673	9.8489	4.5947
48	2304	110592	6.9282	3.6342	98	9604	941192	9.8995	4.6104
49	2401	117649	7.0000	3.6593	99	9801	970299	9.9499	4.6261
50	2500	125000	7.0711	3.6840	100	10000	1000000	10.0000	4.6416

AREAS OF CIRCLES—ADVANCING BY EIGHTHS.

Diam.	0	⅛	¼	⅜	½	⅝	¾	⅞
0	.0	.012	.05	.11	.19	.30	.44	.60
1	.785	.994	1.22	1.48	1.76	2.07	2.40	2.76
2	3.141	3.546	3.97	4.43	4.90	5.41	5.93	6.49
3	7.068	7.669	8.29	8.94	9.62	10.32	11.04	11.79
4	12.56	13.36	14.18	15.03	15.90	16.80	17.72	18.66
5	19.63	20.62	21.64	22.69	23.75	24.85	25.96	27.10
6	28.27	29.46	30.67	31.91	33.18	34.47	35.78	37.12
7	38.48	39.87	41.28	42.71	44.17	45.66	47.17	48.70
8	50.29	51.84	53.45	55.08	56.74	58.42	60.13	61.86
9	63.61	65.39	67.20	69.02	70.88	72.75	74.69	76.58
10	78.54	80.51	82.51	84.54	86.59	88.66	90.76	92.88
11	95.03	97.20	99.40	101.6	103.8	106.1	108.4	110.7
12	113.0	115.4	117.8	120.2	122.7	125.1	127.6	130.1
13	132.7	135.2	137.8	140.5	143.1	145.8	148.4	151.2
14	153.9	156.6	159.4	162.2	165.1	167.9	170.8	173.7
15	176.7	179.6	182.6	185.6	188.6	191.7	194.8	197.9
16	201.0	204.2	207.3	210.5	213.8	217.0	220.3	223.6
17	226.9	230.3	233.7	237.1	240.5	243.9	247.4	250.9
18	254.4	258.0	261.5	265.1	268.8	272.4	276.1	279.8
19	283.5	287.2	291.0	294.8	298.6	302.4	306.3	310.2
20	314.1	318.1	322.0	326.0	330.0	334.1	338.1	342.2
21	346.3	350.4	354.6	358.8	363.0	367.2	371.5	375.8
22	380.1	384.4	388.8	393.2	397.6	402 0	406.4	410.9
23	415.4	420.0	424.5	429.1	433.7	438.3	433.0	447.6
24	452.3	457.1	461.8	466.6	471.4	476.2	481.1	485.9
25	490.8	495.7	500.7	505.7	510.7	515.7	520.7	525.8
26	530.9	536.0	541.1	546.3	551.5	556.7	562.0	567.2
27	572.5	577.8	583.2	588.5	593.9	599.3	604.8	610.2
28	615.7	621.2	626.7	632.3	637.9	643.5	649.1	654.8
29	660.5	666.2	671.9	677.7	683.4	689.2	695.1	700.9
30	706.8	712.7	718.6	724.6	730.6	736.6	742.6	748.6
31	754.8	760.9	767.9	773.1	779.3	785.5	791.7	798.0
32	804.3	810.6	816.9	823.2	829.6	836.0	842.4	848.8
33	855.3	861.8	868.3	874.9	881.4	888.0	894.6	901.3
34	907.9	914.7	921.3	928.1	934.8	941.6	948.4	955.3
35	962.1	969.0	975.9	982.8	989.8	996.8	1003.8	1010.8
36	1017.9	1025.0	1032.1	1039.2	1046.3	1053.5	1060.7	1068.0
37	1075.2	1082.5	1089.8	1097.1	1104.5	1111.8	1119.2	1126.9
38	1134.1	1141.6	1149.1	1156.6	1164.2	1171.7	1179.3	1186.7
39	1194.6	1202.3	1210.0	1217.7	1225.4	1233.2	1241.0	1248.8
40	1256.6	1264.5	1272.4	1280.3	1288.2	1296.2	1304.2	1312.2
41	1320.3	1328.3	1336.4	1344.5	1352.7	1360.8	1369 0	1377.2
42	1385.4	1393.7	1402.0	1410.3	1418.6	1427.0	1435.4	1443.8
43	1452.2	1460.7	1469.1	1477.6	1486.2	1494.7	1503.3	1511.9
44	1520.5	1529.2	1537.9	1546.6	1555.3	1564.0	1572.8	1581.6
45	1590.4	1599.3	1608.2	1617.0	1626.0	1634.9	1643.9	1652.9

EQUIVALENTS OF FRACTIONS OF AN INCH.

Fractions of an Inch.	Decimals of Foot.	Fractions of an Inch.	Decimals of Foot.
⅛	.0104	⅝	.0521
¼	.0208	¾	.0625
⅜	.0313	⅞	.0729
½	.0417	1	.0833

TABLE OF LOGARITHMS.

No.	0	1	2	3	4	5	6	7	8	9	Prop.
0	- 0	00000	30103	47712	60206	69897	77815	84510	90309	95424	
10	00000	00432	00860	01283	01703	02119	02530	02938	03342	03742	415
11	04139	04532	04921	05307	05690	06069	06445	06818	07188	07554	379
12	07918	08278	08636	08990	09342	09691	10037	10380	10721	11059	344
13	11394	11727	12057	12385	12710	13033	13353	13672	13987	14301	323
14	14613	14922	15228	15533	15836	16136	16435	16731	17026	17318	298
15	17609	17897	18184	18469	18752	19033	19312	19590	19865	20139	281
16	20412	20682	20951	21218	21484	21748	22010	22271	22531	22788	264
17	23045	23299	23552	23804	24054	24303	24551	24797	25042	25285	249
18	25527	25767	26007	26245	26481	26717	26951	27184	27415	27646	234
19	27875	28103	28330	28555	28780	29003	29225	29446	29666	29885	222
20	30103	30319	30535	30749	30963	31175	31386	31597	31806	32014	212
21	32222	32428	32633	32838	33041	33243	33445	33646	33845	34044	202
22	34242	34439	34635	34830	35024	35218	35410	35602	35793	35983	193
23	36172	36361	36548	36735	36921	37106	37291	37474	37657	37839	185
24	38021	38201	38381	38560	38739	38916	39093	39269	39445	39619	177
25	39794	39967	40140	40312	40483	40654	40824	40993	41162	41330	170
26	41497	41664	41830	41995	42160	42324	42488	42651	42813	42975	164
27	43136	43297	43457	43616	43775	43933	44091	44248	44404	44560	158
28	44716	44870	45025	45178	45331	45484	45636	45788	45939	46089	153
29	46239	46389	46538	46686	46834	46982	47129	47275	47421	47567	148
30	47712	47856	48000	48144	48287	48430	48572	48713	48855	48995	143
31	49136	49276	49415	49544	49693	49831	49968	50106	50242	50379	138
32	50515	50650	50785	50920	51054	51188	51321	51454	51587	51719	134
33	51851	51982	52113	52244	52374	52504	52634	52763	52891	53020	130
34	53148	53275	53402	53529	53655	53782	53907	54033	54158	54282	126
35	54406	54530	54654	54777	54900	55022	55145	55266	55388	55509	122
36	55630	55750	55870	55990	56110	56229	56348	56466	56584	56702	119
37	56820	56937	57054	57170	57287	57403	57518	57634	57749	57863	116
38	57978	58092	58206	58319	58433	58546	58658	58771	58883	58995	113
39	59106	59217	59328	59439	59549	59659	59769	59879	59988	60097	110
40	60206	60314	60422	60530	60638	60745	60852	60959	61066	61172	107
41	61278	61384	61489	61595	61700	61804	61909	62013	62117	62221	104
42	62325	62428	62531	62634	62736	62838	62941	63042	63144	63245	102
43	63346	63447	63548	63648	63749	63849	63948	64048	64147	64246	99
44	64345	64443	64542	64640	64738	64836	64933	65030	65127	65224	98
45	65321	65417	65513	65609	65705	65801	65896	65991	66086	66181	96
46	66275	66370	66464	66558	66651	66745	66838	66931	67024	67117	95
47	67209	67302	67394	67486	67577	67669	67760	67851	67942	68033	92
48	68124	68214	68304	68394	68484	68574	68663	68752	68842	68931	90
49	69019	69108	69106	69284	69372	69460	69548	69635	69723	69810	88
50	69897	69983	70070	70156	70243	70329	70415	70500	70586	70671	86
51	70757	70842	70927	71011	71096	71180	71265	71349	71433	71516	84
52	71600	71683	71767	71850	71933	72015	72098	72181	72263	72345	82
53	72427	72509	72591	72672	72754	72835	72916	72997	73078	73158	81
54	73239	73319	73399	73480	73560	73639	73719	73798	73878	73957	80
55	74036	74115	74194	74272	74351	74429	74507	74585	74663	74741	78
56	74818	74896	74973	75050	75128	75204	75281	75358	75434	75511	77
57	75587	75663	75739	75815	75891	75960	76042	76117	76192	76267	75
58	76342	76417	76492	76566	76641	76715	76789	76863	76937	77011	74
59	77085	77158	77232	77305	77378	77451	77524	77597	77670	77742	73
60	77815	77887	77959	78031	78103	78175	78247	78318	78390	78461	72

TABLE OF LOGARITHMS.

No.	0	1	2	3	4	5	6	7	8	9	Prop.
61	78533	78604	78675	78746	78816	78887	78958	79028	79098	79169	71
62	79239	79309	79379	79448	79518	79588	79657	79726	79796	79865	70
63	79934	80003	80071	80140	80208	80277	80345	80414	80482	80550	69
64	80618	80685	80753	80821	80888	80956	81023	81090	81157	81224	68
65	81291	81358	81424	81491	81557	81624	81690	81756	81822	81888	67
66	81954	82020	82085	82151	82216	82282	82347	82412	82477	82542	66
67	82607	82672	82737	82801	82866	82930	82994	83058	83123	83187	64
68	83251	83314	83378	83442	83505	83569	83632	83695	83758	83822	63
69	83885	83947	84010	84073	84136	84198	84261	84323	84385	84447	6
70	84509	84571	84633	84695	84757	84819	84880	84942	85003	85064	62
71	85125	85187	85248	85309	85369	85430	85491	85552	85612	85673	61
72	85733	85793	86853	85913	85973	86033	86093	86153	86213	86272	60
73	86332	86391	86451	86510	86569	86628	86687	86746	86805	86864	59
74	86923	86981	87040	87098	87157	87215	87273	87332	87390	87440	58
75	87506	87564	87621	87679	87737	87794	87852	87909	87967	88024	57
76	88081	88138	88195	88252	88309	88366	88422	88479	88536	88592	57
77	88649	88705	88761	88818	88874	88930	88986	89042	89098	89153	56
78	89209	89265	89320	89376	89431	89487	89542	89597	89652	89707	55
79	89762	89817	89872	89927	89982	90036	90091	90145	90200	90254	54
80	90309	90363	90417	90471	90525	90579	90633	90687	90741	90794	54
81	90848	90902	90955	91009	91062	91115	91169	91222	91275	91328	53
82	91381	91434	91487	91540	91592	91645	91698	91750	91803	91855	53
83	91907	91960	92012	92064	92116	92168	92220	92272	92324	92376	52
84	92427	92479	92531	92582	92634	92685	92737	92788	92839	92890	51
85	92942	92993	93044	93095	93145	93196	93247	93298	93348	93399	51
86	93449	93500	93550	93601	93651	93701	93751	93802	93852	93902	50
87	93952	94001	94051	94101	94151	94200	94250	94300	94349	94398	49
88	94448	94497	94546	94596	94645	94694	94743	94792	94841	94890	49
89	94939	94987	95036	95085	95133	95182	95230	95279	95327	95376	48
90	95424	95472	95520	95568	95616	95664	95712	95760	95808	95856	48
91	95904	95951	95999	96047	96094	96142	96189	96237	96284	96331	48
92	96378	96426	96473	96520	96567	96614	96661	96708	96754	96801	47
93	96848	96895	96941	96988	97034	97081	97127	97174	97220	97266	47
94	97312	97359	97405	97451	97497	97543	97589	97635	97680	97726	46
95	97772	97818	97863	97909	97954	98000	98045	98091	98136	98181	46
96	98227	98272	98317	98362	98407	98452	98497	98542	98587	98632	45
97	98677	98721	98766	98811	98855	98900	98945	98989	99034	99078	45
98	99122	99167	99211	99255	99299	99343	99387	99431	99475	99519	44
99	99563	99607	99651	99695	99738	99782	99826	99869	99913	99956	44
No.	0	1	2	3	4	5	6	7	8	9	Prop.

By the use of these tables the logarithm of any number below 10,000 can be found with sufficient accuracy in the manner exemplified on the following page, and for most uses it will be found equally convenient as many much more extensive tables.

The use of the foregoing table is explained by the following rules:

RULES FOR LOGARITHMS.

To multiply by logarithms add the logarithms together and find number of logarithms so found.

To divide by logarithms subtract one from the other.

To extract the roots, divide the logarithms by the index of the root.

To raise a number to any power, multiply the logarithms by the index.

Find Log. of 5065
Log. of 5060 = 3.70415
Prop 86 x Diff. 5 = 430
Log. required = 3.704580

Find Number of Log. 3.771442
Log. of 5900 = 3.770850
Diff. 592 + Prop. 73 = 8 Diff. = 592
No. required 5908

Indices of Logarithms.

Log.	4030	= 3.60530
"	403	= 2.60530
"	40.3	= 1.60530
Log	4.03	= .60530
"	.403	= $\bar{1}$.60530
"	.0403	= $\bar{2}$.60530
"	.00403	= $\bar{3}$.60530

Log. nat. n=2.3026 log. n.

WEIGHTS AND MEASURES.

TROY WEIGHT.

24 grains	1 pennyweight: dwt.
20 pennyweights	1 ounce=480 grains.
12 ounces	1 pound=240 dwts.=5,760 grains.

AVOIRDUPOIS OR COMMERCIAL WEIGHT.

27.34375 grains	1 drachm.
16 drachms	1 ounce=437.5 grains.
16 ounces	1 pound=256 drachms=7,000 grains.
28 pounds	1 quarter=448 ounces.
4 quarters	1 cwt=112 pounds.
20 cwts.	1 ton=80 quarters=2,240 lbs.

APOTHECARIES' WEIGHT.

20 grains	1 scruple.	8 drachms	1 ounce.
3 scruples.	1 drachm.	12 ounces	1 pound.

The grain in each of the foregoing tables is the same.
An avoirdupois pound of pure water has the following volumes.

At 32° F.=.016021 cu. ft. or 27.684 cu. ins.
39.1° " =.016019 " " " 27.680 " "
62° " =.016037 " " " 27.712 " "
212° " =.016770 " " " 28.978 " "

—*D. K. Clark, Rules, Tables and Data.*

LONG MEASURE.

By law the U. S. standards of length and weight are made equal to the British.

12 inches	1 foot.
3 feet	1 yard = 36 ins. = .9143919 metre.
5½ yards	1 rod, pole or perch = 16½ feet.
40 rods	1 furlong.
8 furlongs	1 mile = 5,280 feet = 63,360 ins.
3 miles	1 league.

A palm = 3 ins. A hand = 4 ins. A span = 9 ins.
A fathom = 6 ft. A cable's length = 120 fathoms.

A Gunter's chain is 66 ft. long, and 80 Gunter's chains = 1 mile.
In the U. S. a nautical mile is 1.15157 times a common mile.

INCHES AND THEIR EQUIVALENT DECIMAL VALUES IN PARTS OF A FOOT.

Inches.	Fraction of Foot.	Decimal Part of Foot.
1	1/12	.0833
2	1/6	.1667
3	1/4	.25
4	1/3	.3333
5	5/12	.4167
6	1/2	.5
7	7/12	.5833
8	2/3	.6667
9	3/4	.75
10	5/6	.8333
11	11/12	.9167
12	1	1.0

SQUARE OR LAND MEASURE.

144 sq. ins. = 1 sq. foot.
9 sq. ft. = 1 sq. yard.
30¼ sq. yds. = 1 sq. rod.
40 sq. rods = 1 rood.
4 roods = 1 acre = 43560 sq. ft.

In the United States surveys a SECTION OF LAND is one mile square, or 640 acres.

A square acre is 208.71 feet on each side.

A circular acre is 235.504 feet in diameter.

CUBIC OR SOLID MEASURE.

1,728 cubic inches = 1 cubic foot.
27 cubic feet = 1 cubic yard.

A cord of wood, being 4×4×8 feet, contains 128 cubic feet. A ton, 2,240 pounds of Pennsylvania anthracite coal, in size for domestic use, occupies from 41 to 43 cubic feet; bituminous coal, 44 to 48 cubic feet; coke, 80 cubic feet.

LIQUID MEASURE.

4 gills = 1 pint.
2 pints = 1 quart.
4 quarts = 1 gallon = 231 cubic inches.

A cylinder 3½ inches in diameter and 6 inches high will hold almost exactly one quart, and one 7 inches in diameter and 6 inches high will hold very nearly one gallon.

This United States gallon is only .8333 of the British imperial gallon. A cubic foot contains about 7½ United States gallons.

DRY MEASURE.

2 pints = 1 quart.
8 quarts = 1 peck.
4 pecks = 1 bushel.

Four quarts in dry measure contain 268.8 cubic inches, or .96945 of the British imperial gallon. The flour barrel should contain 3.75 cubic feet and 196 pounds.

THE METRIC STANDARDS OF WEIGHTS AND MEASURES.

The primary metric standards are: The meter, the unit of length, and the kilogramme, the unit of weight, derived from the meter, being the two platinum standards deposited at the Palais des Archives at Paris. This standard meter is alleged to be equal to the one-ten-millionth part of the quadrant of the meridian of the earth.

METRIC MEASURES OF LENGTH.

10 millimetres	= 1 centimetre
10 centimetres	= 1 decimetre
10 decimetres 100 centimetres 1,000 millimetres	= 1 METRE
10 metres	= 1 decametre
10 decametres	= 1 hectometre
10 hectometres	= 1 KILOMETRE
10 kilometres	= 1 myriametre

A table of METRIC MEASURES OF SURFACE is obtained from the foregoing table by squaring the numbers, and placing the word "square" before each of the names; thus, 100 square millimetres= 1 square centimetre. And A TABLE FOR VOLUMES is obtained by cubing the numbers, and placing the word "cubic" before the names; thus, 1,000 cubic millimetres = 1 cubic centimetre.

FOR MEASURES OF CAPACITY the unit is the litre, and the table is—

10 centilitres	= 1 decilitre
10 decilitres	= 1 LITRE
10 litres	= 1 decalitre

and a litre contains 1 cubic decimetre. This portion of the capacity table belongs especially to the measurement of liquids.

FOR DRY MEASURES the table is contained and we have—

	10 litres	= 1 decalitre
10 decalitres or	100 litres	= 1 hectolitre
10 hectolitres or	1,000 litres	= 1 kilolitre = 1 cu. metre.

METRIC MEASURES OF WEIGHT.

10 milligrames	= 1 centigramme
10 centigrammes	= 1 decigramme
10 decigrammes	= 1 GRAMME
10 grammes	= 1 decagramme
10 decagrammes	= 1 hectogramme
10 hectogrammes or 1,000 grammes	= 1 KILOGRAMME
10 kilogrammes	= 1 myriagramme
10 myriagrammes 100 kilogrammes	= 1 quintal metrigue
10 quintaux 1,000 kilogrammes	= 1 millier or tonne.

A millier or tonne is the weight of 1 cubic metre of water at 39.1° F.

APPROXIMATE EQUIVALENTS OF FRENCH AND ENGLISH MEASURES.

1 inch	25 millimeters (exactly 25.4).
1 yard	11-12 meter.
1 kilometer	⅝ mile.
1 mile	1.6 or 1 3-5 kilometers.
1 square yard	6-7 square meter.
1 acre	4,000 square meters.
1 gallon	4½ liters fully.
1 cubic foot	28.3 liters.
1 cubic meter of water	1 ton nearly.
1 gramme	15½ grains nearly.
1 kilogramme	2.2 pounds fully.

SPECIFIC GRAVITY AND WEIGHT OF MATERIALS.

METALS.

	Specific Gravity.	Weight per cu. ft.	Cu. ft. in one ton.
Aluminum	2.6	162	13.3
Antimony, cast, 6.66 to 6.74	6.7	418	5.3
Bismuth, cast and native	9.74	607	3.6
Brass, copper and zinc, cast 7.8 to 8.4	8.1	504	4.4
Brass, rolled	8.4	524	4.2
Bronze, copper, 8, and tin, 1; gun metal, 8.4 to 8.6	8.5	529	4.2
Copper, cast, 8.6 to 8.8	8.7	542	4.1
Copper, rolled, 8.7 to 8.9	8.8	549	4.0

METALS—CONTINUED.

	Specific Gravity.	Weight per cu. ft.	Cu. ft. in one ton.
Gold, cast, pure or 24 carat	19.258	1204	1.86
Iron, cast, 6.9 to 7.4	7.21	450	4.8
" wrought, 7.6 to 7.9	7.77	485	4.6
" large rolled bars..............	7.69	480	4.6
" sheet		485	4.6
Lead......................................	11.4	712	3.15
Mercury at 32° F.	13.62	849	2.6
" " 60° F.......................	13.58	846	2.6
" " 212° F.......................	13.38	836	2.6
Platinum, 21 to 22	21.5	1,342	1.6
Silver..............	10.5	655	3.4
Steel, crucible, average.............	7.842	489	4.5
" cast, "	7.848	489.3	4.5
" Bessemer	7.852	489.6	4.5
Spelter or zinc, 6.8 to 7.2	7.00	437.5	5.1
Tin, cast, 7.2 to 7.5....................	7.35	459.—	4.8
Type metal..............................	10.45	653.—	3.4

WOODS.

	Specific Gravity.	Weight per cu. ft.	Cu. ft. in one ton.
Ash, perfectly dry.................	.752	47.—	1.748
Ash, American white, dry.........	.61	38	1.414
Chestnut, perfectly dry............	.66	41	1.525
Elm " "	.56	35	1.302
Hemlock " "	.40	25.—	.930
Hickory " "	.85	53	1.971
Maple, dry............................	.79	49	
Oak, live, dry.........................	.95	59.3	
" white, dry.........................	.70	44	
" red.....		32 to 45	–
Pine, white............................	.40	25	.930
Pine, yellow, southern.............	.72	45	1.674
Sycamore, perfectly dry........ ...	.59	37	1.370
Spruce, " "	.40	25	.930

STONES AND MINERALS.

	Specific Gravity.	Weight per cu. ft.	Cu. ft. in one ton.
Granite, syenite, gneiss	2.36 to 2.96	147.1 to 184.6	12.1
" gray...................	2.80 to 3.06	174.6 to 190.8	11.8
Graphite..............................	2.20	137.2	16.3
Gypsum, plaster of Paris..........	2.27	141.6	15.8
" in irregular lumps.......		82	
Greenstone, trap, 2.8 to 3.2.........	3.—	187	
Limestones and marbles, 2.4 to 2.86	2.6	164.4	13.6
Limestones and marbles, they are frequently	2.7	168.0	13.3
Quicklime, ground, loose, per struck bushel, 62 to 70 lbs		53	42.2
Quartz, common, finely pulverized, loose		90	24.8
Sand, with its natural moisture and loose.......................		.85 to .90	24.8
Sand, pure, quartz, perfectly dry..	1.7	106	
Sand, perfectly wet, voids full of water		118 to 129	17.3
Sandstones, fit for building, dry, 2.1 to 2.73	2.41	150.—	
Standstones, quarried and piled. One measure, solid, makes 1¾ piled		86	26.
Serpentines	2.81	175.2	12.8
Shales, red or black, 2.4 to 2.8	2.6	162	
" quarried in piles............		92	24.3
Slate, 2.7 to 2.9........................	2.8	175	12.8
Soapstone or steatite, 2.65 to 2.8....	2.73	170	13.1
Air, atmosphere at 60° F., Barom. 30"	.00123	.0765	
Alcohol, pure........................	.793	49.43	
" of commerce.............	.834	52.10	
" proof spirit...............	.916	57.2	
Alabaster, a compact plaster of Paris...........	2.31	144.0	

STONES AND MINERALS—CONTINUED.

	Specific Gravity.	Weight per cu. ft.	Cu. ft. in gross ton.
Anthracite, solid, 1.3 to 1.84, average	1.50	93.5	
Asphaltum	1.4	87.3	25.6
Carbonic acid gas, 1½ times as heavy as air	.00187		
Charcoal of pines and oaks		15 to 30	74.6
Clay, potters', dry, 1.8 to 2.1	1.9	119	18.8
Coke, loose, of good coal		23 to 32	
Cement, English, Portland	1.25 to 1.51	78 to 92	23.8 to 28.7
Cork	.25	15.6	
Cork (comminuted)		6.0	
Earth, common loam, perfectly dry, shaken moderately		82 to 92	
Earth, common loam, more moist, packed		90 to 100	
Earth, common loam, as a soft flowing mud		104 to 112	
Fat	.93	58	
Glass, 2.5 to 3.45	2.98	186	
Gutta percha	.98	61.1	
Hydrogen gas is 14.5 times lighter than air and 16 times lighter than oxygen		.00527	
Ice, at 32° F	.92	57.5	38.9
India rubber	.93	58	
Lard	.95	59.3	
Masonry, of granite or limestones, well dressed		165	13.57
Masonry, of brickwork, pressed brick, fine joints		140	16.—
Masonry, of brickwork, coarse, soft bricks		100	22.4
Mortar, hardened, 1.4 to 1.9	1.65	103	
Naphtha	.848	52.9	
Nitrogen gas is about 1-35 part lighter than air		.0744	
Oils, whale, olive	.92	57.3	
Oxygen gas, a little more than 1-10 heavier than air	.00136	.0846	
Petroleum	.878	54.8	40.87
Pitch	1.15	71.7	
Rosin	1.1	68.6	32.65
Salt, coarse, per struck bushel, Syracuse, N. Y., 56 lbs		45.—	49.77
Salt, coarse, per struck bushel, St. Barts, 84 to 90		70.—	32.—
Salt, coarse, per struck bushel, well dried, W. I. 90 to 96		74	
Sand		90 to 106	
Snow, fresh fallen		5 to 12	
" moistened and compacted by rain		15 to 50	
Sulphur	2	125	
Tallow	.94	58.6	
Tar	1.—	62.4	
*Water, pure rain, or distilled, at 32° F. Barom. 30"		62.416	
60° F. " "	1.—	62.366	35.918
80° F. " "		62.217	
Water, sea, 1.026 to 1.030	1.028	64.08	34.96
Wax, bees	.97	60.5	
Gypsum, plaster of Paris	2.27	141.6	15.8
" in irregular lumps		82.	
Gas (natural)		0.0316	
Limestones and marbles, 2.4 to 2.86.	2.6	164.4	13.6
" " they are frequently	2.7	168.0	13.3
Lime-quick, ground, loose, per struck bushel, 62 to 70 pounds		53.	42.2
Quartz, common, finely pulverized, loose		90.	24.8

TABLE OF CONTENTS IN CUB. FEET AND IN U. S. GALLON,

(From Trautwine.)

Of 231 cubic inches (or 7.4805 gallons to a cubic foot); and for one foot of length of the cylinder. For the contents for a greater diameter than any in the table, take the quantity opposite one-half said diameter and multiply it by 4. Thus, the number of cubic feet in one foot length of a pipe eighty inches in diameter is equal to 8.728×4=34.912 cub. ft. So also with gallons and areas.

Diameter in Inches.	Diameter in decimals of a foot.	For 1 foot in length.		Diameter in Inches.	Diameter in decimals of a foot.	For 1 foot in length.	
		Cubic feet. Also area in square feet.	Gallons of 231 cub. in.			Cubic feet. Also area in square feet.	Gallons of 231 cub. in.
¼......	.0208	.0003	.0026	¾......	.5625	.2485	1.859
5-16......	.0260	.0005	.0040	7.	.5833	.2673	1.999
⅜......	.0313	.0008	.9057	¼......	.6042	.2868	2.144
7-16......	.0365	.0010	.0078	½......	.6250	.3068	2.295
½......	.0417	.0014	.0102	¾......	.6458	.3275	2.450
9-16......	.0469	.0017	.0129	8.	.6667	.3490	2.611
⅝......	.0521	.0021	.9159	¼......	.6875	.3713	2.777
11-16......	.0573	.0026	.0193	½......	.7083	.3940	2.948
¾......	.0625	.0031	.0230	¾......	.7292	.4175	3.125
13-16......	.0677	.0036	.0270	9.	.7500	.4418	3.305
⅞......	.0729	.0042	.0312	¼......	.7708	.4668	3.492
15-16......	.0781	.0048	.0359	½......	.7917	.4923	3.682
1.	.0833	.0055	.0408	¾......	.8125	.5185	3.879
¼......	.1042	.0085	.0638	10.	.8333	.5455	4.081
½......	.1250	.0123	.0918	¼......	.8542	.5730	4.286
¾......	.1458	.0168	.1250	½......	.8750	.6013	4.498
2.	.1667	.0218	.1632	¾......	.8958	.6303	4.714
¼......	.1875	.0276	.2066	11.	.9167	.6600	4.937
½......	.2083	.0341	.2550	¼......	.9375	.6903	5.163
¾......	.2292	.0413	.3085	½......	.9583	.7213	5.395
3.	.2500	.0491	.3673	¾......	.9792	.7530	5.633
¼......	.2708	.0576	.4310	12.	1 Foot	.7854	5.876
½......	.2917	.0668	.4998	½......	1.042	.8523	6.375
¾......	.3125	.0767	.5738	13.	1.083	.9218	6.895
4.	.3333	.0873	.6528	½......	1.125	.9940	7.435
¼......	.3542	.0985	.7370	14.	1.167	1.069	7.997
½......	.3750	.1105	.8263	½......	1.208	1.147	8.578
¾......	.3958	.1231	.9205	15.	1.250	1.227	9.180
5.	.4167	.1364	1.020	½......	1.292	1.310	9.801
¼......	.4375	.1503	1.124	16.	1.333	1.396	10.44
½......	.4583	.1650	1.234	½......	1.375	1.485	11.11
¾......	.4792	.1803	1.349	17.	1.417	1.576	11.79
6.	.5000	.1963	1.469	½......	1.458	1.670	12.50
¼......	.5208	.2130	1.594	18.	1.500	1.767	13.22
½......	.5417	.2305	1.724	½......	1.542	1.867	13.97

TABLE OF GALLONS.

	Cubic inch. in a gallon.	Weight of a gallon in pounds avoirdupois.	Gallons in a cubic foot.	Weight of a cubic foot of water, English standard, 62.3210286 lbs. avoirdupois.
United States.	231.	8.33111	7.480519	
New York....	231.81918	8.00	7.901285	
Imperial......	277.274	10.00	6.232102	

COMPARISON OF WEIGHTS AND MEASURES.

METRIC SYSTEM.	U. S. STANDARD.
LENGTH.	LENGTH.
1 millimeter = .0394 inches.	1 inch = 2.5309 centimeters.
1 centimeter = .3937 inches.	1 foot = 30.4794 centimeters.
1 METER = 39.3708 inches.	1 yard = .9143 meters.
1 kilometer = .6214 miles.	1 mile = 1.6093 kilometers.
SQUARE.	SQUARE.
1 sq. centimeter = .1549 sq. in.	1 sq. in. = 6.4513 sq. centimeters.
1 sq. meter = 10.7631 sq. ft.	1 sq. ft. = .0929 sq. meters.
1 ARE = 119.5894 sq.yds.	1 sq. yd. = .8361 sq. meters.
1 hectare = 2.4711 acres.	1 acre = .4047 hectares.
CUBIC.	CUBIC.
1 CUBIC METER = 35.3166 cubic ft.	1 cubic foot = .02831 cubic meters
WEIGHT.	WEIGHT.
1 gram = 15.4323 grains.	1 lb. = .4536 kilos.
1 KILOGRAM = 2.2046 lbs.	1 cwt. = 50.8024 kilos.
1 tonneau = 2204.55 lbs.	1 ton = 1016.0483 kilos.
DRY MEASURE.	DRY MEASURE.
1 centiliter = .0181 pints.	1 pint = 55.0661 centiliters.
1 LITER = .908 quarts.	1 quart = 1.1013 liters.
1 hectoliter = 2.837 bushels.	1 bushel = 35.2416 liters.
LIQUID MEASURE.	LIQUID MEASURE.
1 centiliter = .0211 pints.	1 pint = 47.3171 centiliters.
1 LITER = 1.0567 quarts.	1 quart = .9563 liters.
1 hectoliter = 26.4176 gallons.	1 gallon = 3.7854 liters.

COMPARISON OF ALCOHOLOMETERS.

In the absence of a specific gravity or Beaumé scale, an alcoholometer may also be used for ascertaining the strength of ammonia liquor. The accompanying table is to be used in connection with the table on page 97 for this purpose.

Specific gravity.	Per cent Tralles (by volume).	Per cent Richter (by weight).	Per cent Gendar United States.
0.793	100	100	100
0.815	95	91.5	90
0.832	90	85	80
0.848	85	79.1	70
0.863	80	74.2	60
0.876	75	68.4	50
0.889	70	62.5	40
0.901	65	57.3	30
0.912	60	51.7	20
0.923	55	46.5	10
0.933	50	42.0	P
0.942	45	37.7	10
0.951	40	33.0	20
0.958	35	28.7	30
0 964	30	24.4	40
0.970	25	20.2	50
0.975	20	16.4	60
0.980	15	13.0	70
0.985	10	10 4	80
0.991	5	6.0	90
0.999	0	1.0	100

P in the last column stands for proof spirits. Percentage over proof U. S. gendar scale can be converted into per cent Tralles by dividing by two and adding fifty. Degrees below proof are converted by dividing by two and subtracting from fifty.

HORSE POWER OF BELTING.

TABLE FOR SINGLE LEATHER, 4-PLY RUBBER AND 4-PLY COTTON BELTING, BELTS NOT OVERLOADED. (ONE INCH WIDE, 800 FEET PER MINUTE = 1-HORSE POWER.)

Speed in Ft. Per Minute.	Width of Belts in Inches.											
	2	3	4	5	6	8	10	12	14	16	18	20
	h.p	h.p	h.p	h.p	h.p	h.p	h.p	h.p	h.p	h.p	h. p.	h. p.
400	1	1½	2	2½	3	4	5	6	7	8	9	10
600	1½	2¼	3	3¾	4½	6	7½	9	10½	12	13½	15
800	2½	3	4	5	6	8	10	12	14	16	18	20
1,000	2	3¾	5	6¼	7½	10	12½	15	17½	20	22½	25
1,200	3	4½	6	7½	9	12	15	18	21	24	27	30
1,500	3¾	5¾	7½	9½	11½	15	18¾	22½	26½	30	33¾	37½
1,800	4½	6¾	9	11¼	13½	18	22½	27	31½	36	40½	45
2,000	5	7½	10	12½	15	20	25	30	35	40	45	50
2,400	6	9	12	15	18	24	30	36	42	48	54	60
2,800	7	10½	14	17½	21	28	35	42	49	56	63	70
3,000	7½	11¼	15	18¾	22½	30	37½	45	52½	60	67½	75
3,500	8¾	13	17½	22	26	35	44	52½	61	70	79	88
4,000	10	15	20	25	30	40	50	60	70	80	90	100
4,500	11¼	17	22½	28	34	45	57	69	78	90	102	114
5,000	12½	19	25	31	37½	50	62½	75	87½	100	112	125

Double leather, 6-ply rubber or 6-ply cotton belting will transmit 50 to 75 per cent more power than is shown in this table.

A simple rule for ascertaining transmitting power of belting, without first computing speed per minute that it travels, is as follows: Multiply diameter of pulley in inches by its number of revolutions per minute, and this product by width of the belt in inches; divide this product by 3,300 for single belting, or by 2,100 for double belting, and the quotient will be the amount of horse power that can be safely transmitted.

HORSE POWER OF SHAFTING.

Diameter of Shaft in Inches.	Revolutions per Minute.				
	100	125	150	175	200
	h. p.	h. p.	h. p.	h. p.	h. p.
15-16	1.2	1.4	1.7	2.1	2.4
1 3-16	2.4	3.1	3.7	4.3	4.9
1 7-16	4.3	5.3	6.4	7.4	8.5
1 11-16	6.7	8.4	10.1	11.7	13.4
1 15-16	10.0	12.5	15.0	17.5	20.0
2 3-16	14.3	17.8	21.4	24.9	28.5
2 7-16	19.5	24.4	29.3	34.1	39.0
2 11-16	26.0	32.5	39.0	43.5	52.0
2 15-16	33.8	42.2	50.6	59.1	67.5
3 3-16	43.0	53.6	64.4	75.1	85.8
3 7-16	53.6	67.0	79.4	93.8	107.2
3 11-16	65.9	82.4	97.9	115.4	121.8
3 15-16	80.0	100.0	120.0	140.0	160.0
4 7-16	113.9	142.4	170.8	199.8	227.8
4 15-16	156.3	195.3	234.4	273.4	312.5

CAPACITIES OF TANKS IN BARRELS OF 31 GALLONS.

DIAMETER IN FEET.

DEPTH.	5 feet	6 feet	7 feet	8 feet	9 feet	10 feet	11 feet	12 feet	13 feet	14 feet	15 feet	16 feet	17 feet	18 feet	19 feet	20 feet
5 feet	23.3	33.6	45.7	59.7	75.5	93.2	112.8	134.3	157.6	182.8	209.8	238.7	269.5	302.1	336.6	373.0
6 feet	28.0	40.3	54.8	71.7	90.6	111.0	135.4	161.1	189.1	219.3	251.8	286.5	323.4	362.6	404.0	447.6
7 feet	32.7	47.0	64.0	83.6	105.7	130.6	158.0	188.0	220.6	255.9	293.7	334.2	377.3	423.0	471.3	522.2
8 feet	37.3	53.7	73.1	95.5	120.9	149.1	180.5	214.8	252.1	292.4	335.7	382.0	431.2	483.4	538.6	596.8
9 feet	42.0	60.4	82.2	107.4	136.0	167.9	203.1	241.7	283.7	329.0	377.7	429.7	485.1	543.8	605.9	671.4
10 feet	46.7	67.1	91.4	119.4	151.1	186.5	225.7	268.6	315.2	365.5	419.6	477.4	539.0	604.3	673.3	746.0
11 feet	51.3	73.9	100.5	131.3	166.2	205.1	248.2	295.4	346.7	402.1	461.6	525.2	592.9	667.7	740.6	820.6
12 feet	56.0	80.6	109.7	143.2	181.3	223.8	270.8	322.3	378.2	438.6	503.5	572.0	646.8	725.1	807.9	895.2
13 feet	60.7	87.3	118.8	155.2	196.4	242.4	293.4	349.1	409.7	475.2	545.5	620.7	700.7	785.5	875.2	969.8
14 feet	65.3	94.0	127.9	167.1	211.5	261.1	315.9	376.0	441.3	511.0	587.5	668.2	754.6	846.0	942.6	1044.4
15 feet	70.0	100.7	137.1	179.0	226.6	289.8	338.5	402.8	472.8	548.3	629.4	716.2	808.5	906.4	1009.9	1119.0
16 feet	74.7	107.4	146.2	191.0	241.7	298.4	361.1	429.7	504.3	584.9	671.4	773.9	862.4	966.8	1077.2	1193.6
17 feet	79.3	114.1	155.4	202.9	256.8	317.0	383.6	456.6	535.8	621.4	713.4	811.6	916.3	1027.2	1144.6	1268.2
18 feet	84.0	120.9	164.5	211.8	272.0	335.7	406.2	483.4	567.3	658.0	755.3	859.4	970.2	1087.7	1211.9	1342.8
19 feet	88.7	127.6	173.6	226.8	287.0	354.3	428.8	510.3	598.0	694.5	797.3	907.1	1024.1	1148.1	1279.2	1417.4
20 feet	93.3	134.3	182.8	238.7	302.1	373.0	451.3	537.1	634.4	731.1	839.3	954.9	1078.0	1208.5	1346.5	1492.0

TABLE FOR CONVERTING FEET HEAD OF WATER INTO PRESSURE PER SQUARE INCH.

Feet. Head.	Pounds per square inch.	Feet. Head.	Pounds per square inch.	Feet. Head.	Pounds per square inch.
1	.43	55	23.82	190	82.29
2	.87	60	25.99	200	86.62
3	1.30	65	28.15	225	97.45
4	1.73	70	30.32	250	108.27
5	2.17	75	32.48	275	119.10
6	2.60	80	34.65	300	129.93
7	3.03	85	36.81	325	140.75
8	3.40	90	38.98	350	151.58
9	3.90	95	41.14	375	162.41
10	4.33	100	43.31	400	173.24
15	6.50	110	47.64	500	216.55
20	8.66	120	51.97	600	259.85
25	10.83	130	56.30	700	303.16
30	12.99	140	60.63	800	346.47
35	15.16	150	64.96	900	389.78
40	17.32	160	69.29	1000	433.09
45	19.49	170	73.63	....	......
50	21.65	180	77.96	....	......

1 lb. pressure	per square inch is equivalent to a head of water of................	2.3093 feet.
2 lb. "		27.71 inches.
14.7 lbs. or 1 atmosphere,		33.947 feet.
14.7 lbs. "		10.347 meters.
0.433 lbs. "		1 foot.
43.3 lbs. "		100 feet.

TABLE OF THEORETICAL HORSE POWER REQUIRED TO RAISE WATER TO DIFFERENT HEIGHTS.

Feet.	5	10	15	20	25	30	35	40	45	50	60
Gals. per Minute.											
5	.006	.012	.019	.025	.031	.037	.044	.05	.06	.06	.07
10	.012	.025	.037	.050	.062	.075	.087	.10	.11	.12	.15
15	.019	.037	.056	.075	.094	.112	.131	.15	.17	.19	.22
20	.025	.050	.075	.100	.125	.150	.175	.20	.22	.25	.30
25	.031	.062	.093	.125	.156	.187	.219	.25	.28	.31	.37
30	.037	.075	.112	.150	.187	.225	.262	.30	.34	.37	.45
35	.043	.087	.131	.175	.219	.262	.306	.35	.39	.44	.52
40	.050	.100	.150	.200	.250	.300	.350	.40	.45	.50	.60
45	.056	.112	.168	.225	.281	.337	.394	.45	.51	.56	.67
50	.062	.125	.187	.250	.312	.375	.437	.50	.56	.62	.75
60	.075	.150	.225	.300	.375	.450	.525	.60	.67	.75	.90
75	.093	.187	.281	.375	.469	.562	.656	.75	.84	.94	1.12
90	.112	.225	.337	.450	.562	.675	.787	.90	1.01	1.12	1.35
100	.125	.250	.375	.500	.625	.750	.875	1.00	1.12	1.25	1.50
125	.156	.312	.469	.625	.781	.937	1.094	1.25	1.41	1.56	1.87
150	.187	.375	.562	.750	.937	1.125	1.312	1.50	1.69	1.87	2.25
175	.219	.437	.656	.875	1.093	1.312	1.531	1.75	1.97	2.19	2.62
200	.250	.500	.750	1.000	1.250	1.500	1.750	2.00	2.25	2.50	3.00
250	.312	.625	.937	1.250	1.562	1.875	2.187	2.50	2.81	3.12	3.75
300	.375	.750	1.125	1.500	1.875	2.250	2.625	3.00	3.37	3.75	4.50
350	.437	.875	1.312	1.750	2.187	2.625	3.062	3.50	3.94	4.37	5.25
400	.500	1.000	1.500	2.000	2.500	3.000	3.500	4.00	4.50	5.00	6.00
500	.625	1.250	1.875	2.500	3.125	3.750	4.375	5.00	5.62	6.25	7.50

TABLE SHOWING LOSS OF PRESSURE IN POUNDS PER SQUARE INCH BY FRICTION IN CLEAR IRON PIPES FOR EACH 100 FEET IN LENGTH WHILE DISCHARGING GIVEN QUANTITIES OF WATER. ALSO VELOCITY OF FLOW IN FEET PER SECOND.

Gallons Discharged per Minute.	½ Inch.		¾ Inch.		1 Inch.		1¼ Inch.		1½ Inch.		2 Inch.		2½ Inch.		3 Inch.		4 Inch.		6 Inch.		Gallons Discharged per Minute.
	Veloc. in Pipe per Second.	Friction Loss in Pounds.	Veloc. in Pipe per Second.	Friction Loss in Pounds.	Veloc. in Pipe per Second.	Friction Loss in Pounds.	Veloc. in Pipe per Second.	Friction Loss in Pounds.	Veloc. in Pipe per Second.	Friction Loss in Pounds.	Veloc. in Pipe per Second.	Friction Loss in Pounds.	Veloc. in Pipe per Second.	Friction Loss in Pounds.	Veloc. in Pipe per Second.	Friction Loss in Pounds.	Veloc. in Pipe per Second.	Friction Loss in Pounds.	Veloc. in Pipe per Second.	Friction Loss in Pounds.	
5	8.17	24.6	3.63	3.3	2.04	0.84	1.31	0.31	0.91	0.12											5
10	16.3	96.0	7.25	13.0	4.08	3.16	2.61	1.05	1.82	0.47	1.02	0.12									10
15			10.9	28.7	6.13	6.98	3.92	2.38	2.73	0.97											15
20			14.5	50.4	8.17	12.3	5.22	4.07	3.63	1.66	2.04	0.42									20
25			18.1	78.0	10.2	19.0	6.53	6.40	4.54	2.62			1.63	0.21	1.13	0.10					25
30					12.8	27.5	7.84	9.15	5.45	3.75	3.06	0.91									30
35					14.3	37.0	9.14	12.04	6.36	5.05											35
40					16.3	48.0	10.4	16.1	7.26	6.52	4.09	1.60									40
45							11.7	20.2	8.17	8.15											45
50							13.1	24.9	9.08	10.0	5.11	2.44	3.26	0.81	2.27	0.35	1.28	0.09			50
75							19.6	56.1	13.6	22.4	7.66	5.82	4.90	1.80	3.40	0.74					75
100									18.2	39.0	10.2	9.46	6.53	3.20	4.54	1.31	2.55	0.33	1.13	0.05	100
125											12.8	14.9	8.16	4.89	5.67	1.99					125
150											15.3	21.2	9.80	7.00	6.81	2.85	3.83	0.69	1.70	0.10	150
175											17.1	28.1	11.4	9.46	7.94	3.85					175
200											20.4	37.5	13.1	12.47	9.08	5.02	5.11	1.22	2.27	0.17	200
250													16.3	19.66	11.3	7.76	6.39	1.89	2.84	0.26	250
300													19.6	28.06	13.6	11.2	7.66	2.66	3.40	0.37	300
350															15.9	15.2	8.94	3.65	3.97	0.50	350
400															18.2	19.5	10.2	4.73	4.54	0.65	400
450															20.4	25.0	11.5	6.01	5.11	0.81	450
500															22.7	30.8	12.8	7.43	5.67	0.96	500
750																			8.51	2.21	750
1,000																			11.3	3.88	1,000

FLOW OF STEAM THROUGH PIPES.

Initial Pressure per Square Inch.	Diameter of Pipe in inches. Length of each Pipe, 240 Diameters.						
	¾	1	1¼	2	2½	3	4
	Weight of Steam per Minute in Pounds, with One Pound Fall of Pressure.						
Lbs.	Lbs.	Lbs.	Lbs.	Lbs.	Lbs.	Lbs.	Lbs.
1	1.16	2.07	5.7	10.27	15.45	25.38	46.85
10	1.44	2.57	7.1	12.72	19.15	31.45	48.05
20	1.70	3.02	8.3	14.94	22.49	36.94	68.20
30	1.91	3.40	9.4	16.84	25.35	41.63	76.84
40	2.10	3.74	10.3	18.51	27.87	45.77	84.49
50	2.27	4.04	11.2	20.01	30.13	49.48	91.34
60	2.43	4.32	11.9	21.38	32.19	52.87	97.60
70	2.57	4.58	12.6	22.65	34.10	56.00	103.37
80	2.71	4.82	13.3	23.82	35.87	58.91	108.74
90	2.83	5.04	13.9	24.92	37.52	61.02	113.74
100	2.95	5.25	14.5	25.96	39.07	64.18	118.47
120	3.16	5.63	15.5	27.85	41.93	68.87	127.12
150	3.45	6.14	17.0	30.37	45.72	75.09	138.61

For any other given length of pipe divide 240 by the given length in diameters and multiply the tubular values by the square root of the quotient, to give the flow for one pound fall of pressure.

For any other given fall of pressure multiply the tubular weight by the square root of the given fall of pressure.

HORSE POWER OF BOILERS.

Thirty pounds of water evaporated at seventy pounds steam pressure per hour from feed water at 100°=1 horse power. In calculating horse power of steam boilers consider for—

Tubular boilers, fifteen square feet of heating surface equivalent to one horse power.

Flue boilers, twelve square feet of heating surface =1 horse power.

Cylinder boilers, ten square feet of heating surface =1 horse power.

Doubling the diameter of a pipe increases its capacity four times; friction of liquids increases as the square of velocity.

To find the pressure, in square inches, of a column of water: Multiply the height of the column in feet by .434 approximately. Every foot elevation is equal to half pound pressure per square inch; this allows for ordinary friction.

WOOD'S TABLE OF SATURATED AMMONIA.*

Recalculated by GEORGE DAVIDSON, M. E.

Temperature.		Pressure, Absolute.		Gauge Pressure, Pound per Sq. Inch.	Heat of Vaporization, Thermal Units. h.	Volume of Vapor per Pound, Cubic Feet. v.	Volume of Liquid per Pound, Cubic Feet. v_1	Weight of Vapor in Pounds per Cubic Foot. w.	Weight of Liquid in Pounds per Cubic Foot. w_1	Temperature.
Degrees F. t.	Absolute. T.	Pounds per Sq. Foot. P.	Pounds per Sq. Inch. p.							Degrees F.
—40	420.66	1539.90	10.69	—4.01	579.67	24.388	.02348	.0410	42.589	—40
39	1	1584.43	11.00	—3.70	579.07	23.735	.02351	.0421	42.535	39
38	2	1630.03	11.32	—3.38	578.42	23.102	.02354	.0433	42.483	38
37	3	1676.71	11.64	—3.06	577.88	22.488	.02357	.0444	42.427	37
36	4	1724.51	11.98	—2.72	577.27	21.895	.02359	.0457	42.391	36
—35	425.66	1773.43	12.31	—2.39	576.68	21.321	.02362	.0469	42.337	—35
34	6	1823.50	12.66	—2.04	576.08	20.763	.02364	.0482	42.301	34
33	7	1874.73	13.02	—1.68	575.48	20.221	.02366	.0495	42.265	33
32	8	1927.17	13.38	—1.32	574.89	19.708	.02368	.0507	42.213	32
31	9	1980.78	13.75	—0.95	574.39	19.204	.02371	.0521	42.176	31
—30	430.66	2035.69	14.13	—0.57	573.69	18.693	.02374	.0535	42.123	—30
29	1	2091.83	14.53	—0.17	573.08	18.225	.02378	.0549	42.052	29
28	2	2149.23	14.92	+0.22	572.48	17.759	.02381	.0563	42,000	28
27	3	2207.94	15.33	+0.63	571.89	17.307	.02384	.0577	41.946	27
26	4	2267.97	15.75	+1.05	571.28	16.869	.02387	.0593	41.893	26
—25	435.66	2329.34	16.17	+1.47	570.68	16.446	.02389	.0608	41.858	—25
24	6	2392.09	16.61	1.91	570.08	16.034	.02392	.0624	41.806	24
23	7	2456.23	17.05	2.35	569.48	15.633	.02395	.0640	41.754	23
22	8	2520.45	17.50	2.8	568.88	15.252	.02398	.0656	41.701	22
21	9	2588.77	17.97	3.27	568.27	14.875	.02401	.0672	41.649	21
—20	440.66	2657.23	18.45	+3.75	567.67	14.507	.02403	.0689	41.615	—20
19	1	2727.17	18.94	4.24	567.06	14.153	.02406	.0706	41.563	19
18	2	2798.62	19.43	4.73	566.43	13.807	.02409	.0725	41.511	18
17	3	2871.61	19.94	5.24	565.85	13.475	.02411	.0742	41.480	17
16	4	2946.17	20.46	5.76	565.25	13.150	.02414	.0760	41.425	16
—15	445.66	3022.31	20.99	+6.29	564.64	12.834	.02417	.0779	41.374	—15
14	6	3100.07	21.53	6.83	564.04	12.527	.02420	.0798	41.322	14
13	7	3179.45	22.08	7.38	563.43	12.230	.02423	.0818	41.271	13
12	8	3260.52	22.64	7.94	562.82	11.939	.02425	.0838	41.237	12
11	9	3343.29	23.22	8.52	562.21	11.659	.02428	.0858	41.186	11
—10	450.66	3427.75	23.80	+9.10	561.61	11.385	.02431	.0878	41.135	—10
9	1	3513.97	24.40	9.70	560.99	11.117	.02434	.0899	41.084	9
8	2	3601.97	25.01	10.31	560.39	10.860	.02437	.0921	41.034	8
7	3	3691.75	25.64	10.94	559.78	10.604	.02439	.0943	41.000	7
6	4	3783.37	26.27	11.57	559.17	10.362	.02442	.0965	40.950	6
—5	455.66	3876.85	26.92	+12.22	558.56	10.125	.02445	.0988	40.900	—5
4	6	3972.62	27.59	12.89	557.94	9.894	.02448	.1011	40.845	4
3	7	4069.48	28.26	13.56	557.33	9.669	.02451	.1034	40.799	3
2	8	4168.70	28.95	14.25	556.73	9.449	.02454	.1058	40.749	2
1	9	4269.90	29.65	14.95	556.11	9.234	.02457	.1083	40.700	1
+0	460.66	4373.10	30.37	+15.67	555.50	9.028	.02461	.1107	40.650	+0
1	1	4478.32	31.10	16.40	554.88	8.825	.02463	.1133	40.601	1
2	2	4485.60	31.84	17.14	554.27	8.630	.02466	.1159	40.551	2
3	3	4694.96	32.60	17.90	553.65	8.436	.02469	.1186	40.502	3
4	4	4806.46	33.38	18.68	553.04	8.250	.02472	.1212	40.453	4

* For values at temperatures higher than 100° F. see Wood's table on page 92.

WOOD'S TABLE OF SATURATED AMMONIA—Continued.

Temperature.		Pressure, Absolute.		Gauge Pressure, Pound per Sq. Inch.	Heat of Vaporization, Thermal Units. h.	Volume of Vapor per Pound, Cubic Feet. v.	Volume of Liquid per Pound, Cubic Feet. v_1	Weight of Vapor in Pounds per Cubic Foot. w.	Weight of Liquid in Pounds per Cubic Foot. w_1	Temperature.
Degrees F. t.	Absolute. T.	Pounds per Sq. Foot. P.	Pounds per Sq. Inch. p.							Degrees F.
+5	465.66	4920.11	34.16	+19.46	552.43	8.070	.02475	.1240	40.404	+5
6	6	5035.95	34.97	20.27	551.81	7.892	.02478	.1267	40.355	6
7	7	5153.99	35.79	21.09	551.19	7.717	.02480	.1296	40.322	7
8	8	5274.28	36.63	21.93	550.58	7.553	.02483	.1324	40.274	8
9	9	5396.83	37.48	22.78	549.96	7.388	.02486	.1353	40.225	9
+10	470.66	5521.71	38.34	+23.64	549.35	7.229	.02490	.1383	40.160	+10
11	1	5649.48	39.23	24.53	548.73	7.075	.02493	.1413	40.112	11
12	2	5778.50	40.13	25.43	548.11	6.924	.02496	.1444	40.064	12
13	3	5910.52	41.04	26.34	547.49	6.786	.02499	.1474	40.016	13
14	4	6044.96	41.98	27.28	546.88	6.632	.02502	.1507	39.968	14
+15	475.66	6182.00	42.94	+28.24	546.26	6.491	.02505	.1541	39.920	+15
16	6	6321.24	43.90	29.20	545.63	6.355	.02508	.1573	39.872	16
17	7	6463.24	44.88	30.18	545.01	6.222	.02511	.1607	39.872	17
18	8	6607.77	45.89	31.19	544.39	6.093	.02514	.1641	39.777	18
19	9	6754.90	46.91	32.21	543.74	5.966	.02517	.1676	39.729	19
+20	480.66	6904.68	47.95	33.25	543.15	5.843	.02520	.1711	39.682	+20
21	1	7057.15	49.01	34.31	542.53	5.722	.02523	.1748	39.635	21
22	2	7211.33	50.09	35.39	541.90	5.605	.02527	.1784	39.572	22
23	3	7370.27	51.18	36.48	541.28	5.488	.02529	.1822	39.541	23
24	4	7530.96	52.30	37.60	540.66	5.378	.02533	.1860	39.479	24
+25	485.66	7694.52	53.43	+38.73	540.03	5.270	.02536	.1897	39.432	+25
26	6	7860.89	54.59	39.89	539.41	5.163	.02539	.1937	39.386	26
27	7	8030.16	55.76	41.06	538.78	5.058	.02542	.1977	39.339	27
28	8	8202.38	56.96	42.26	538.16	4.900	.02545	.2016	39.292	28
29	9	8377.56	58.17	43.47	537.53	4.858	.02548	.2059	39.246	29
+30	490.66	8555.74	59.42	+44.72	536.91	4.763	.02551	.2099	39.200	+30
31	1	8736.96	60.67	45.97	536.28	4.668	.02554	.2142	39.115	31
32	2	8921.26	61.95	47.25	535.66	4.577	.02557	.2185	39.108	32
33	3	9108.71	63.25	48.55	535.03	4.486	.02561	.2229	39.047	33
34	4	9299.32	64.58	49.88	534.40	4.400	.02564	.2273	39.001	34
+35	495.66	9493.07	65.92	+51.22	533.78	4.314	.02568	.2318	38.940	+35
36	6	9690.04	67.29	52.59	533.13	4.234	.02571	.2362	38.894	36
37	7	9890.75	68.68	53.98	532.52	4.157	.02574	.2413	38.850	37
38	8	10093.91	70.09	55.39	531.89	4.068	.02578	.2458	38.789	38
39	9	10300.88	71.53	56.83	531.26	3.989	.02582	.2507	38.729	39
+40	500.66	10511.16	72.99	+58.29	530.63	3.915	.02585	.2554	38.684	+40
41	1	10724.95	74.48	59.78	529.99	3.839	.02588	.2605	38.639	41
42	2	10942.18	75.99	61.29	529.36	3.766	.02591	.2655	38.595	42
43	3	11162.93	77.52	62.82	528.73	3.695	.02594	.2706	38.550	43
44	4	11387.21	79.08	64.38	528.10	3.627	.02597	.2757	38.499	44
+45	505.66	11615.12	80.66	+65.96	527.47	3.559	.02600	.2809	38.461	+45
46	6	11846.64	82.27	67.57	526.83	3.493	.02603	.2863	38.417	46
47	7	12081.80	83.90	69.20	526.20	3.428	.02606	.2917	38.373	47
48	8	12320.71	85.56	70.86	525.57	3.362	.02609	.2974	38.328	48
49	9	12563.36	87.25	72.55	524.93	3.303	.02612	.3027	38.284	49
+50	510.66	12809.91	88.96	+74.26	524.30	3.242	.02616	.3084	38.226	+50
51	1	13080.21	90.70	76.00	523.66	3.182	.02620	.3143	38.167	51
52	2	13314.43	92.46	77.76	523.03	3.124	.02623	.3201	38.124	52
53	3	13572.52	94.25	79.55	522.39	3.069	.02626	.3258	38.080	53
54	4	13834.64	96.07	81.37	521.76	3.012	.02629	.3320	38.037	54

WOOD'S TABLE OF SATURATED AMMONIA—Continued.

Temperature.		Pressure, Absolute.								
Degrees F. t.	Absolute. T	Pounds per Sq. Foot. P.	Pounds per Sq. Inch. p.	Gauge Pressure, Pound per Sq. Inch.	Heat of Vaporization, Thermal Units. h.	Volume of Vapor per Pound, Cubic Feet. v.	Volume of Liquid per Pound, Cubic Feet. v_1	Weight of Vapor in Pounds per Cubic Foot. w.	Weight of Liquid in Pounds per Cubic Foot. w_1	Temperature. Degrees F.
+55	515.66	14100.74	97.92	+83.22	521 12	2.958	.02632	.3380	37.994	+55
56	6	14370.92	99.80	85.10	520.48	2.905	.02636	.3442	37.936	56
57	7	14645.18	101.70	87.00	519.84	2.853	.02639	.3505	37.893	57
58	8	14923.98	103.64	88.94	519.20	2.802	.02643	.3568	37.835	58
59	9	15206.28	105.60	90.90	518.57	2.753	.02646	.3632	37.793	59
+60	520.66	15493.09	107.59	+92.89	517.93	2.705	.02651	.3697	37.736	+60
61	1	15784.23	109.61	94.91	517.29	2.658	.02654	.3762	37.678	61
62	2	16079.67	111.66	96.96	516.65	2.610	.02658	.3831	37.622	62
63	3	16379.51	113.75	99.05	516.01	2.565	.02661	.3898	37.579	63
64	4	16683.75	115.86	101.16	515.37	2.520	.02665	.3968	37.523	64
+65	525.66	16992.50	118.08	+103.33	514.73	2.476	.02668	.4039	37.481	+65
66	6	17305.70	120.18	105.48	514.09	2.433	.02671	.4110	37.439	66
67	7	17623.45	122.38	107.68	513.45	2.389	.02675	.4189	37.383	67
68	8	17945.89	124.62	109.92	512.81	2.351	.02678	.4254	37.341	68
69	9	18272.81	126.89	112.19	512.16	2.310	.02682	.4329	37.285	69
+70	530.66	18604.53	129.19	+114.49	511.52	2.272	.02686	.4401	37.230	+70
71	1	18941.00	131.54	116.84	510.87	2.233	.02689	.4479	37.188	71
72	2	19282.21	133.90	119.20	510.22	2.194	.02693	.4558	37.133	72
73	3	19628.32	136.31	121.61	509.58	2.153	.02697	.4645	37.079	73
74	4	19979.22	138.74	124.04	508.93	2.122	.02700	.4712	37.037	74
+75	535.66	20335.16	141.22	+126.52	508.29	2.087	.02703	.4791	36.995	+75
76	6	20696.00	143.72	129.02	507.64	2.052	.02706	.4873	36.954	76
77	7	21061.85	146.26	131.56	506.99	2.017	.02710	.4957	36.900	77
78	8	21432.82	148.84	134.14	506.34	1.995	.02714	.5012	36.845	78
79	9	21808.85	151.45	136.75	505.69	1.952	.02717	.5123	36.805	79
+80	540.66	22190.15	154.10	+139.40	505.05	1.921	.02721	.5205	36.751	+80
81	1	22576.51	156.78	142.08	504.40	1.889	.02725	.5294	36.696	81
82	2	22968.88	159.50	144.80	503.75	1.858	.02728	.5382	36.657	82
83	3	23365.38	162.26	147.56	503.10	1.827	.02732	.5473	36.603	83
84	4	23767.81	165.05	150.35	502.45	1.799	.02736	.5558	36.549	84
+85	545.66	24175.61	167.88	+153.18	501.81	1.770	.02739	.5649	36.509	+85
86	6	24588.92	170.75	156.05	501.15	1.741	.02743	.5744	36.456	86
87	7	25007.80	173.66	158.96	500.50	1.714	.02747	.5834	36.407	87
88	8	25432.16	176.61	161.91	499.85	1.687	.02751	.5927	36.350	88
89	9	25862.14	179.59	164.89	499.20	1.660	.02754	.6024	36.311	89
+90	550.66	26297.88	182.62	+167.92	498.55	1.634	.02758	.6120	36.258	+90
91	1	26739.88	185.69	170.99	497.89	1.608	.02761	.6219	36.219	91
92	2	27186.56	188.79	174.09	497.24	1.583	.02765	.6317	36.166	92
93	3	27639.43	191.94	177.24	496.59	1.558	.02769	.6418	36.114	93
94	4	28098.26	195.13	180.43	495.94	1.534	.02772	.6518	36.075	94
+95	555.66	28563.00	198.35	+183.65	495.29	1.510	.02776	.6622	36.023	+95
96	6	29033.86	201.62	186.92	494.63	1.486	.02780	.6729	35.971	96
97	7	29510.69	204.94	190.24	493.97	1.463	.02784	.6835	35.919	97
98	8	29993.52	208.29	193.59	493.32	1.442	.02787	.6934	35.881	98
99	9	30482.52	211.68	196.98	492.66	1.419	.02791	.7047	35.829	99
+100	560.66	30977.78	215.12	+200.42	492.01	1.398	.02795	.7153	35.778	+100

TABLE OF HUMIDITY IN AIR.

Temperature Cels.	Greatest Tension of Vapor in Millimeter.	Weight of Vapor in One Cubic Meter of Air in Grams.	Temperature Cels.	Greatest Tension of Vapor in Millimeter.	Weight of Vapor in One Cubic Meter of Air in Grams.
−10	2.1	2.3	+13	11.2	11.4
−9	2.3	2.5	+14	11.9	12.1
−8	2.5	2.7	+15	12.7	12.9
−7	2.7	2.9	+16	13.5	13.6
−6	2.9	3.2	+17	14.4	14.5
−5	3.1	3.4	+18	15.4	15.4
−4	3.4	3.7	+19	16.3	16.3
−3	3.7	4.0	+20	17.4	17.3
−2	4.0	4.3	+21	18.5	18.4
−1	4.3	4.6	+22	19.7	19.4
0	4.6	4.9	+23	20.9	20.6
+1	5.0	5.3	+24	22.2	21.8
+2	5.3	5.6	+25	23.6	23.1
+3	5.7	6.0	+26	25.0	24.4
+4	6.1	6.4	+27	26.5	25.8
+5	6.5	6.8	+28	28.1	27.2
+6	7.0	7.3	+29	29.8	28.8
+7	7.5	7.8	+30	31.5	30.4
+8	8.0	8.3	+31	33.4	32.1
+9	8.6	8.9	+32	35.4	33.8
+10	9.2	9.4	+33	37.4	35.7
+11	9.8	10.1	+34	39.3	37.6
+12	10.5	10.7	+35	41.5	39.3

TABLE SHOWING AMOUNT OF MOISTURE TO 100 LBS. OF DRY AIR WHEN SATURATED AT DIFFERENT TEMPERATURES.

Temperature. Fahr. Degrees.	Weight of Vapor in lbs.	Temperature. Fahr. Degrees.	Weight of Vapor in lbs.	Temperature. Fahr. Degrees.	Weight of Vapor in lbs.
−20	0.0350	62	1.179	142	16.170
−10	0.0574	72	1.680	152	22.465
0	0.0918	82	2.361	162	31.713
+10	0.1418	92	3.289	172	46.338
20	0.2265	102	4.547	182	71.300
32	0.379	112	6.253	192	122.643
42	0.561	122	8.584	202	280.230
52	0.819	132	11.771	212	Infinite.

LATENT UNITS OF HEAT OF FUSION AND VOLATILIZATION PER POUND OF SUBSTANCE.

Solids Melted to Liquids.	Latent Heat B. T. Units	Liquids Converted to Vapor.	Latent Heat B. T. Units
Ice to water..........	142	Water to steam	966
Tin....................	25.6	Ammonia	495
Zinc....................	50.6	Alcohol, pure........	372
Sulphur	17.0	Carbonic acid	298
Lead....................	9.72	Bisulphite of carbon.	212
Mercury................	5.00	Ether, sulphuric.....	174
Beeswax.........	175	Essence of turpentine	137
Bismuth..............	550	Oil of turpentine	184
Cast iron	233	Mercury	157
Spermaceti....... ...	46.4	Chimogene	175

COLD STORAGE RATES.

The charges for cold storage and rates for freezing must depend greatly upon various conditions, such as capacity of house, demand and supply, competition to be met and other local conditions. For general use and as a basis for figuring, the following rates, which are those now in force in the principal cold storage points and which are generally adhered to, will be found useful:

COLD STORAGE RATES PER MONTH.

Goods and Quantity.	First Month.	Each Succeeding Month.	In Large Quantities, per Month.	Season Rate per Bbl. or 100 Lbs.	Season Ends.
Apples, per bbl	$0.15	$0.12½	$0.12½	$0.45	May 1.
Bananas, per bunch	.15	.10	.10		
Beef, mutton, pork and fresh meats, per lb	.00¾	.00½	.00⅜		
Beer and ale, per bbl	.25	.25			
Beer and ale, per ½ bbl	.15	.15			
Beer and ale, per ¼ or ⅛ bbl	.10	.10			
Beer, bottled, per case	.10	.10			
Beer, bottled, per bbl	.20	.20			
Berries, fresh, of all kinds, per quart	.00½	.00⅕	.00⅕		
Berries, fresh, of all kinds, per stand	.10				
Butter and butterine, per lb. (See also butter freezing rate.)	.00⅓	.00¼	.00⅛	.50–75	Jan. 1.
Buckwheat flour, per bbl	.15	.12½	.10	.50	Oct. 1.
Cabbage, per bbl	.25	.25	.20		
Cabbage, per crate	.10	.10	.08		
Calves (per day), each	.10				
Calves, per lb	.00¾	.00½	.00⅜		
Canned and bottled goods, per lb	.00¼	.00⅛	.00⅛		
Celery, per case	.15	.10	.10		
Cheese, per lb	.00⅓	.00¼	.00⅛	.50–60	Jan. 1.
Cherries, per quart	.00½	.00⅓	.00⅓		
Cider, per bbl	.25	.15	.15		
Cigars, per lb	.00⅕	.00¼	.00⅛		
Cranberries, per bbl	.25	.20	.15		
Cranberries, per case	.10				
Corn meal, per bbl	.15	.12½	.10		
Dried and boneless fish, etc., per lb	.00 1-5	.00⅛	.00⅛	.50	Nov. 1.
Dried corn, per bbl	.12½	.10	.10		
Dried and evaporated apples, per lb	.00⅛	.00 1-10		.50	Nov. 1.
Dried fruit, per lb	.00 1-6	.00⅛	.00⅛	.40–50	Nov. 1.
Eggs, per case	.15	.12½	.10	.50–60	Jan. 1.
Figs, per lb	.00⅛	.00⅛	.00 1-10		
Fish, per bbl	.20	.18	.15	.75	Oct. 1.
Fish, per tierce (See also fish freezing rates.)	.15	.13	.12½	.50	Oct. 1.
Fruits, fresh, per bbl	.25	.20	.20		
Fruits, fresh, per crate	.10	.08	.08		
Furs, undressed, hydraulic pressed, per lb	.00½	.00¼	.00¼	1.00	Oct. 1.
Furs, dressed, per lb	.03	.02½	.02	8.00	Oct. 1.
Ginger ale, bottled, per bbl	.20	.15	.15		
Grapes, per lb	.00½	.00⅓	.00¼	2.00	May 1.
Grapes, per basket	.03	.02	.01		

COLD STORAGE RATES PER MONTH—Continued.

GOODS AND QUANTITY.	First Month.	Each Succeeding Month.	In Large Quantities, per Month.	Season Rate per Bbl. or 100 Lbs.	Season Ends.
Grapes, Malaga, etc., per keg.	.15	.12½	.12½	....	....
Hops, per lb	.00⅕	.00¼	.00⅛	....	....
Lard, per tierce	.25	.20	.20	1.00	Nov. 1.
Lard oil, per cask	.25	.20	.20	1.00	Nov. 1.
Lemons, per box	.15	.12½	.10	.50	Nov. 1.
Macaroni, per bbl	.20	.15	.12½	....	....
Maple sugar, per lb	.00⅕	.00¼	.00⅛	.40–50	Nov. 1.
Maple syrup, per gallon	.01½	.01¼	.01	....	....
Meats, fresh, per lb	.00¾	.00½	.00⅜	....	....
Nuts of all kinds, per lb	.00¼	.00 1-5	.00⅛	.40–50	Nov. 1.
Oatmeal, per bbl	.20	.15	.12½	....	....
Oil, per cask	.25	.20	....	....	....
Oil, per hhd	1.00	.80	....	....	....
Oleomargarine, per lb	.00⅓	.00¼	.00⅛	....	....
Onions, per bbl	.15	.12½	.10	.50–60	May 1.
Onions, per box	.12½	10	....	....	....
Oranges, per box	.15	.12½	.10	.50	Nov. 1.
Oysters, in tubs, per gal	.05	.04	....	....	....
Oysters, in shell, per bbl	.50	.40	.30	....	....
Peaches, per basket	.10	.08	.07	2.00	Jan. 1.
Pears, per box	.20	.15	....	.60	May 1.
Pears, per bbl	.40	.30	....	1.20	May 1.
Pigs' feet, per lb	.00⅕	.00¼	.00¼	1.00	Nov. 1.
Pork, per tierce	.20	.15	.15	....	....
Potatoes, per bbl	.25	.20	.20	....	....
Preserves, jellies, jams, etc., per lb	.00¼	.00⅛	.00⅛	....	....
Provisions, per bbl	.25	.20	.20	....	....
Rice flour, per bbl	.20	.15	.12½	....	....
Sauerkraut, per cask	.25	.20	.15	.60–75	Nov. 1.
Sauerkraut, per ½ bbl	.15	.12½	.10	....	....
Syrup, per bbl	.30	.25	.20	1.00	Oct. 1.
Tobacco, per lb	.00⅓	.00¼	.00⅛	....	....
Vegetables, fresh, per bbl	.25	.20	.15	....	....
Vegetables, fresh, per case	.15	.10	.08	....	....
Wine, in wood, per bbl	.25	.25	....	....	....
Wine, in bottles, per case	.10	.10	....	....	....

RATES FOR FREEZING POULTRY, GAME, FISH, MEATS, BUTTER, EGGS, ETC.

The rates for freezing goods, or for storing goods at a freezing temperature when they are already frozen, as follows:

POULTRY, GAME, ETC., IN UNBROKEN PACKAGES.

Poultry, including turkeys, fowl, chickens, geese, etc., and rabbits, squirrels and ducks when picked.

Four rates, A, B, C and D, for storing poultry, and the rate to be charged will be determined by the amount of such goods as may be frozen and stored during a season of six months, usually from October or November 1 to April or May 1.

RATE A.—For customers storing fifty (50) or more tons of poultry, the rate to be one-third cent per pound for the first month stored, and one-fourth cent per pound for each month or fraction of a month, including the first month, if stored for more than one month.

RATE B.—For customers storing five or more, but less than fifty tons of poultry, the rate to be one-third cent per pound for the first month stored, and one-fourth cent per pound for each month or fraction of a month thereafter.

RATE C.—For customers storing one or more, but less than five tons of poultry, the rate to be three-eighths cent per pound for the first month stored, and one-fourth cent per pound for each month or fraction of a month thereafter.

RATE D.—For customers storing less than one ton of poultry, the rate to be one-half cent per pound for the first month stored, and three-eighths cent per pound for each month or fraction of a month thereafter.

Venison, etc., and ducks when unpicked, one to one-half cent per pound per month, according to quantity and length of time stored.

Grouse and partridges, three cents to five cents per pair per month. Woodcock, one cent to two cents per pair per month.

Squabs and pigeons, four cents to six cents per dozen per month. Quail, plover, snipe, etc., three cents to five cents per dozen per month.

When a portion of the goods is removed from a package, storage to be charged for the whole package as it was received until the balance of the package is removed from the freezer.

For goods received loose, when to be taken out of the packages in which they are received, or when to be laid out, the following rates to be charged:

Poultry, including turkeys, chickens, geese, etc., and rabbits and squirrels, one-half cent to one-fourth cent per pound extra, according to quantity and length of time stored.

Grouse, partridges, woodcock, squabs, pigeons, quail, plover and snipe, 50 per cent more than the rates as above specified.

Ducks weighing less than two pounds each, two cents to three cents each per month. Ducks weighing two pounds or more each, three cents to four cents each per month.

For all kinds of poultry and birds not herein specified, the rate from one cent to one-half cent per pound per month, according to quantity and length of time stored.

SUMMER FREEZING RATES.

Freezing rates for the summer months, 50 per cent more than the specified winter rates for the first month stored, and the same as the winter rates for the second and succeeding months.

STORING UNFROZEN POULTRY, ETC.

For holding poultry, game, etc., which are not frozen, at a temperature which shall be about 30° F., the rate to be one-fifth cent to two-fifths cent per pound, according to quantity, for any time not exceeding two weeks.

FREEZER RATES FOR FISH AND MEATS.

Salmon, blue fish and other fresh fish in packages, one-half cent per pound for the first month stored, three-eighths cent per pound per month thereafter.

Fresh fish of all kinds when to be hung up or laid out, three-fourths cent per pound for the first month stored, one-half cent per pound per month thereafter.

Fish in small quantities, 50 per cent more than the above rates.

Special rates for large lots of large fish.

Scallops, three-fourths cent per pound, gross, per month.

Sweetbreads and lamb fries, one cent per pound, gross, per month.

Beef, mutton, lamb, pork, veal, tongues, etc., three-fourths cent to one-half cent per pound, net, for the first month stored, one-fourth cent to three-eighths cent per pound per month thereafter.

BUTTER FREEZING RATES.

For freezing and storing butter in a temperature of 20° F. or lower, the rate to be charged will be determined by the amount of such goods that may be frozen and stored during the season of eight months, from April 1 to December 1, or from May 1 to January 1. There will be three rates, A, B and C.

RATE A.—For customers storing thirty-five (35) or more tons of butter, the rate to be fifteen cents per 100 pounds, net, per month.

RATE B.—For customers storing five or more, but less than thirty-five tons of butter, the rate to be eighteen cents per 100 pounds, net, per month.

RATE C.—For customers storing less than five tons of butter, the rate to be twenty-five cents per 100 pounds, net, per month.

EGG FREEZING RATES.

For freezing broken eggs in cans, the charge to be one-half cent per pound, net weight, per month, and for a season of eight months the rate to be one and one-half cents per pound, net weight.

RENT OF ROOMS.

For freezing temperatures, four cents to five cents per cubic foot per month.

TERMS OF PAYMENT OF COLD STORAGE AND FREEZING RATES.

All the above rates are the charges for each month, or fraction of a month, unless otherwise specified; and in all cases, fractions of months to be charged as full months.

Charges to be computed in all cases when possible upon the marked weights and numbers of all goods at the time they are received.

All storage bills are due and payable upon the delivery of a whole lot, or balance of a lot of goods, or every three months, when goods are stored more than three months.

Unless special instructions regarding insurance accompany each lot of goods, they are held at owner's risk.

DESCRIPTION OF TWO-FLUE BOILERS.

NUMBER.	1	2	3	4	5	6
Heating Surface, square feet.	105	152	201	249	356	508
Horse power at 10 square feet.	10	15	20	25	36	51
Diameter, inches.	30	32	36	40	44	50
Length, feet....	10	14	16	18	22	28
Diameter of Flues, inches	10	10	12	13	15	18
Thickness of Shell, inches....	¼	¼	¼	¼	¼	5-16
Thickness of Head, inches	⅜	⅜	⅜	⅜	⅜	7-16
Size of Dome, inches....	15 x 15	15 x 15	18 x 18	20 x 20	24 x 24	30 x 30
Width of Grate Bars, inches..	30	32	36	40	44	50
Length of Grate Bars, feet. ..	3	3½	4	4	4½	5
Number of Wall Binding Bars	4	6	6	6	6	8
Length Wall Binding Bars, ft.	6	6	6	6	7	7
Diameter of Blow-off Cock, ins	1½	1½	1½	1½	1½	2
Diameter of Safety Valve, ins	1½	2	2	2	2½	3
Diameter of Smoke Stack, ins	15	15	18	18	24	26
Length of Stack, feet.....	30	30	30	35	40	50
Number of Iron in Stack.....	16	16	16	16	16	16
Approximate Weight of Boiler	2100	2580	3300	4250	5225	10000
Total Weight..........	4400	5100	6400	7350	8800	15000

USEFUL NUMBERS FOR RAPID APPROXIMATION.

Feet	×	.00019	=miles.
Yards	×	.0006	=miles.
Links	×	.22	=yards.
Links	×	.66	=feet.
Feet	×	1.5	=links.
Square inches	×	.007	=square feet.
Circular inches	×	.00546	=square feet.
Square feet	×	.111	=square yards.
Acres	×	4840.	=square yards.
Square yards	×	.0002066	=acres.
Cubic feet	×	.04	=cubic yards.
Cubic inches	×	.00058	=cubic feet.
U. S. bushels	×	.046	=cubic yards.
U. S. bushels	×	1.244	=cubic feet.
U. S. bushels	×	2150.42	=cubic inches.
Cubic feet	×	.8036	=U. S. bushels.
Cubic inches	×	.000466	=U. S. bushels.
U. S. gallons	×	.13368	=cubic feet.
U. S. gallons	×	231.	=cubic inches.
Cubic feet	×	7.48	=U. S. gallons.
Cylindrical feet	×	5.878	=U. S. gallons.
Cubic inches	×	.004329	=U. S. gallons.
Cylindrical inches	×	.0034	=U. S. gallons.
Pounds	×	.009	=cwt. (112 lbs.)
Pounds	×	.00045	=tons (2,240 lbs.)
Cubic feet water	×	62.5	=lbs. avdps.
Cubic inches water	×	.03617	=lbs. avdps.
Cylindrical feet of water	×	49.1	=lbs. avdps.
Cylindrical inches of water	×	.02842	=lbs. avdps.
U. S. gallons of water	÷	13.44	=cwt. (112 lbs.)
U. S. gallons of water	÷	268.8	=tons.
Cubic feet water	÷	1.8	=cwt. (112 lbs.)
Cubic feet water	÷	35.88	=tons.
Cylindrical feet of water	÷	5.875	=U. S. gallons.
Col. of water 12 in. high, 1 in. diam			=.34 lbs.
183 346 circular inches			=1 square foot.
2,200 cyclindrical inches			=1 cubic foot.
French meters	×	3.281	=feet.
Kilogrammes	×	2.205	=avdps. lbs.
Grammes	×	.0022	=avdps lbs.

---

12 × wt. of pine pattern = iron casting.
13 × wt. of pine pattern = brass casting.
19 × wt. of pine pattern = lead casting.
12.2 × wt. of pine pattern = tin casting.
11.4 × wt. of pine pattern = zinc casting.
1 cubic foot anthracite coal = 54 lbs.
40—43 cubic feet anthracite coal = 1 ton.
49 cubic feet bituminous coal = 1 ton.
537 lbs. per cubic foot = wt. of copper.
450 lbs. per cubic foot = wt. of cast iron.
485 lbs. per cubic foot = wt. of wrought iron.
708 lbs. per cubic foot = wt. of cast lead.
490 lbs. per cubic foot = wt. of steel.
1 gallon water = 8⅓ lbs. = 231 cubic inches.
1 cubic foot water = 62½ lbs. = 7½ gallons.
1 lb. water = 27.8 cubic inches = 1 pint.

The friction of water in pipes is as the square of its velocity.

Doubling the diameter of a pipe increases its capacity four times.

In tubular boilers, 15 square feet of heating surface are equivalent to one horse power; in flue boilers, 12 square feet of heating surface are equivalent to one horse power; in cylinder boilers, 10 square feet of heating surface are equivalent to one horse power.

One square foot of grate will consume, on an average, 12 lbs. of coal per hour.

Consumption of coal averages 7½ lbs. of coal, or 15 lbs. of dry pine wood, for every cubic foot of water evaporated.

The ordinary speed to run steam pumps is at the rate of 100 feet piston travel per minute.

SOLUBILITY OF GASES IN WATER AT ATMOSPHERIC PRESSURE.

1 Vol. Water dissolves Vols. Gas.	32° Fahr.	39.2° Fahr.	50° Fahr.	60° Fahr.	70° Fahr.
Air	.0247	.0224	.0195	.0179	.0171
Ammonia	1049.6	941.9	812.8	727.2	654.0
Carbon di-oxide	1.7987	1.5126	1.1847	1.0020	.9014
Sulphur di-oxide	79.789	69.828	56.647	47.276	39.374
March gas	.0545	.0499	.0437	.0391	.0350
Nitrogen	.0204	.0184	.0161	.0148	.0140
Hydrogen	.0193	.0193	.0191	.0193	.0193
Oxygen	.0411	.0372	.0325	.0299	.0284

DOUBLE EXTRA STRONG PIPE.—Table of Standard Dimensions.

Diameter.			Thickness. Inches.	Circumference.		Transverse Areas.			Length of Pipe per Square Foot of External Surface. Feet.	Nominal Weight per Foot. Pounds.
Nominal Internal. Inches.	Actual External. Inches.	Actual Internal. Inches.		External. Inches.	Internal. Inches.	External. Sq. Inch.	Internal. Sq. Inch.	Metal. Sq. Inch.		
½	.84	.244	.298	2.630	.766	.554	.047	.507	4.547	1.7
¾	1.05	.422	.314	3.299	1.326	.866	.139	.727	3.637	2.44
1	1.315	.587	.364	4.131	1.844	1.358	.271	1.087	2.904	3.65
1¼	1.66	.885	.388	5.215	2.78	2.164	.615	1.549	2.304	5.2
1½	1.9	1.088	.406	5.969	3.418	2.835	.93	1.905	2.01	6.4
2	2.375	1.491	.442	7.461	4.684	4.43	1.744	2.686	1.608	9.02
2½	2.875	1.755	.560	9.032	5.513	6.492	2.419	4.073	1.328	13.68
3	3.5	2.284	.608	10.996	7.175	9.621	4.097	5.524	1.091	18.56
3½	4.	2.716	.642	12.566	8.533	12.566	5.794	6.772	.955	22.75
4	4.5	3.136	.682	14.137	9.852	15.904	7.724	8.18	.849	27.48
5	5.563	4.063	.75	17.477	12.764	24.306	12.965	11.34	.687	38.12
6	6.625	4.875	.875	20.813	15.315	34.472	18.666	15.896	.577	53.11

SIZES AND DIMENSIONS OF STANDARD CORLISS ENGINES.

Size of Cylinder: Diameter.	Size of Cylinder: Stroke.	No. of Revolutions.	Indicated Horse Power, 80 lbs. Pressure: 1-5 Cut-off.	80 lbs.: 1-4 Cut-off.	80 lbs.: 1-3 Cut-off.	90 lbs. Pressure: 1-5 Cut-off.	90 lbs.: 1-4 Cut-off.	90 lbs.: 1-3 Cut-off.	100 lbs. Pressure: 1-5 Cut-off.	100 lbs.: 1-4 Cut-off.	100 lbs.: 1-3 Cut-off.	Fly Wheel: Diameter.	Fly Wheel: Face.	Fly Wheel: Weight in Pounds.	Crank Shaft: Diameter.	Crank Shaft: Length.	Crank Shaft: Center Shaft above Found.	Crank Shaft: Center Shaft to Back of Cylinder Head. (Ft.)	Center Shaft to Back of Cylinder Head. (In.)	Steam Pipes: Diameter Steam.	Steam Pipes: Diameter Exhaust.	Approx. Shipping Weight.
In.	In.											Ft.	In.		In.	Ft.	In.	Ft.	In.	In.	In.	Pounds.
10	24	90	30	36	43	34	41	48	39	49	60	8	13	3,500	5	6	25	10	11	3	4	10,700
10	30	90	37	45	53	42	51	60	48	58	68	8	13	4,300	5	6	25	13		3	4	11,750
12	30	90	54	65	76	62	74	87	69	83	97	9	15	5,700	6	6½	25	13	4	3½	4½	15,550
12	36	85	61	73	86	70	84	98	78	94	110	10	17	6,300	6	7	25	15	5	3½	4½	16,900
14	36	85	83	100	118	95	114	134	107	128	150	10	19	8,900	7	7½	25½	15	6	4	5	22,300
14	42	82	93	112	132	107	128	140	120	144	168	12	19	8,900	7	7½	25½	17	7	4	6	22,700
16	36	82	105	126	140	120	144	167	135	162	187	12	21	10,500	8	8	27	15	9	4½	6	26,610
16	42	78	116	139	163	133	159	188	150	179	210	12	23	11,800	8	8	27	17	10	4½	6	28,930
18	36	80	129	155	182	148	177	208	166	199	233	12	25	12,600	9	9	30	15	11	5	7	30,000
18	42	78	147	176	208	168	202	237	189	227	266	14	25	13,300	9	9	30	18		5	7	33,800
18	48	75	162	194	227	185	222	259	208	249	291	15	25	15,300	9	9	30	20	1	5	7	39,570
20	42	75	175	210	246	200	240	280	225	270	314	15	29	16,600	10	9	30	18	4	6	8	43,450
20	48	72	192	230	268	219	263	307	246	296	346	16	29	18,700	10	9	30	20	5	6	8	47,612
22	42	75	211	254	298	242	290	340	271	326	382	16	31	21,000	11	10	32	18	5	6	8	50,600
22	48	72	232	278	308	265	318	368	298	358	410	16	31	23,100	11	10	32	20	6	6	8	56,220
24	48	70	268	322	380	307	368	432	345	414	486	18	33	24,400	12	11	32	20	11	7	9	65,140
24	60	65	311	374	438	356	427	500	401	481	561	18	37	30,500	12	11	32	25	2	7	9	75,740
26	48	70	315	378	445	360	432	507	405	486	570	18	37	29,000	13	11½	32	21		7	9	77,240
26	60	65	366	439	512	418	502	586	470	564	658	18	37	31,000	13	11½	32	25	3	7	10	80,000
28	48	68	355	426	503	406	487	568	457	548	680	18	37	31,000	14	12	34	21	3	8	12	77,920
28	54	68	400	477	554	461	546	631	523	616	709	18	37	31,000	14	12	34	23	4	8	12	83,000
28	60	65	424	509	594	485	582	655	545	654	763	18	37	31,000	14	12	34	25	6	8	12	89,000
30	48	68	407	489	571	466	559	652	524	629	734	22	46	33,000	15	12½	36	21	8	8	12	93,000
30	60	62	464	557	650	531	637	743	597	717	837	24	52	38,500	15	12½	36	25	11	8	12	104,000
30	72	55	494	593	693	565	678	791	635	762	890	24	60	52,000	15	13	36	30	2			127,000
32	48	65	443	532	622	507	608	710	570	684	800	24	48	34,500	16	13	36	21	11			
32	60	62	529	634	739	604	725	846	680	816	952	24	60	43,900	16	13	36	26	2			
32	72	55	563	675	787	643	772	902	723	868	1013	24	60	58,200	16	14	36	30	5			

TABLE OF MEAN TEMPERATURE OF DIFFERENT LOCALITIES, DEGREES FAHR.

LOCATION.	Year.	Spring.	Summer.	Autumn.	Winter.
Algiers	63.0	63.0	74.5	70.5	50.4
Berlin	47.5	46.4	63.1	47.8	30.6
Berne	46.0	45.8	60.4	47.3	30.4
Boston	49.0	48	66	53	28
Baltimore	54.9	60.0	83.0	64.6	43.5
Buenos Ayres	62.5	59.4	73.0	64.6	52.5
Cairo	72.3	71.6	84.6	74.3	58.5
Calcutta	78.4	82.6	83.3	80.0	67.8
Canton	69.8	69.8	82.0	72.9	54.8
Christiania	41.7	39.2	59.5	42.4	25.2
Cape of Good Hope	66.4	63.5	74.1	66.9	58.6
Constantinople	56.7	51.8	73.4	60.4	40.6
Copenhagen	46.8	43.7	63.0	48.7	31.3
Chicago	45.9	52.8	74.6	61.2	38.4
Cincinnati	54.7	63.2	81.8	66.4	46.6
Edinburgh	47.5	45.7	57.9	48.0	38.5
Jerusalem	62.2	60.6	72.6	66.3	49.6
Jamaica (Kingston)	79	78.3	81.3	80	76.3
Lima (Peru)	66.2	63.0	73.2	69.6	59.0
Lisbon	61.5	59.9	71.1	62.6	52.3
London	50.7	49.1	62.8	51.3	39.6
Madeira (Funchal)	65.7	63.5	70	67.6	61.3
Madrid	57.6	57.6	74.1	56.7	42.1
Mexico City	60.5	53.6	63.4	65.2	60.1
Montreal	43.7	44.2	69.1	47.1	17.5
Moscow	38.5	43.3	62.6	34.9	13.5
Naples	61.5	59.4	74.8	62.2	49.6
New Orleans	72	73	84	72	58
New York	53	50	72	56	33
New Zealand	59.6	60.1	66.7	58.0	53.5
Nice	60.1	55.9	72.5	63.0	48.7
Nicolaief (Russia)	48.7	49 3	71.2	50	25.9
Paramatto (Australia)	64.6	66.6	73.9	64.8	54.5
Palermo	63	59.0	74.3	66.2	52.5
Pekin (China)	52.6	56.6	77.8	54.9	29
Paris	51.4	50.5	61.6	52.2	37.9
Philadelphia	55	52	76	57	34
Quito (Equador)	60.1	60.3	60.1	62.5	59.7
Quebec	40.3	....	....	....	....
Rio Janeiro	73.6	72.5	79.0	74.5	68.5
Rome	59.7	57.4	73.2	61.7	46.6
San Francisco	57.5	58	59	60	53
St. Louis	55.0	84.6	67.8	44.6	46.0
St. Petersburg	38.3	35.1	60.3	40.5	16.6
Stockholm	42.1	38.3	61.0	43 7	25.5
Trieste	55.8	53.8	71.5	56.7	39.4
Turin	53.1	53.1	71.6	53.8	33.4
Vienna	50.7	49.1	62.8	51.3	39.6
Warsaw	45.5	44 6	63.5	46.4	27.5
Washington	59	69	79	58	38

USEFUL DATA ABOUT LIQUIDS.

A gallon of water contains 231 cubic inches, and weighs 8⅓ pounds (U. S. standard).

A cubic foot of water contains 7½ gallons, and weighs 62½ pounds.

One U. S. gallon=.133 cubic feet; .83 imperial gallon; 3.8 liters.

An imperial gallon contains 277.274 cubic inches. .16 cubic feet; 10.00 pounds; 1.2 U. S. gallons; 4.537 liters.

A cubic inch of water=.03607 pound; .003607 imperial gallons; .004329 U. S. gallon.

A cubic foot of water =6.23 imperial gallons; 7.48 U. S. gallons; 28.375 liters; .0283 cubic meters; 62.35 pounds; 557 cwt.; .028 ton.

A pound of water = 27.72 cubic inches; .10 imperial gallon; .083 U. S. gallon; .4537 kilos.

One cwt. of water = 11.2 imperial gallons; 13.44 U. S. gallons; 1.8 cubic foot.

A ton of water = 35.9 cubic feet; 224 imperial gallons; 298.8 gallons; 1,000 liters (about); 1 cubic meter (about).

A liter of water = .220 imperial gallon: .264 U. S. gallon; 61 cubic inches; .0353 cubic foot.

A cubic meter of water = 220 imperial gallons; 264 U. S. gallons; 1.308 cubic yards; 61,028 cubic inches; 35.31 cubic feet; 1,000 kilos; 1 ton (nearly); 1,000 liters.

A kilo of water = 2.204 pounds.

A vedros of water = 2.7 imperial gallons.

An eimer of water = 2.7 imperial gallons.

A pood of water = 3.6 imperial gallons.

A Russian fathom = 7 feet.

One atmosphere = 1.054 kilos per square inch.

One ton of petroleum =275 imperial gallons (nearly).

One ton of petroleum = 360 U. S. gallons (nearly).

A column of water 1 foot in height = .434 pound pressure per square inch.

A column of water 1 meter in height = 1.43 pounds pressure per square inch.

One pound pressure per square inch = 2.31 feet of water in height.

One U. S. gallon of crude petroleum = 6.5 pounds (nearly).

One wine gallon, or U. S. gallon, is equal to 8.331 pounds=3,785 cubic centimeters=58,318 grains.

One imperial gallon (English gallon) is equal to about ten pounds=4.543 cubic-centimeters=70,000 grains.

One grain=0.0649 grams—one gram=15.36 grains.

One barrel=1.192 hectoliters—one hectoliter=0.843 barrels.

One English quarter=eight bushels=290.78 liters.

One English bushel=36.35 liters=0.3635 hectoliters.

One English barrel=36 gallons. One American barrel=31 gals.

One bushel malt (English), 40 pounds; American, 34 pounds (32 pounds cleaned); one bushel barley (American), 38 pounds.

One kilogram square centimeter equal to 14.2 pounds inch pressure (equal to about one atmosphere).

Four B. T. units equal to about one calorie.

TEMPERATURES—FAHRENHEIT AND CENTIGRADE.

°F.	°C.	°F.	°C.	°F.	°C.	°F.	°C.	°F.	°C.	°F.	°C.
330	165.6	267	130.6	206	96.7	143	61.7	80	26.7	19	-7.2
329	165.	266	130.	205	96.1	142	61.1	79	26.1	18	-7.8
328	164.4	265	129.4	204	95.6	141	60.6	78	25.6	17	-8.3
327	163.9	264	128.9	203	95	140	60	77	25.	16	-8.9
326	163.3	263	128.3	202	94.4	139	59.4	76	24.4	15	-9.4
325	162.8	262	127.8	201	93.9	138	58.9	75	23.9	14	-10.
324	162.2	261	127.2	200	93.3	137	58.3	74	23.3	13	-10.6
323	161.7	260	126.7	199	92.8	136	57.8	73	22.8	12	-11.1
322	161.1	259	126.1	198	92.2	135	57.2	72	22.2	11	-11.7
321	160.6	258	125.6	197	91.7	134	56.7	71	21.7	10	-12.2
320	160.	257	125.	196	91.1	133	56.1	70	21.1	9	-12.8
319	159.4	256	124.4	195	90.6	132	55.6	69	20.6	8	-13.3
318	158.9	255	123.9	194	90.	131	55.	68	20.	7	-13.9
317	158.3	254	123.3	193	89.4	130	54.4	67	19.4	6	-14.4
316	157.8	253	122.8	192	88.9	129	53.9	66	18.9	5	-15.
315	157.2	252	122.2	191	88.3	128	53.3	65	18.3	4	-15.6
314	156.7	251	121.7	190	87.8	127	52.8	64	17.8	3	-16.1
313	156.1	250	121.1	189	87.2	126	52.2	63	17.2	2	-16.7
312	155.6	249	120.6	188	86.7	125	51.7	62	16.7	1	-17.2
311	155.	248	120.	187	86.1	124	51.1	61	16.1	0	-17.8
310	154.4	247	119.4	186	85.6	123	50.6	60	15.6	-1	-18.3
309	153.9	246	118.9	185	85.	122	50.	59	15	-2	-18.9
308	153.3	245	118.3	184	84.4	121	49.4	58	14.4	-3	-19.4
307	152.8	244	117.8	183	83.9	120	48.9	57	13.9	-4	-20
306	152.2	243	117.2	182	83.3	119	48.3	56	13.3	-5	-20.6
305	151.7	242	116.7	181	82.8	118	47.8	55	12.8	-6	-21.1
304	151.1	241	116.1	180	82.2	117	47.2	54	12.2	-7	-21.7
303	150.6	240	115.6	179	81.7	116	46.7	53	11.7	-8	-22.2
302	150.	239	115.	178	81.1	115	46.1	52	11.1	-9	-22.8
301	149.4	238	114.4	177	80.6	114	45.6	51	10.6	-10	-23.3
300	148.9	237	113.9	176	80.	113	45.	50	10.	-11	-23.9
299	148.3	236	113.3	175	79.4	112	44.4	49	9.4	-12	-24.4
298	147.8	235	112.8	174	78.9	111	43.9	48	8.9	-13	-25.
297	147.2	234	112.2	173	78.3	110	43.3	47	8.3	-14	-25.6
296	146.7	233	111.7	172	77.8	109	42.8	46	7.8	-15	-26.1
295	146.1	232	111.1	171	77.2	108	42.2	45	7.2	-16	-26.7
294	145.6	231	110.6	170	76.7	107	41.7	44	6.7	-17	-27.2
293	145.	230	110.	169	76.1	106	41.1	43	6.1	-18	-27.8
292	144.4	229	109.4	168	75.6	105	40.6	42	5.6	-19	-28.3
291	143.9	228	108.9	167	75.	104	40.	41	5	-20	-28.9
290	143.3	227	108.3	166	74.4	103	39.4	40	4.4	-21	-29.4
289	142.8	226	107.8	165	73.9	102	38.9	39	3.9	-22	-30.
288	142.2	225	107.2	164	73.3	101	38.3	38	3.3	-23	-30.6
287	141.7	224	106.7	163	72.8	100	37.8	37	2.8	-24	-31.1
286	141.1	223	106.1	162	72.2	99	37.2	36	2.2	-25	-31.7
285	140.6	222	105.6	161	71.7	98	36.7	35	1.7	-26	-32.2
284	140.	221	105.	160	71.1	97	36.1	34	1.1	-27	-32.8
283	139.4	220	104.4	159	70.6	96	35.6	33	0.6	-28	-33.3
282	138.9	219	103.9	158	70.	95	35.	Water freezes		-29	-33.9
281	138.3	218	103.3	157	69.4	94	34.4	32	0.	-30	-34.4
280	137.8	217	102.8	156	68.9	93	33.9			-31	-35.
279	137.2	216	102.2	155	68.3	92	33.3	31	-0.6	-32	-35.6
278	136.7	215	101.7	154	67.8	91	32.8	30	-1.1	-33	-36.1
277	136.1	214	101.1	153	67.2	90	32.2	29	-1.7	-34	-36.7
276	135.6	213	100.6	152	66.7	89	31.7	28	-2.2	-35	-37.2
275	135.			151	66.1	88	31.1	27	-2.8	-36	-37.8
274	134.4	Water boils		150	65.6	87	30.6	26	-3.3	-37	-38.3
273	133.9	212	100.	149	65.	86	30.	25	-3.9	-38	-38.9
272	133.3	211	99.4	148	64.4	85	29.4	24	-4.4	-39	-39.4
271	132.8	210	98.9	147	63.9	84	28.9	23	-5		
270	132.2	209	98.3	146	63.3	83	28.3	22	-5.6		
269	131.7	208	97.8	145	62.8	82	27.8	21	-6.1	Mercury freezes	
268	131.1	207	97.2	144	62.2	81	27.2	20	-6.7	-40	-40

SPECIFIC GRAVITY TABLE (BEAUMÉ).

The meaning of the degrees of the Beaumé scale for liquids heavier than water has been defined somewhat differently by the manufacturing chemists of the United States. Accordingly the specific gravity for any given degree Beaumé is found after the formula:

$$\text{Specific gravity} = \frac{145}{145 - \text{deg. Beaumé.}}$$

The following table is calculated after this formula by Clapp:

Degrees.	Specific Gravity.	Degrees.	Specific Gravity.	Degrees.	Specific Gravity.
0	1.000	18	1.142	45	1.450
1	1.007	19	1.151	50	1.526
2	1.014	20	1.160	55	1.611
3	1.021	21	1.169	60	1.706
4	1.028	22	1.179	65	1.812
5	1.036	23	1.188	70	1.933
6	1.043	24	1.198		
7	1.051	25	1.208	66 Used by sulphuric acid manufacturers.	1.835
8	1.058	26	1.218		
9	1.066	27	1.229		
10	1.074	28	1.239		
11	1.082	29	1.250		
12	1.090	30	1.261		
13	1.098	32	1.283		
14	1.107	34	1.295		
15	1.115	36	1.306		
16	1.124	38	1.318		
17	1.133	40	1.381		

TABLE ON SOLUTIONS OF CHLORIDE OF CALCIUM.

Specific Gravity at 64° F.	Degree Beaumé at 64° F.	Degree Salometer at 64° F.	Per Cent of Chloride of Calcium.	Freezing Point. Deg. F.	Ammonia Gauge. Pounds per Square Inch at Freezing Point.
1.007	1	4	0.943	+31.20	46
1.014	2	8	1.886	+30.40	45
1.021	3	12	2.829	+29.60	44
1.028	4	16	3.772	+28.80	43
1.035	5	20	4.715	+28.00	42
1.043	6	24	5.658	+26.89	41
1.050	7	28	6.601	+25.78	40
1.058	8	32	7.544	+24.67	38
1.065	9	34	8.487	+23.56	37
1.073	10	40	9.430	+22.09	35.5
1.081	11	44	10.373	+20.62	34
1.089	12	48	11.316	+19.14	32.5
1.097	13	52	12.259	+17.67	30.5
1.105	14	56	13.202	+15.75	29
1.114	15	60	14.145	+13.82	27
1.112	16	64	15.088	+11.89	25
1.131	17	68	16.031	+ 9.96	23.5
1.140	18	72	16.974	+ 7.68	21.5
1.149	19	76	17.917	+ 5.40	20
1.158	20	80	18.860	+ 3.12	18
1.167	21	84	19.803	− 0.84	15
1.176	22	88	20.746	− 4.44	12.5
1.186	23	92	21.689	− 8.03	10.5
1.196	24	96	22.632	−11.63	8
1.205	25	100	23.575	−15.23	6
1.215	26	104	24.518	−19.56	4
1.225	27	108	25.461	−24.43	1.5
1.236	28	112	26.404	−29.29	1 " Vacuum
1.246	29	116	27.347	−35.30	5 "
1.257	30	120	28.290	−41.32	8.5 "
1.268	31	....	29.233	−47.66	12 "
1.279	32	....	30.176	−54.00	15 "
1.290	33	....	31.119	−44.32	10 "
1.302	34	....	32.062	−34.66	4 "
1.313	35	....	33	−25.00	1.5 lbs.

This table, which has been published by a manufacturer of chloride of calcium, gives the freezing points much lower in some cases than the small table on page 142.

FRICTION OF WATER IN PIPES.

Frictional loss in pounds pressure for each 100 feet in length of cast iron pipe discharging the stated quantities per minute:

Imperial Gallons.	Sizes of Pipes, Inside Diameter.															U.S. Gallons.
	¾"	1"	1¼"	1½"	2"	2½"	3"	4"	6"	8"	10"	12"	14"	16"	18"	
4	3.3	0.84	.31	.12												5
8	13.	3.16	1.05	.47	.12											10
12	28.7	6.98	2.38	.97	.27											15
16	50.4	12.30	4.07	1.66	.42											20
20	78.	19.00	6.40	2.62	.67	.21	.10									25
25		27.5	9.15	3.75	.91	.30	.12									30
29		37	13.4	5.05	1.26	.42	.14									35
33		48	16.1	6.52	1.60	.51	.17									40
37			20.2	8.15	2.02	.62	.27									45
41			24.9	10.00	2.44	.81	.35	.09								50
62			56.1	22.40	5.32	1.80	.74	.21								75
83				39	9.46	3.20	1.31	.33	.05							100
103				48.1	14.9	4.89	1.99	.51	.07							125
124					21.2	7.00	2.85	.69	.10	.02						150
145					28.1	9.46	3.85	.95	.14	.03						175
166					37.5	12.47	5.02	1.22	.17	.05	.01					200
207					47.7	19.66	7.76	1.89	.26	.07	.03					250
249						28.06	11.20	2.66	.37	.09	.04	.005				300
290						33.41	15.20	3.65	.50	.11	.05	.007				350
332						42.95	19.50	4.73	.65	.15	.06	.01				400
373							25.00	6.01	.81	.20	.08	.02				450
415							30.80	7.43	.96	.25	.09	.04	.017	.009	.005	500
622								14.32	2.21	.53	.18	.08	.036	.019	.011	750
830									3.88	.94	.32	.13	.062	.036	.020	1000
1037										1.46	.49	.20	.091	.049	.028	1250
1245										2.09	.70	.29	.135	.071	.040	1500
1450											.95	.38	.181	.095	.054	1750
1660											1.23	.49	.234	.123	.071	2000
1867												.63	.297	.153	.086	2250
2075												.77	.362	.188	.107	2500
2490												1.11	.515	.267	.150	3000
2905													.697	.365	.204	3500
3320													.920	.472	.263	4000
3735														.593	.333	4500
4150														.730	.408	5000
4980															.585	6000

The frictional loss is greatly increased by bends or irregularities in the pipes.

COMPARISON OF UNITS OF ENERGY (HERING.)

Acceleration of gravity	=	981.000	centimeters per second.
Acceleration of gravity	=	32.186	feet per second.
1 dyne . . .	=	.015731	grain.
1 dyne . . .	=	.0010194	gram.
1 grain . . .	=	63.668	dynes.
1 gram . . .	=	981.	dynes.
1 pound avdp. . .	=	444976.	dynes.
1 foot pound . .	=	.0012953	pound Fah., heat unit.
1 foot pound . .	=	.0[illegible]7196	pound C., heat unit.
1 foot pound . .	=	.[illegible]03264	kilogr.-C., heat unit.
1 metric horsepower hour	=	1952940.	foot pounds.
1 metric horsepower hour	=	270000.	kilogram meters.
1 metric horsepower hour	=	2529.7	pound Fah., heat units
1 metric horsepower hour	=	1405.4	pound C., heat units.
1 metric horsepower hour	=	.98634	horsepower hour.
1 horsepower hour .	=	2685400.	joules.
1 horsepower hour .	=	1980000.	foot pounds.
1 horsepower hour .	=	2564.8	pound Fah., heat units
1 horsepower hour .	=	1424.9	pound C., heat units.
1 horsepower hour .	=	646.31	kilogr.-C., heat units.

COMPARISON OF UNITS OF ENERGY (HERING).

1 pound Fahrenheit .	=	1047.03	joules.
1 pound Fahrenheit .	=	772.	foot pounds.
1 pound Fahrenheit .	=	106.731	kilogram meter.
1 pound Fahrenheit .	=	.55556	pound Centigrade.
1 pound Fahrenheit .	=	.25200	kilogram Centigrade.
1 pound Fahrenheit .	=	.29084	watt-hour.
1 pound Fahrenheit .	=	1	Brit. therm. unit (B.T.U.)
1 pound Centigrade	=	1884.66	joules.
1 pound Centigrade	=	1389.6	foot pounds.
1 pound Centigrade	=	192.116	kilogram meters.
1 pound Centigrade	=	1.8	pound Fahrenheit.
1 pound Centigrade	=	.52352	watt-hour.
1 pound Centigrade	=	.0007018	horsepower hour.
1 kilogram Centigrade	=	3063.5	foot pounds.
1 kilogram Centigrade	=	423.54	kilogram meters.
1 kilogram Centigrade	=	3.9683	pound Fahrenheit.
1 kilogram Centigrade	=	1.1542	watt-hours.
1 kilogram Centigrade	=	.0015472	horsepower hour.
1 watt-hour	=	3600.	joules.
1 watt-hour	=	2654.4	foot pounds.
1 watt-hour	=	3.4383	
1 erg	=	1.	dyne-centimeter.
1 erg	=	.0000001	joules.
1 gram centimeter	=	981.00	ergs.
1 joule	=	10000000.	ergs.
1 volt-coulomb	=	.737324	foot pound.
1 watt during every second		.101937	kilogram meter.
1 volt ampere during every second	=	.0013406	horsepower for one sec.
1 volt ampere during every second	=	.0009551	pound F. heat unit.
1 foot-pound	=	13562600.	ergs.
1 foot-pound	=	1.35626	joules.
1 foot-pound	=	13825.	kilogram meter.
1 foot-pound	=	.0018434	metric horsepower for one second.
1 horsepower .	=	745.941	watts.
1 horsepower .	=	33000.	foot pounds per minute.
1 horsepower .	=	42.746	lb. F., heat unit per min.
1 horsepower .	=	1.01385	metric horsepowers.
1 lb. F. heat unit per min	=	17.4505	watts.
1 lb. F. heat unit per min	=	.023718	metric horsepower.
1 lb. F. heat unit per min	=	.023394	horsepower.
1 lb. Ct. heat unit per min	=	.042109	horsepower.
1 k. Ct. heat unit per min	=	.092835	horsepower.
1 Pferdekraft	=	10.625	klg. cent.
1 erg per second	=	.0000001	watt.
1 watt	=	10000000.	ergs per second.
1 volt ampere	=	442394.	foot pounds per min.
1 volt coulomb per sec.	=	.0573048	lb. F. heat unit per min.
1 volt coulomb per sec.	=	.0013406	horsepower
1 foot pound per min.	=	.0226043	watt:
1 foot pound per min.	=	.00003072	metric horsepower.
1 foot pound per min.	=	.000030303	horsepower.
1 metric horsepower	=	735.75x107	ergs per second.
1 French horsepower	=	32549.0	foot pounds per minute.
1 cheval vapeur	=	42.162	lb. F. heat units per min.
1 force de cheval	=	23.423	lb. Ct. heat unit per min.
1 horsepower	=	33000.	foot pounds per minute.
1 horsepower	=	1980000.	foot pounds per hour.
1 horsepower	=	2565.	H. units per hour (B.T. units).
1 horsepower	=	42.75	B. T. units per minute.
1 ton of refrig. capacity	=	284000.	B. T. units.
1 ton of refrig. capacity	=	to about ¼-ton ice making capacity.	
1 ton of ref. cap. per day	=	to about 12000	B. T. units per hour.
1 ton of ref. cap. per day	=	to about 200	B. T. units per minute.

In these tables the mechanical equivalent of heat is taken at 772. Many engineers prefer the more recent figure, 778.

TABLE OF MEAN EFFECTIVE PRESSURES.

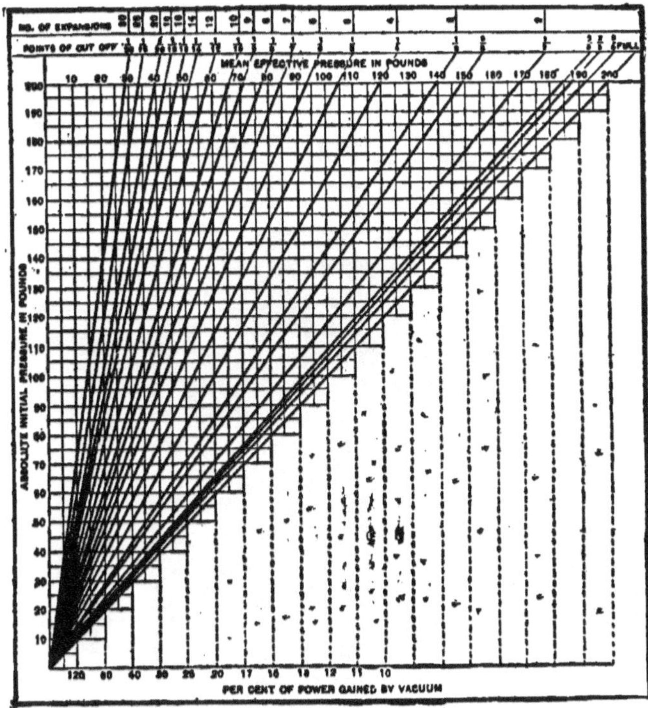

The above graphical table will be found of assistance to the engineer by affording a ready, and, at the same time, comprehensive means of ascertaining the mean effective pressure of steam in an engine cylinder, when the initial steam pressure and the point of cut-off or the number of expansions of the steam are known.

AMMONIA COMPRESSION UNDER DIFFERENT CONDITIONS.

	Wet Gas.		Dry Gas.	
Condenser pressure	113.8	116.7	147.3	161.3
Suction pressure	15.6	27.2	13.	27.5
Condenser temperature	69.2'	70.5'	82.7'	87.7'
Suction temperature	.5'	14.3'	—8.2'	14.5'
Horse power (indicated of steam cylinder)	16.8	18.	73.6	88.6
Refrigeration (tons per 24 hours)	13.3	19.5	46.5	74.4
M. E. P. in compressor	50.3	53.4	59.9	70.5
Refrigerating capacity per horse power (tons per 24 hours)	.792	1.083	.632	.840
Economy of high over low evaporating pressure	....	36.7%	....	82.9%

MEAN EFFECTIVE PRESSURE OF DIAGRAM OF STEAM CYLINDER.

Cut-off at		1/10	1/9	1/8	1/7	1/6	1/5	1/4	3/10	1/3	2/5	1/2	3/5	2/3	7/10	3/4
Apparent Ratio of Expansion.		10	9	8	7	6	5	4	3.33	3	2.5	2	1.67	1.5	1.43	1.33
M. E. P. per Lb. Initial.		.330	.355	.385	.421	.465	.523	.596	.661	.699	.776	.846	.906	.937	.949	.964
Initial Pressure.																
Gauge	Absolute															
25	39.7	13.11	14.09	15.28	16.71	18.46	20.72	23.68	26.24	27.77	30.43	33.61	35.96	37.20	37.69	38.27
30	44.7	14.76	15.87	17.21	18.82	20.79	23.33	26.67	29.54	31.27	34.26	37.84	40.49	41.88	42.44	43.00
35	49.7	16.41	17.64	19.13	20.92	23.11	25.94	29.65	32.85	34.77	38.10	42.07	45.02	46.57	47.19	47.91
40	54.7	18.07	19.42	21.06	23.03	25.44	28.55	32.63	36.15	38.28	41.93	46.31	49.55	51.25	51.94	52.73
45	59.7	19.72	21.19	22.98	25.12	27.76	31.16	35.62	39.46	41.76	45.76	50.54	54.08	55.94	56.68	57.55
50	64.7	21.37	22.97	24.91	27.23	30.09	33.77	38.60	42.76	45.26	49.59	54.77	58.61	60.62	61.43	62.07
55	69.7	23.02	24.74	26.83	29.34	32.31	36.38	41.58	46.07	48.76	53.43	59.00	63.15	65.31	66.18	67.19
60	74.7	24.67	26.52	28.75	31.45	34.74	38.99	44.56	49.37	52.26	57.26	63.24	67.67	69.99	70.92	72.01
65	79.7	26.32	28.29	30.68	33.55	37.06	41.59	47.55	52.67	55.75	61.09	67.47	72.20	74.68	75.67	76.83
70	84.7	27.97	30.07	32.60	35.66	39.39	44.20	50.53	55.98	59.25	64.92	71.70	76.73	79.36	80.42	81.64
75	89.7	29.62	31.84	34.53	37.76	41.71	46.81	53.51	59.28	62.75	68.76	75.94	81.26	84.05	85.17	86.27
80	94.7	31.28	33.62	36.45	39.87	44.04	49.42	56.50	62.59	66.25	72.59	80.17	85.79	88.73	89.91	91.28
85	99.7	32.93	35.39	38.38	41.97	46.36	52.03	59.48	65.89	69.74	76.42	84.40	90.32	93.42	94.66	96.10
90	104.7	34.58	37.17	40.30	44.08	48.69	54.64	62.46	69.20	73.24	80.26	88.63	94.84	98.10	99.41	100.92
95	109.7	36.23	38.94	42.23	46.18	51.01	57.25	65.44	72.50	76.74	84.09	92.87	99.37	102.79	104.15	105.74
100	114.7	37.88	40.72	44.15	48.20	53.34	59.86	68.43	75.81	80.24	87.92	97.10	103.90	107.47	108.90	110.56
110	124.7	41.18	44.27	48.00	52.50	57.98	65.08	74.39	82.41	87.23	95.59	105.26	112.96	116.84	118.39	120.20
120	134.7	44.49	47.82	51.85	56.71	62.64	70.30	80.36	89.02	94.23	103.25	114.04	122.02	126.21	127.89	129.84
130	144.7	47.79	51.37	55.70	60.92	67.29	75.52	86.32	95.63	101.22	110.91	122.50	131.08	135.58	137.38	139.48
140	154.7	51.09	54.92	59.55	65.13	71.94	80.74	92.29	102.24	108.22	118.58	130.96	140.14	144.95	146.88	149.12
150	164.7	54.39	58.47	63.40	69.34	76.59	85.96	98.26	108.85	115.21	126.26	139.43	149.20	154.32	156.37	158.76
160	174.7	57.70	61.02	67.25	73.55	81.24	91.17	104.22	115.46	122.21	133.91	147.89	158.25	163.69	165.87	168.40
170	184.7	61.00	65.57	71.10	77.76	85.89	96.39	110.19	122.07	129.20	141.90	156.36	167.31	173.06	175.37	178.04
180	194.7	64.30	69.12	74.95	81.97	90.54	101.61	116.15	128.68	136.20	149.24	164.82	176.37	182.43	184.86	187.67
190	204.7	67.60	72.67	78.79	86.18	95.19	106.83	122.12	135.29	143.19	156.91	173.29	185.43	191.80	194.35	197.32
200	214.7	70.91	76.22	82.64	90.39	99.84	112.05	128.08	141.90	150.19	164.57	181.85	194.94	200.71	203.85	206.95
210	224.7	74.21	79.78	86.49	94.60	104.49	117.27	134.05	148.51	157.19	172.24	190.22	203.55	210.54	213.34	216.59
220	234.7	77.51	83.32	90.34	98.81	109.14	122.48	140.02	155.12	164.18	179.90	198.68	212.61	219.91	222.83	226.23
230	244.7	80.81	86.87	94.19	103.02	113.79	127.70	145.98	161.72	171.18	187.57	207.15	221.66	229.28	232.33	235.87

The M. E. P. for any initial pressure not given in the table can be found by multiplying the (absolute) given pressure by the M. E. P. per pound of initial, as given in the third horizontal line of the table.

NOTE.—This table is reprinted from "Indicating the Refrigerating Machine," published by H. S. Rich & Co., Chicago.

RELATIVE EFFICIENCY OF FUELS.

One cord of air dried hickory or hard maple weighs about 4,500 pounds and is equal to about 2,000 pounds of coal.

One cord of air dried white oak weighs about 3,850 pounds and is equal to about 1,715 pounds of coal.

One cord of air dried beech, red oak or black oak weighs about 3,250 pounds and is equal to about 1,450 pounds of coal.

One cord of air dried poplar (whitewood), chestnut or elm weighs about 2,350 pounds, and is equal to about 1,050 pounds of coal.

One cord of air dried average pine weighs about 2,000 pounds, and is equal to about 625 pounds of coal.

From the above it is safe to assume that two and one-quarter pounds of dry wood is equal to one pound average quality of soft coal, and that the full value of the same weight of different wood is very nearly the same. That is, a pound of hickory is worth no more for fuel than a pound of pine, assuming both to be dry. It is important that the wood be dry, as each 10 per cent of water or moisture in wood will detract about 12 per cent from its value as fuel.

TABLE SHOWING TENSION OF WATER VAPOR AT DIFFERENT TEMPERATURES IN ABSOLUTE PRESSURE, AND CORRESPONDING VACUUM IN INCHES OF MERCURY.

Temperature. Deg. F.	Absolute Pressure.		Vacuum, Inches.
	Atmospheres.	Inch of Mercury.	
212	1.	30.	0.
158	0.307	9.270	20.730
140	0.196	5.880	24.120
122	0.121	3.630	26.370
113	0.094	2.820	27.180
104	0.0722	2.166	27.834
95	0.0550	1.650	28.350
86	0.0415	1.245	28.755
77	0.0310	0.930	29.070
68	0.0229	0.687	29.313
59	0.0167	0.501	29.499
50	0.0121	0.363	29.637
41	0.0086	0.258	29.742
32	0.0061	0.183	29.817
14	0.0026	0.078	29.922
−4	0.0012	0.036	29.964

BOILING POINTS UNDER ATMOSPHERIC PRESSURE.

Liquids.	Fahr. deg.	Cent. deg.	Liquids.	Fahr. deg.	Cent. deg.
Wrought iron (?).	5000	2760	Alcohol...........	173	78
Cast iron (?)......	3300	1815	Ether.............	96	35
Mercury..........	675	352	Carbon, bi-sulphurated..	116	47
Whale oil.........	630	332	Water, distilled..	212	100
Oil of linseed	600	316	Salt, sea water....	213	101
Oil of turpentine	357	180	Water, 20% salt...	218	103
Sulphuric acid...	593	312	Water, 30% salt...	222	105
Sulphur..........	570	300	Water, 40% saturated	227	108
Phosphorus......	557	292	Ammonia, liquid.	140	60
Sweet oil........	412	211	Water, in vacuo.	98	36
Naphtha..........	320	160	Chimogene......	+38	33
Nitric acid........	220	104	Carbonic acid....	−112	−80
Milk of cows.....	213	101	Ammonia........	−30	−34
Petroleum, rectified	316	158	Benzine..........	187	86

COMPOSITION OF COMMON WATER CONSTITUENTS.

Chloride of sodium contains........	Na	39.3	and Cl	60.6
Chloride of magnesium contains...	Mg	25.28	and Cl	74.72
Chloride of calcium contains	Ca	36.06	and Cl	63.94
Chloride of potassium contains	K	52.45	and Cl	47.55
Carbonate of soda contains	Na O	58.5	and CO_2	41.5
Carbonate of magnesia contains...	Mg O	47.62	and CO_2	52.38
Carbonate of lime contains	Ca O	56.0	and CO_2	44.0
Carbonate of potassa contains	K O	68.17	and CO_2	31.83
Sulphate of soda contains	Na O	43.66	and SO_3	56.34
Sulphate of magnesia contains	Mg O	33.33	and SO_3	66.67
Sulphate of lime contains	Ca O	41.18	and SO_3	58.82
Sulphate of potassa contains.......	K O	54.08	and SO_3	45.92

Carbonate of lime multiplied by 0.56=lime.
Sulphate of baryta multiplied by 0.343=sulphuric acid.
Phosphate of magnesia multiplied by 0.036=magnesia.
Magnesia multiplied by 0.6=magnesium.
Magnesium multiplied by 1.66=magnesia.
Cubic centimeter carbonic acid multiplied by 0.002=carbonic acid in grams.
C. C. nitrate of silver solution multiplied by 0.0035=chlorine in grams.
Chloride of sodium multiplied by 0.39=sodium.
Carbonate of soda multiplied by 0.58=soda.
Chloride of potassium and platinum multiplied by 0.16=potassium.

In the construction of water analysis from constituents it is advisable, as most consistent with practical requirements to combine chlorine with magnesium (balance of chlorine with sodium or balance of magnesia with sulphuric acid).

Carbonic acid combines with lime, balance of lime with sulphuric acid, balance of sulphuric acid with soda (or balance of carbonic acid with magnesia).

When alkaline carbonates are present all the chlorine is to be combined with sodium. Magnesium carbonate and calcium sulphate are supposed not to coexist.

MILLIGRAMS PER LITER TO GRAINS PER U. S. GALLON.				GRAINS PER U. S. GALLON TO MILLIGRAMS PER LITER.			
Milligrams per Liter.	Grains per U. S. Gal.	Milligrams per Liter.	Grains per U. S. Gal.	Grains per U. S. Gal.	Milligrams per Liter.	Grains per U. S. Gal.	Milligrams per Liter.
1	0.058	26	1.519	1	17.1	26	444.9
2	0.117	27	1.578	2	34.2	27	462.0
3	0.175	28	1.636	3	51.3	28	479.1
4	0.234	29	1.695	4	68.4	29	496.2
5	0.292	30	1.753	5	85.6	30	513.4
6	0.351	31	1.812	6	102.7	31	530.5
7	0.409	32	1.870	7	119.8	32	547.6
8	0.468	83	1.929	8	136.9	33	564.7
9	0.526	34	1.987	9	154.0	34	581.8
10	0.584	35	2.045	10	171.1	85	598.9
11	0.643	36	2.104	11	188.2	36	616.0
12	0.701	37	2.162	12	205.3	87	633.1
13	0.760	38	2.221	13	222.5	38	650.3
14	0.818	39	2.279	14	239.6	39	667.4
15	0.877	40	2.338	15	256.7	40	684.5
16	0.935	41	2.396	16	273.8	41	701.6
17	0.993	42	2.454	17	290.9	42	718.7
18	1.052	43	2.513	18	308.0	48	735.8
19	1.110	44	2.571	19	325.1	44	752.9
20	1.169	45	2.630	20	342.2	45	770.0
21	1.227	46	2.688	21	359.4	46	787.2
22	1.286	47	2.747	22	376.5	47	804.3
23	1.344	48	2.805	23	393.6	48	821.4
24	1.403	40	2.864	24	410.7	49	838.5
25	1.461	50	2.922	25	427.8	50	855.6

EXPERIMENTS IN WORT COOLING.

The following tabulated experiments of the performance of a tubular refrigerator for wort cooling are gleaned from *Engineering*. The water and wort are moved in opposite directions, the former through thin metallic tubes, which are surrounded by the wort to be cooled:

Area of Cooling Surface of Refrigerator.	Wort.					Water.			
	Specific Gravity.	Quantity Passed through Per Hour	Initial Tempera-ture.	Final Tempera-ture.	Cooled Down.	Quantity Passed through Per Hour.	Initial Tempera-ture.	Final Tempera-ture.	Warmed up.
Sq. Feet.		Bbls.	Fahr	Fahr	Fahr	Bbls.	Fahr	Fahr	Fahr
1. 881	......	23.9	212°	72°	140°	61 1	65°	169°	104°
2. 514	1.104	36.1	155	59	96	75 5	54	100	46
3. 514	1.188	36.6	191	59	132	99.5	54	100	46
4. 514	1.035	47.3	193	59	134	90.7	54	100	46
5. 514	1.018	48.0	178	59	119	102.0	54	100	46

NOTE 1.—A barrel contains thirty-six gallons, or 360 pounds of water.

2.—The temperature of the air in Nos. 2 and 4 was 44° F. and in Nos. 3 and 5, 40°.

DIMENSIONS OF EXTRA STRONG PIPE.

Nominal Inside Diameter.	Actual Inside Diameter.	Actual Outside Diameter.	Thickness.	Internal Circumference.	External Circumference.	Length of Pipe per Sq. Ft. of Outside Surface	Internal Area.	External Area.	Length of Pipe Containing One Cubic Foot.	Weight per Foot of Length.
In.	In.	In.	In.	In.	In.	Ft.	In.	In.	Ft.	Lbs.
⅛	0.205	0.465	0.100	0.644	1.461	8.21	0.0329	0.1694	4377	0.29
¼	0.294	0.54	0.123	0.924	1.697	7.07	0.0678	0.2290	2124	0.54
⅜	0.421	0.675	0.127	1.323	2.121	5.66	0.1394	0.3573	1033	0.74
½	0.542	0.84	0.149	1.703	2.639	4.65	0.2307	0.5542	624.2	1.09
¾	0.736	1.05	0.157	2.312	3.299	3.67	0.4254	0.8650	338.7	1.39
1	0.951	1.315	0.182	2.988	4.131	2.90	0.7103	1.3582	202.7	2.17
1¼	1.272	1.66	0.194	3.990	5.215	2.30	1.2707	2.1642	113.3	3.00
1½	1.494	1.90	0.203	4.695	5.969	2.01	1.7530	2.8353	82.15	3.63
2	1.933	2.375	0.221	6.075	7.461	1.61	2.9345	4.4302	49.72	5.02
2½	2.315	2.875	0.280	7.304	9.032	1.33	4.1989	6.4918	34.28	7.67
3	2.892	3.5	0.304	9.085	10.996	1.09	6.5688	9.6211	21.91	10.25
3½	3.358	4.0	0.321	10.550	12.566	0.931	8.7561	12.5664	16.23	12.47
4	3.818	4.5	0.341	11.995	14.137	0.849	11.4608	15.9043	12.56	14.97
5	4.813	5.563	0.375	15.121	17.477	0.687	18.193	24.3010	7.915	20.54
6	5.750	6.625	0.437	18.064	20.813	0.576	25.967	34.4496	5.542	28.58

APPENDIX II.—PRACTICAL EXAMPLES.

INTRODUCTORY REMARKS.

The following practical examples, problems and questions have been discussed for a two-fold purpose. In the first place their object is to give to those not accustomed to the use of books an idea as to how the Compend may be utilized, and to show them in particular that the formulæ may be referred to by any man of ordinary acquaintance with the rules of common arithmetic; and to also show them how most questions can be answered without the use of such formulæ, by referring to more convenient rules or tables in the book or appendix of tables.

In the second place these problems are calculated to answer such questions as frequently occur in the refrigerating practice, and to discuss certain questions in a more direct way than it was practicable to do in the body of the book.

By carrying out the formulæ in numerical quantities in this appendix it was also intended to please those who profess a great preference in favor of formulæ written altogether in figures, and not with figures and letters of the alphabet mixed. It is also probable that by studying the solutions in this appendix more carefully they will discover the reasons why formulæ are thus written, viz.: In order to make the necessary distinction between constant numerical quantities which never change, and which therefore are given their constant numerical value in the formula and between the quantities which change with every example and which therefore are given in letters of the alphabet, for which the different values are to be inserted in every different example.

FORTIFYING AMMONIA CHARGE.

Q.—How many pounds (x) anhydrous ammonia should be added to 600 pounds of ammonia liquor in absorption machine showing 20° Beaumé (scale showing 10° in pure water) to make it 26° Beaumé?

From table on page 97 we find 20° Beaumé to correspond to 17 per cent, and 26° Beaumé to correspond to about 28 per cent of ammonia; hence in formula on page 285, m is equal to 600, $a = 20$, $b = 28$, and therefore $x =$

$$\frac{600\,(28 - 17)}{100 - 28} = \frac{600 \times 11}{72} = 97.6 \text{ pounds.}$$

EXAMPLES ON SPECIFIC HEAT.

QUESTION.—What amount of heat must be abstracted from 1,000 pounds of beer wort of 14 per cent to reduce its temperature from 70 to 40° F. ?

Specific heat c of wort from page 158 = 0.902 = 0.9, according to page 16: $S = c \times t \times w = 0.9 \times (70-40) \times 1000 = 0.9 \times 30 \times 1000 = 27000$ units.

Q.—What will be the final temperature T, if 1,000 pounds of beer wort of 14 per cent and of a temperature of 180° degrees are mixed with 1,200 pounds of water of 60° F.?

In accordance with page 17 we find—

$$T = \frac{w\,t\,s + w_1\,t_1\,s_1}{w\,s + w_1\,s_1} = \frac{1200 \times 60 \times 1 + 1000 \times 180 \times 0.9}{1200 \times 1 \times 1000 \times 0.9} = \frac{23400}{2100} = 111.4^\circ \text{ F.}$$

EVAPORATIVE POWER OF COAL.

Q.—If a lignite contains 60 per cent of carbon (C) and 5 per cent of hydrogen (H), what will be its evaporative power (e) expressed in pounds of water?

From page 37 we find—

$e = .15\,(C + 4.29\,H) = 0.15\,(60 + 5 \times 4.29) = 12.21$ pounds.

CAPACITY OF FREEZING MIXTURE.

Q.—How many pounds of ammonia nitrate must be dissolved in so many pounds of water to obtain a theoretical refrigerating effect equal to one ton ice melting capacity = 284,000 uuits?

On page 32 we find that 1 pound will reduce the temperature from 40° to 4°, which is equivalent to a refrigerating effect of $2 \times (40-4) = 72$ units if we assume the specific heat of the solution equal to that of water = 1. Hence $\frac{284000}{72}$ = about 4,000 pounds of the salt must be dissolved in 4,000 pounds of water to obtain the required effect theoretically; practically it would take a great deal more.

EXAMPLES ON PERMANENT GASES.

Q.—If the volume V of a gas is ten cubic feet at a pressure of eighteen atmospheres and a temperature of 40°, what will be its volume V_1 if expanded to a pressure of one atmosphere and a temperature of —80°?

Examples of this kind occur quite frequently, and their study will be found very instructive and profitable.

The formula at bottom of page 48 gives—

$$V_1 = V\frac{p\,(t_1+461)}{p_1\,(t+461)} = 10\,\frac{18\,(-80+461)}{1\,(40+461)} = 10\,\frac{18\times381}{1\times501} = 137\text{c.ft}$$

Q.—What volume x in cubic feet is occupied by 180 cubic feet of a permanent gas if its temperature is reduced from 40° to —80° F.?

According to page 55 the volume of a gas is proportional in its *absolute* temperature. Hence we have—

$$180 : x = (40+461) : (-80+461) = 501 : 381$$

$$\text{or } x = \frac{180\times381}{501} = 137 \text{ cubic feet.}$$

EXAMPLES SHOWING USE OF GAS EQUATION.

Q.—What will be the pressure of a confined volume of air at a temperature 45° if its pressure at 32° F. is equal to one atmosphere?

According to the equation for perfect gases, page 55, is:

$$p\,v = \frac{T}{493} = \frac{45+461}{493} \text{ or } v \text{ remaining unit.}$$

$$p = \frac{506}{493} = 1.03 \text{ atmospheres or thereabouts.}$$

Q.—What will be the volume of one cubic foot of air if heated at constant pressure from 32° to 45° F., its pressure at the former temperature being one atmosphere?

According to the same equation we find—

$$p\,v = \frac{T}{493} = \frac{45+461}{493} \text{ or } p \text{ remaining unit.}$$

$$v = \frac{506}{493} = 1.03 \text{ cubic feet.}$$

Q.—If the volume of a confined body of a permanent gas be one cubic foot at the temperature of 32° and at a pressure of one atmosphere, what will have to be its temperature T in order that it may occupy a volume of one-half cubic foot at a pressure of four atmospheres?

The same equation answers the question, viz.:

$$p\,v = \frac{T}{493} \text{ or } T = 493\,p\,v = 493\times4\times\tfrac{1}{2} = 986° \text{ F. absolute.}$$

or 986 — 461 = 525° F.

WORK REQUIRED TO LIFT HEAT.

Q.—What amount of work must be expended theoretically by a perfect refrigerating machine to withdraw 284,000 units of heat (one ton refrigeration) from a refrigerator at temperature of 10° if the temperature in the condenser is 90°?

From the equation, page 71,

$$\frac{H}{W} = \frac{T}{T_1 - T_0} \text{ we find—}$$

$$W = \frac{H(T_1 - T_0)}{T_0} = \frac{284000\ (90-10)}{(461+10)} = 49000 \text{ units.}$$

The work is here expressed in heat units, which are equivalent to:

49,000 × 772 = 37,830,000 foot-pounds (page 346) or to

$$\frac{49000}{2565 \cdot \times 24} = \frac{49000}{61560} = 0.8\text{-horse power per day (page 346).}$$

REFRIGERATING EFFECT OF SULPHUROUS ACID.

Q.—What is the theoretical refrigerating effect r of one pound and of one cubic foot of sulphurous acid if used in a compression machine, the temperature in refrigerator coils being 5° and in condenser coils 95° F.?

The equation $r = h_1 - (t - t_1) s$, on page 115, applies also for sulphurous acid, for which we find $h_1 = 171$ units (page 250) and $s = 0.41$ (page 250); hence—

$$r = 161 - (95-5)\,0.41\ 161 - 37 = 124 \text{ units.}$$

From same table we find the weight of one cubic foot of sulphurous acid at 5° equal to 0.153 pounds; hence the refrigeration of one cubic foot is—

$$124 \times 0.153 = 18.97 \text{ units.}$$

REFRIGERATING CAPACITY OF A COMPRESSOR.

Q. What is the refrigerating capacity of a double-acting compressor, 70 revolutions per minute, diameter 9⅞ inches and stroke 16½ inches, temperature in refrigerator coil 5° and in condenser coil 85° F.?

The compressor volume C per minute after formula on page 303 is—

$$C = d^2 \times l \times m \times 0.785 = 9⅞^2 \times 16½ \times 0.785 \times 70.$$

From table on page 314 we find $9⅞^2 \times 0.785 = 76.58$.

Hence $C = 76.58 \times 16.5 \times 70 = 88410$ cubic inches.

The compressor being double-acting, this is equal to

$$\frac{88410 \times 2}{1728} = 102 \text{ cubic feet.}$$

From table on page 125 we find that 3.34 cubic feet of ammonia must be pumped per minute, at above named condenser and compressor temperature to produce a refrigerating effect of one ton in twenty-four hours, hence the above compressor represents a theoretical refrigerating efficiency of—

$$\frac{102}{3.34} = 30.5 \text{ tons.}$$

SECOND METHOD OF CALCULATION.

The actual refrigeration will be from 15 to 20 per cent less, or equivalent to about 25 tons (commercial capacity ?); see table page 302, according to which the nominal daily refrigerating capacity is—

$$\frac{Vm}{4} = \frac{102}{4} = 25.5 \text{ tons.}$$

The agreement between this amount and the amount found by the first calculation holds good only for the temperature selected; otherwise the last rule affords only a crude approximation.

THIRD METHOD OF CALCULATION.

The theoretical refrigerating effect R of this compressor can also be calculated after the formula on page 118—

$$R = \frac{C \times 60 \times r}{v}$$

We find, according to formula on page 115 and table on page 94, $r = h, - (t - t_1)\, s = 552.43 - (85 - 5)\, 1 = 552{-}80 = 472$ units, and $v = 8.06$ (page 94)—

$$R = \frac{102 \times 60 \times 472}{8.06} \text{ units or in tons per day—}$$

$$R = \frac{102 \times 60 \times 472 \times 24}{8 \times 284.000.} = 29.8 \text{ tons.}$$

Or, again, the actual refrigerating capacity will be about 15 to 20 per cent less, or equivalent to about twenty-five tons, and the actual ice making capacity will be about thirteen tons per twenty-four hours. The last method of calculation will answer also for other refrigerating media if r and v are found and inserted accordingly.

COOLING WORT.

Q.—A direct expansion ammonia refrigerating machine is applied to the cooling of beer wort, and reduces the temperature of 300 barrels of wort from 70° to 40° F. in four hours. What is the refrigerating capacity U of the machine if the weight of the wort is 14° Balling?

From table on page 202 we find the specific gravity corresponding to 14° equal to 1.0572 (1.06), and from table on page 197 we find the specific heat 0.895 (0.9), hence in accordance with formula on page 198—

$$U\ B \times 259 \times g \times s\,(70 - 40) \quad 300 \times 259 \times 1.06 \times 0.9 \times 30 \quad 222300 \text{ units.}$$

This is the refrigeration in three hours; expressed for twenty-four hours, and in tons of refrigeration, it is equal to—

$$\frac{2223000 \times 8}{284000} \quad \text{about 60 tons.}$$

The actual refrigeration required to cool the wort is only one-eighth of that (for three hours), *i. e.*, 7½ tons, which is about one ton for every forty barrels. The rule on page 199 allows one ton for every thirty-eight barrels.

HEAT BY ABSORPTION OF AMMONIA.

Q.—What is the heat H_n developed when one pound of ammonia vapor is absorbed by enough of a 20 per cent solution of ammonia in order to produce a 33 per cent solution of ammonia?

On page 226 we find—

$$H_n = 925 - \frac{248 + 142\,b}{n} \text{ units.}$$

In accordance with the definitions given on page 226 we find n, that is the number of pounds of water present to one pound of ammonia in a 20 per cent solution, $= \frac{80}{20} = 4$, and the number of pounds of ammonia ($b + 1$ pounds) which are present for every four pounds of water in a 33 per cent solution $\frac{4 \times 33}{67}$ 2 pounds.

$b + 1$ being $= 2$, it follows $b = 1$.

We now insert these values in the above equation for H_n:

$$H_n = 925 - \frac{284 + 142 \times 1}{4} = 819 \text{ units.}$$

RICH LIQUOR TO BE CIRCULATED.

Q.—How many pounds P_2 of rich liquor of 33 per cent strength must be circulated in an ammonia absorption plant if the poor liquor enters the absorber at 20 per cent strength?

We find this in accordance with equation on page 224:

$$P_2 = \frac{(100 - a)\ 100}{(100 - a)\,c - (100 - c)\,a} = \frac{(100 - 20)\ 100}{(100 - 20)\,33 - (100 - 33)\,20}$$

$$= \frac{8000}{1300} = 6.1 \text{ pounds.}$$

CAPACITY OF ABSORPTION MACHINE.

Q.—What should be the theoretical refrigerating capacity R of an ammonia absorption machine if the rich liquor is 33 per cent and the poor liquor 20 per cent strong, and if the ammonia pump makes fifty revolutions per minute and each stroke is seven inches and the diameter of pump piston three inches?

From page 139 we find the capacity C of this pump to be 2.14 gallons at ten strokes per minute, hence at fifty strokes it is expressed in pounds—

$$C = 2.14 \times 5 \times 8.3 = 90 \text{ pounds in round figures.}$$

By calculating as before from table on page 226 we find $P_2 = 6.1$, and according to formula on page 230 we find—

$$R = \frac{C \times r}{P_2} = \frac{90 \times 453}{6.1} = 6700 \text{ units per minute.}$$

The value for r is found after the rule on page 115, which, assuming the temperature in refrigerator coils to be 4° and the pressure in condenser to 210 pounds, equivalent to an absolute pressure of 225 pounds and to a temperature of 104° F. (see table on page 94), reads—

$$r = h_1 - (t - t_1)\, s = 553 - (104 - 4)\, 1 = 453 \text{ units.}$$

$$h_1 = 553 \text{ units from table on page 94.}$$

One ton of refrigerating capacity per day being equal to 284000 units, one ton per hour is equal to about 12000 units, and one ton of refrigerating capacity per minute is equivalent to 200 units per minute, and therefore the above refrigeration is equivalent to—

$$\frac{6700}{200} = 33.5 \text{ tons per day (theoretical capacity).}$$

If we allow 25 per cent for losses all around slips of pumps, radiation, etc., we find the actual refrigerating capacity $33.5 - 8.4 = 25.1$ tons, and the actual ice making capacity being about half of that = twelve tons per day.

HEAT AND STEAM REQUIRED.

What is the theoretical amount of heat W_1 and of steam P_6 required in still of above plant?

From page 230 we find—

$W_1 = H_n - (h_2 - h) = 819 - (495 - 489) = 813$ units, h_2 at a temperature of 90° in absorber being 495 units, h and at a temperature of 104° in condenser being 489 units, after table on page 94. *H* 819, from table on page 226. The amount of steam P_s in pounds required per hour to run this plant would be (see page 229):

$$P_s \quad \frac{W_1 \times m \times 28400}{24 \times r \times h_s} \quad \frac{813 \times 25 \times 284000}{24 \times 453 \times 686} \quad 770$$

pounds of steam per hour.

$h_s = 886$ (at pressure in boiler 100 absolute or 85 pounds gauge pressure). To this should be added about $\frac{1}{6}$ to allow for steam to run the ammonia pump, so that the whole would amount to about 900 pounds per hour.

COLD STORAGE EXAMPLES.

Q.—What is the refrigeration *R* required for a local storage room 40×50×10 if each day about 30,000 pounds of fresh meat (about 120 hogs) are placed in the same at a temperature of 95° to be chilled to a temperature of 35, if the temperature of atmosphere is to be 85° F.?

METHOD OF CALCULATION.

The side walls of room 2×50×10+2×40×10=1,800 sq.ft.
The ceiling and floors 2×40×50 =4,000 "

Total..............................5,800 sq.ft.

If we take *n* as 3 all around (assuming an average degree of insulation (see page 181), we have—

$$R = \frac{f\,n\,(t-t)}{284000} = \frac{5800 \times 3\,(85-35)}{284000} = 3.1 \text{ tons of re-}$$

frigeration per day to keep the room at the desired temperature.

The additional refrigeration to chill the meat, assuming its specific heat to be 0.8, we find (page 183)—

$$R = \frac{P\,(t-t_1)}{355000} = \frac{30000\,(95-35)}{355000} = 5 \text{ tons,}$$

which makes a total refrigeration of 8.1 tons required. For closer estimates the rules on pages 181 and 182 may be used.

APPROXIMATE ESTIMATE.

The cubic contents of the room are equal to 20,000 cubic feet, and in accordance with the rules on page 173 from fifty to 100 units (say seventy-five units in this case) are allowed per cubic foot, and in addition to that about 50 per cent more for chilling, which amounts to about 110 units in all per cubic feet, or a daily refrigeration of

$$\frac{20000 \times 110}{284000} = 8 \text{ tons in round figures.}$$

For opening doors, for windows, etc., about 10 to 15 per cent extra refrigeration may be allowed, making the total about nine tons refrigerating capacity per day. See also rules on page 212 and 213.

PIPING REQUIRED.

Q.—What will be the amount of 2-inch pipe direct circulation required for the above room and purpose?

In accordance with rule on page 128 we assume that one square foot of pipe will convey about 2,500 units of refrigeration; this is equal to 1.6 foot of 2-inch pipe (table on page 129), hence to distribute nine tons in twenty-four hours the pipe required will be—

$$\frac{9 \times 284000 \times \cdot 1.6}{2500} = 1600 \text{ feet of 2-inch pipe.}$$

According to another rule, given on page 212, one running foot of 2-inch pipe is allowed for thirteen cubic feet chilling room capacity, in accordance with which $\frac{20,000}{13} = 1,540$ feet or thereabouts of 2-inch pipe would be required.

After still another rule, given on page 212, we find that six feet 2-inch pipe are allowed per hog slaughtered in chilling room; according to this rule we would only require 720 feet of 2-inch pipe for above room, but the rule from which this result is obtained applies to large installations having over a hundred times the capacity contemplated in the example as given and calculated above.

EXAMPLES ON NATURAL GAS.

Q.—What amount of refrigeration and work can be produced by natural gas expanding adiabatically from a pressure of 255 pounds (seventeen atmospheres) to a pressure of fifteen pounds (one atmosphere) absolute pressure, and to a volume of 1,000,000 cubic feet at the ordinary temperature and pressure?

TEMPERATURE AFTER EXPANSION.

If we assume the initial temperature of the gas to be $70^\circ = 70 + 461 = 531^\circ$ absolute we find the temperature T_2 of the gas after expansion after the rule on page 257, viz:

$$T_2 - T\left(\frac{P_2}{P_1}\right)^{\frac{k-1}{k}} = 531\left(\frac{1}{17}\right)^{\frac{1.41-1}{1.41}} = 531\left(\frac{1}{17}\right)^{0.291} \text{ or}$$

$\log. T_2 = \log. 531 \times 0.291 (\log. 1 - \log. 17) = 2.7251 - 0.6432 = 2.0819.$

$T_2 =$ num. log. $2.0819 = 121^\circ$ absolute $= 461 - 121 = -340^\circ$ F.

REFRIGERATING CAPACITY.

The theoretical refrigeration H produced by 1,000,000 cubic feet expanded in this manner if the gas leaves the refrigerator at the temperature T_0 of $5^\circ = 466^\circ$ absolute is found after the formula on page 257.

$$H = mkc\,(T_0 - T_2) = mkc\,(466 - 121)$$

$c = 0.468$ (page 47)

$m = 0.0316$ pounds (page 233 coal gas) hence

$H = 0.0316 \times 0.468 \times 1.41 \times 345 = 7.0$ units per cubic foot or 7,000,000 units per 1,000,000 cubic feet, which is equivalent to a theoretical refrigerating capacity of about twenty-five tons. The actual ice making capacity would probably be less than ten tons per day.

WORK DONE BY EXPANSION.

The amount of work, Wm, that can be obtained theoretically by the adiabatic expansion to 1,000,000 cubic feet of the gas is expressed by the formula—

$$Wm = \frac{mkc}{A}(T - T_2) = 0.0316 \times 0.468 \times 1.41\,(531 - 121) \times 772 = \text{about } 6600 \text{ foot-pounds}$$

per cubic foot, or for 1,000,000 cubic feet per day

$$\frac{6600 \times 1000000}{1980000 \times 24} = \text{about 147 horse powers per day.}$$

According to this calculation the power to be gained would be of considerably more consequence than the ice, but it must not be forgotten that these are theoretical calculations which are naturally greatly reduced in practical working, not to speak of possible difficulties connected with the same.

SIZE OF EXPANDING ENGINE.

As the expanded gas leaves the expanding engine at the temperature of 121° absolute, its volume x is less in the following proportion—

$$1000000 : x = 531 : 121 \text{ (page 55)}$$

$$x = \frac{121 \times 1000000}{531} = 228000 \text{ cubic feet.}$$

This is the volume over which the piston of expanding engine must sweep in one day. If it is double-acting and makes fifty revolutions a minute the size of the cylinder must be—

$$\frac{228000}{50 \times 2 \times 24 \times 60} = \frac{228}{144} = 1.6 \text{ cubic feet.}$$

If the stroke be two feet the area of piston must be 0.8 square feet.

EXPANSION WITHOUT DOING WORK.

Q.—What amount of refrigeration can be produced by natural gas expanding from a pressure of 255 pounds, absolute, to a volume of 1,000,000 cubic feet at the atmospheric pressure without doing work?

REFRIGERATION OBTAINABLE BY EXPANSION ALONE.

For the sake of simplicity we neglect the contraction of the gas due to reduction of temperature, and allow the theoretical refrigeration to be equivalent to the external work, E, done by the expanding gas, which can be found by the formula for steam on page 106—

$$E = \frac{P\,(v - v_1)}{J}.$$

v representing the final volume and v_1 the original volume of the expanding gas, and calculated for one cubic foot; hence—

$$E = \frac{15\,(17 - 1)}{772} = 0.31 \text{ units per cubic foot.}$$

or 310,000 units for 1,000,000 cubic feet of gas, of which only a fraction could be utilized for ice production, which would probably be less than one-third ton per day.

CALCULATION OF REFRIGERATING DUTY.

Q. A machine is required to cool water from 55° F. to 40° F. during part of the day, and to keep a cold storage at 15° F. at other part of the day. What indicated horse power steam engine will be required to work compressors to extract 3,000,000 B. T. U. per hour from the water at above temperatures, and what size compressors, with number of revolutions per minute? What B. T. U. per hour would same machine extract at same speed when working on the cold storage, and what would then be its indicated horse power? Condensing water at 60° and leaving condenser at 70° F.

If we assume that you work by direct expansion, the temperature of the expanding ammonia would have to be about 10° lower than the water after it is cooled, *i. e.*, 30°; consequently by using the latent heat of vaporization at that temperature, as we find it in table on page 94, viz., 536, and formula on page 115 of Compend we find—

$$r = 536 - (70-30)\ 1 = 496 \text{ units,}$$

which is the refrigerating effect of one pound of ammonia, when the temperature of the refrigerator is 30° and that of the condenser 70°, the specific heat of the ammonia being 1.

The amount of ammonia to be evaporated per minute is, therefore—

$$\frac{3{,}000{,}000}{496 \times 60} = 101 \text{ pounds.}$$

From same table on page 94, we find volume of one pound of ammonia vapor at 30° = 4.75 cubic feet, consequently the compressor capacity per minute will have to be—

$$101 \times 4.75 = 480 \text{ cubic feet in round numbers.}$$

If we add to this 20 per cent for clearance losses by radiation, etc., we require an actual compressor capacity of 576 cubic feet per minute. If we assume that the work is to be done by one double-acting compressor, making, say, seventy revolutions per minute, we require a compressor having the cubic capacity of—

$$\frac{576}{72 \times 2} = 4.2 \text{ cubic feet.}$$

If we distribute this capacity over two compressor cylinders each one has to have a volume of 2.1 cubic feet.

Taking the diameter of each of them at fifteen inches the area is ($1.25^2 \times 0.785$) 1.227 cubic feet, and the stroke will have to be—

$$\frac{2.1}{1.227} = 17.12 \text{ inches.}$$

If we start from a different given stroke and number of revolutions, as we probably shall, the diameter changes accordingly, after the foregoing simple rule.

If a single double-acting compressor making fifty revolutions were to do the work, its dimensions, calculated on the same basis as above, would be twenty inches diameter by 31¼-inch stroke.

The work of the compressor is found after the formula on page 119:

$$W = 0.0234\ WK \text{ horse power};$$

or $W_4 = 0.0234 \frac{t - t_1}{T'} h_1 \times K = 0.0234 \times \frac{70-30}{490} \times 536 \times 101 = 104$ horse power.

And the horse power of engine, after rule on page 121 of Compend, is found to be—

$$104 \times 1.4 = 145.6 \text{ horse power.}$$

The same two compressors, if required to do duty in a cold storage plant, would probably have to run with a temperature of 5° F. in refrigerator. In this case (their cubic capacity being 576 cubic feet per minute) their refrigerating capacity in tons per day is found by the formula on page 303 of Compend, viz.:

$R = \frac{C(h_1 - t + t_1)}{200v} = \frac{576(546 - 70 + 5)}{200 \times 6.49} = 212$ tons in round figures (h_1 and v being found from table on page 94). Or in thermal units per hour—

$$\frac{212 \times 284000}{24} = 2500000 \text{ units.}$$

This is the theoretical capacity; to bring it on a practical basis, we have to subtract 20 per cent, as we did in the case of water cooling before this yields 2500000—500000 = 2000000 units per hour actual refrigerating capacity for cold storage.

To find the horse power of the compressor in this case we find the amount of ammonia to be circulated in a minute, as before, viz.:

$$r = 546 - (70 - 15) = 491, \text{ and}$$

$$K = \frac{2000000}{491 \times 60} = 68 \text{ pounds.}$$

Placing this value in the equation from page 119, as before, we find—

$$W4 = 0.0234 \times \frac{70 - 5}{475} \times 546 \times 68 = 118.7 \text{ horse power};$$

and the horse power of engine—

$$118.7 \times 1.4 = 166 \text{ horse power.}$$

These horse power are calculated from the amount of ammonia theoretically required, and about 15 to 25 per cent should be added to bring them within practical range. We have also assumed that the temperature in condenser is that of outflowing condenser water, when in fact it should be taken 5° higher.

CALCULATING ICE MAKING CAPACITY.

Q.—What is the ice making capacity of two single-acting compressors 7×12 inches, 100 revolutions per minute?

The capacity in cubic feet, C, for each compressor per minute, according to formula on page 117 of Compend, is—

$$C = r^2 \times 3.145 \times 6 \times m$$

$$\frac{7^2 \times 3.145 \times 12 \times 100}{1728} = 26.7 \text{ cubic feet,}$$

or for both compressors, $26.7 \times 2 = 53.4$ cubic feet, which under general conditions, when no back pressure, etc., is mentioned, has been calculated to be equivalent to $\frac{53.4}{4}$ 13.35 tons of refrigerating capacity in twenty-four hours (see page 118 of Compend), and of this from $\frac{4}{10}$ to $\frac{6}{10}$ is available actual ice making capacity, which accordingly is about seven tons per day (more or less; see page 144 of Compend).

VOLUME OF CARBONIC ACID GAS.

Q.—What is the volume of one pound of carbonic acid gas at a pressure of thirty pounds and at a temperature of 50°?

The formula that applies here is given on page 48 of the Compend, viz.:

$$V^1 \quad \frac{Vp\,(t_1 + 461)}{p_1\,(t + 461)}$$

If in this formula we insert for V the volume of one pound of carbonic acid gas at the atmospheric pressure, viz., 8.5 cubic feet, and for p the pressure of the atmosphere, viz., 14.7 pounds, and for t the temperature of 32° F. this formula becomes:

$$V^1 = \frac{8.5\,(461 + t^1)\,14.7}{493\,p^1} \quad \frac{115 + 0.25\,t_1}{p_1}$$

Hence the volume V^1 of one pound of carbonic acid gas at any given temperature and pressure, say at an absolute pressure of thirty pounds. and a temperature of 50°, is found by inserting these quantities in the foregoing formula:

$$V^1 = \frac{115 + 0.25 \times 50}{30} = \frac{127.5}{30} = 4.25 \text{ cubic feet.}$$

For apparent reasons the numerical results of above examples have been rounded off in most cases.

HORSE POWER OF STEAM ENGINE.

Q.—What is the horse power of a steam engine the piston of which has a diameter of 12 inches, a stroke of 30 inches at 90 revolutions a minute if the gauge pressure of the steam is 80 pounds, cut-off $\frac{1}{8}$?

To calculate the horse power in this case we have to find the mean effective pressure by means of an indicator diagram, as shown on page 297. If it is impracticable to obtain a diagram we take the mean average pressure as we find it in table on page 349, which is 49.4 pounds, or 50 pounds in round numbers, in this case. Multiply the same by the area of the piston in square inches and the speed of the piston in feet per minute, and divide the product by 33,000 (foot-pounds of horse power per minute. See table on page 347).

The area of piston in square inches we find, according to rule given on page 309, equal to—

$$12^2 \times 0.7854 = 144 \times 0.785 = 113.0,$$

which is also given direct in table on page 314.

The piston speed is—

$$\frac{30 \times 90 \times 2}{12} = 450 \text{ feet per minute.}$$

Hence the horse power—

$$\frac{113 \times 450 \times 50}{33,000} = 77 \text{ horse power.}$$

This is the indicated horse power, the net effective horse power being the indicated horse power less the friction of the engine.

The table on Corliss engine, on page 340, gives the indicated horse power of an engine of above description at 54, this difference being probably due to a difference in the mean effective pressure and to an allowance for piston space having been made in the latter case.

WORK OF COMPRESSOR.

Q.—What is the work of compression done by a double-acting ammonia compressor 9 inches in diameter, 15 inches stroke at 70 revolutions per minute? The back pressure is 28 pounds and the condenser pressure 115 pounds.

This problem is calculated on the same principles as the foregoing example; but, as in that case, the proper way is to obtain the mean effective pressure from an indicator diagram. If we use the table on page 298 instead we find the mean pressure in this case at 52.6

pounds. The area of piston, by table on page 314, is 63.6 square inches, and its travel per minute equal to—

$$\frac{2\times70\times15}{12}=116 \text{ feet;}$$

hence the work done by the compressor is equal to—

$$\frac{63.6\times116\times52.6}{33,000}=11.6 \text{ horse power.}$$

This is the indicated horse power of the work done by the compressor. In order to find the indicated horse power (of an engine) required to do this work we must add to the above the work required to overcome the friction in the compressor as well as in the engine itself.

CALCULATION OF PUMP.

Q.—How many revolutions must be made by a single-acting pump having a piston of 4 inches in diameter and 12 inches stroke in order to force 400 gallons of water 60 feet above the level of pump per hour, and what will be the power required to do this work?

According to table on page 322 the displacement by this pump for each stroke is 0.653 gallons; hence—

$$\frac{400}{.653}=605 \text{ strokes;}$$

or, in round numbers, 600 strokes must be made per hour; and as the pump is single-acting this corresponds to 600 revolutions per hour, or ten per minute. The work done by this pump in lifting the water may be calculated the same way as the work done by a compressor, by simply inserting, instead of the mean average pressure, the pressure corresponding to a water column of 60 feet in height, viz.: 26 pounds in round numbers, as per table on page 326.

WATER POWER.

Q.—1. What is the power of a water fall twenty feet high and 300,000 cubic feet of water per minute? 2. What amount of coal and steam respectively would give the same power during twenty-four hours?

In accordance with page 108 one cubic foot of water weighs 62.5 pounds; hence by using rule on water power given on page 43, we find the theoretical power of the water fall in question equal to—

$$\frac{300,000\times62\frac{1}{2}\times20}{33,000}=12,400 \text{ horse power.}$$

Of this theoretical effect may be utilized 30 to 75 per cent by water motors, according to construction, etc.; 50 to

75 per cent by turbines; 70 to 80 per cent by water pressure engines (generally not used for falls less than fifty feet in height). Taking 50 per cent as a safe basis, the actual work that can be expected from the fall would be equal to—

$$\frac{12,400}{2} = 6,200 \text{ horse power}$$

This power would of course still be correspondingly reduced if the mechanical power of the water motor had to be converted into electricity, to be transmitted to a distant locality, there to be converted into mechanical power again. Leaving this out of the question, and assuming that electricity was the form of energy wanted, we find, from page 108 of the Compend, that from fifteen to thirteen pounds of steam will produce a horse power per hour, and that a pound of average coal will make about eight pounds of steam; hence a horse power will require not over two pounds of coal per hour with a good engine, and therefore 6,200 horse power may be estimated equivalent to 6,200×24×2=297,600 pounds of coal in twenty-four hours.

Allowing fifteen pounds of steam per horse power, the actual power of the water fall would be represented by 6,200×15=93,000 pounds of steam per hour.

With first-class machinery it would take less steam and coal.

MOTIVE POWER OF LIQUID AIR.

Q.—What is the amount of work expressed in foot-pounds and in horse power that can be done by one pound of liquid air while expanding or volatilizing at the constant temperature of 70°, this being the average atmospheric temperature?

According to page 260, we find the work, W_1, in foot-pounds, which can be done theoretically by the isothermal expansion of V_1, cubic feet of liquid air to the volume of V cubic feet and the pressure P (in pounds per square foot) after the formula—

$$W_1 = P\,V \times 2.3026 \text{ by } \frac{V}{V_1}$$

In the problem on hand we have $P = 2,117$ pounds. V_1, the volume of one pound of liquid air, is not exactly known but $\frac{V}{V_1}$, the ratio of the volume after and before expansion, is about 800, and V_1 the final volume of one pound of air in cubic feet at 70° F. and at atmospheric pressure is equal to about 13.34 cubic feet. Hence the formula developes into—

$$W_1 = 2{,}117 \times 13.34 \times 2.3020 \text{ log. } 800 = 188{,}800 \text{ foot-pounds.}$$

(Log. 800 being equal to 2.9031. See table on page 316.)

In order to express this effect in horse power, the time in which the pound of air is to expand should be stated also. Assuming that it takes place at the rate of one pound of liquid air expanding per minute, the horse power would be—

$$\frac{188{,}800}{33{,}000} = 5.72 \text{ horse power.}$$

This is the theoretical figure; practically, a reduction would have to be made for friction, etc.

MOISTURE IN COLD STORAGE.

Q.—Assuming that 34° is the proper temperature for an egg storage room, what is the proper percentage of moisture which it should contain, and how should the wet bulb thermometer of a hygrometer or sling psychrometer stand in order to indicate that percentage of moisture?

According to Cooper the percentage of moisture for cold storage rooms, especially for eggs, should vary with the temperature as follows:

Temperature in Degrees F.	Relative Humidity, Per Cent.	Temperature in Degrees F.	Rel'tive Humidity, Per Cent.
28	80	35	65
29	78	36	62
30	76	37	60
31	74	38	58
32	71	39	56
33	69	40	53
34	67		

Therefore for a storage temperature of 34° the moisture, or relative humidity, should be 67 per cent (100 per cent corresponding to air saturated with moisture), and by referring to table on page 112 we find that this corresponds to a difference between the dry and the wet bulb of 3.5°. Hence the wet bulb thermometer should show 34—3.5=30.5°.

CARBONIC ACID VS. AMMONIA.

Q.—We would like to ask you for some information on ice machines, as to how the carbonic anhydride ice machines are in comparison with the ammonia ice machines. The carbonic anhydride machine people claim their machine far superior to the ammonia machine. They also claim that carbonic anhydride has more freezing power than ammonia. Is this in accordance with your statement in *Ice and Refrigeration* (see same, page 247 of Compend) or not?

This question, which was directed to the author of the Compend personally, would indicate that his statements with reference to this matter were misunderstood, or at least apt to misconstruction. The superiority of the carbonic acid machine would of course tally with 4,300 and 3,700 calories per horse power; but these figures were quoted by the author as phenomenal, in fact as mere claims, unsupported, so far at least, by any authentical tests. The author of the Compend has taken great pains to find any tests supporting such claims, or to find a carbonic anhydride machine which would give some such results in actual practice, but so far has failed to find any. On the contrary, we have come to the conclusion that the results of the practical comparative tests given in the tables on page 247 of Compend have not been materially exceeded so far, at least not with machines without expansion cylinder, and only such are in the market at present, as far as we know.

As a result of the present status of the theoretical aspect of the questions it appears that at temperatures of 70° before the expansion valve and 20° in refrigeration coils it will take 1.2 horse power in a carbonic anhydride machine to produce the same refrigeration as one horse power in an ammonia refrigerating machine. Hence the advantages of the carbonic acid machine must be looked for in other directions rather than in that of greater efficiency.

APPENDIX III.—LITERATURE ON THERMODYNAMICS, ETC.

a.—BOOKS.

ATKINSON, E.—Ganot's Elementary Treatise on Physics Experimental and Applied; New York, 1883.

BERTHELOT, E.—Mecanique Chimique, two vols.; Paris, 1880.

BEHREND, GOTTLIEB.—Eis und Kälteerzeugungs Maschinen; Halle a.S., 1888,

CARNOT, N. L. S.—Reflections on the Motive Power of Heat; translated by Thurston; New York, 1890.

CLAUSIUS, R.—Die Mechanische Wärmetheorie, three vols.; Braunschweig, 1891.

CLARK D. KINNEAR.—The Mechanical Engineer's Pocket Book; New York, 1892.

COOPER, MADISON.—Eggs in Cold Storage; Chicago, 1899.

DUEHRING, E.—Principien der Mechanic; Leipzig, 1877.

EWING, S. A.—The Steam Engine and Other Heat Engines; Cambridge, 1884.

EDDY, HENRY T.—Thermodynamics; New York, 1879.

FARADAY, M.—Conservation of Force; London, 1857.

FISHER, FERDINAND, DR.—Das Wasser; Berlin, 1891.

GRASHOF, F.—Hydraulik Nebst Mechanische Waermetheorie; Leipzig, 1875.

GAGE, ALFRED P.—A Text Book on the Element of Physics; Boston, 1885.

GIBBS, WILLARD J.—Thermodynamisches Studien, translated by W. Ostwald; Leipzig, 1892.

HELM, G.—Energetik Der Chemischen Erscheinungen; Leipzig, 1894.

HELM, GEORGE.—Die Lehre von der Energie; Leipzig, 1887.

HELMHOLTZ, H.—Erhaltung der Kraft; Berlin, 1847.

HELMHOLTZ, H.—Wechselwirkung der Naturkraefte; Koenigsberg 1854.

HERING, C.—Principles of Dynamo Electric Machines; New York, 1890.

HIRN, G. A.—Equivalent Mecanique de la Chaleur; Paris, 1858.

HIRN, G. A.—Théorie Mécanique de la Chaleur; Paris, 1876.

HOFF, J. H. VAN'T.—Chemische Dynamik; Amsterdam, 1884.

JOULE, J. P.—Scientific Papers; London, 1884.

JEUFFRET, E.—Introduction a la Théorie de l'Energie; Paris, 1883.

KIMBALL, ARTHUR L.—The Physical Properties of Gases; Boston and New York, 1890.

KENNEDY, ALEX. C.—Compressed Air; New York, 1892.

LEDOUX, M.—Ice Making Machines; New York, 1879.

LEAR, VAN J. J.—Die Thermodynamik in der Chemie; Leipzig, 1893

LEASK, A. R.—Refrigerating Machinery; London, 1895.

LEDOUX, M.—Ice Making Machines, with Additions by Messrs. Denton, Jacobus and Riesenberger; New York, 1892.

LORENZ, HANS.—Neuere Kuehlmaschinen; Muenchen und Leipzig; 1899.

MARCHENA, R. E. DE.—Machines frigorifiques a gas liquifiable; Paris, 1894.

MAYER, J. R.—The Forces of Inorganic Nature, 1842. Translated by Tyndall.

MAYER, J. R.—Mechanik der Waerme; Stuttgart, 1847.

MAYER, J. R.—Bemerkungen ueber das Mechanische Equivalent der Waerme; Heilbronn und Leipzig, 1851.

MAXWELL, CLERK J.—The Theory of Heat; London, 1891.

NYSTROM'S Pocket Book of Mechanics and Engineering; Philadelphia, 1895.

OSTWALD, W.—Die Energie und ihre Wandlungen; Leipzig, 1888.

OSTWALD, W.—Lehrbuch der allgemeinen Chemie, vom Standpunkt der Thermodynamik, 3 Vols; Leipzig, 1891-94.

PLANCK, MAX.—Ueber der Zweiten Hauptsatz der Mechanischen Waermetheorie; Muenchen, 1879.

PLANCK, MAX.—Grundriss der Thermochemie; Breslau, 1893.

PLANCK, Max.—Erhaltung der Energie; Leipzig, 1887.

PARKER, J.—Thermo-Dynamics; Treated with Elementary Mathematics; London, 1894.

PECLET, E.—Traité de la Chaleur, two vols.; Paris, 1843.

PICTET, RAOUL.—Synthèse de la Chaleur. Genève, 1879.

PEABODY, C. H.—Tables on Saturated Steam and Other Vapors; New York, 1888.

PUPIN, M. T.—Thermodynamics; New York, 1894.

REDWOOD, J.—Theoretical and Practical Ammonia Refrigeration; New York, 1895.

RICHMOND, GEO.—Notes on the Refrigerating Process and its plac- in Thermodynamics; New York, 1892.

RÖNTGEN, ROBT.—Principles of Thermodynamics; translated by Du Bois; New York. 1889.

RUHLMANN, RICHARD.—Handbuch der Mechanischen Waerme Theorie, two vols.; Braunschweig, 1876.

SCHWACKHOEFER, FRANZ.—Vol. II, des Officiellen Berichts der K. K. Osterr. Central Commission für die Weltausstellung in Chicago, im Jahre 1893; Wien, 1894.

SCHWARZ, ALOIS.—Die Eis und Kuehlmaschinen; Muenchen und Leipzig, 1888.

SKINKLE, EUGENE T.—Practical Ice Making and Refrigerating; Chicago, 1897.

TAIT, P. G.—Sketch of Thermodynamics; Edinburgh, 1877.

TAIT, P. G.—Vorlesungen ueber einige neuere Fortschritte in der Physik; Braunschweig, 1877.

THURSTON, R. H.—The Animal as a Machine and a Prime Motor and the Laws of Energetics.

THURSTON, R. H.—Engine and Boiler Trials and of the Indicator and Prory Brake; New York, 1890.

THURSTON, ROBT. H.—Heat as a Form of Energy; Boston and New York, 1890.

THOMSEN, I.—Thermochemische Untersuchungen, three vols.; Leipzig, 1883.

THOMSON, SIR W.—Lectures on Molecular Dynamics; Baltimore, 1884.

TYNDALL, J.—Heat Considered as a Mode of Motion; London, 1883.

VERDET, E.—Théorie mécanique de la Chaleur; Paris, 1872.

VORHEES, GARDNER T.—Indicating the Refrigerating Machine; Chicago, 1899.

WALD, F.—Die Energie und ihre Entwerting; Leipzig, 1889.

WALLIS-TAYLOR, A. J.—Refrigerating and Ice-Making Machinery; London, 1896.

WOOD, DE VOLSON.—Thermodynamics, Heat, Motors and Refrigerating Machines; New York, 1896.

WAALS, VAN DER.—Die Continuität des Gasförmigen und Flüssigen Zustandes; Leipzig, 1881.

ZENNER, GUSTAVE.—Technische Thermodynamik, two vols.; Leipzig, 1890.

b.—CATALOGUES.

American Insulating Material Manufacturing Co. (Granite Rock Wool and Insulating Materials), St. Louis, Mo.

Arctic Machine Manufacturing Co. (Ice Making and Refrigerating Machinery, Ammonia compression system), Cleveland, Ohio.

Austin Separator Co. (Oil Separators), Detroit, Mich.

Barber, A. H., Manufacturing Co. (Ice Making and Refrigerating Machinery, Ammonia compression system), Chicago, Ill.

Buffalo Refrigerating Machine Co. (Ice Making and Refrigerating Machinery, Ammonia compression system), Buffalo, N. Y.

Carbondale Machine Co. (Ice Making and Refrigerating Machinery, Ammonia absorption system), Carbondale, Pa.

Case Refrigerating Machine Co. (Ice Making and Refrigerating Machinery, Ammonia compression system), Buffalo, N. Y.

Challoner's, Geo., Sons Co. (Ice Making and Refrigerating Machinery, Ammonia compression system), Oshkosh, Wis.

Cochran Company (Ice Making and Refrigerating Machinery, Carbonic anhydride system), Lorain, Ohio.

De La Vergne Refrigerating Machine Co. (Ice Making and Refrigerating Machinery, Ammonia compression system), New York City, N. Y.

Direct Separator Co. (Water and Oil Separator), Syracuse, N. Y.

Farrell & Rempe Co. (Wrought Iron Coils and Ammonia Fittings), Chicago, Ill.

Featherstone Foundry and Machine Co. (Ice Making and Refrigerating Machinery, Ammonia compression system, and Corliss Engines), Chicago, Ill.

Frick Co. (Ice Making and Refrigerating Machinery, Ammonia compression system, and Corliss Engines), Waynesboro, Pa.

Gifford Bros. (Ice Elevating, Conveying and Lowering Machinery), Hudson, N. Y.

Gloekler, Bernard (Cold Storage Doors and Fasteners), Pittsburg, Pa.

Hall, J. & E., Limited (Ice Making and Refrigerating Machinery, Carbonic anhydride system), London, E. C., England.

Harrisburg Pipe and Pipe Bending Co., Limited (Coils and Bends, and Ammonia Fittings and Feed-water Heaters), Harrisburg, Pa.

Haslam Foundry and Engineering Co. (Ice Making and Refrigerating Machinery, Ammonia absorption system), Derby, England.

Hohmann & Maurer Manufacturing Co. (Thermometers), Rochester, N. Y.

Hoppes Manufacturing Co. (Water Purifiers and Heaters), Springfield, Ohio.

Hoppes Manufacturing Co. (Steam Separators and Oil Illuminators), Springfield, Ohio.

Kilbourn Refrigerating Machine Co., Limited (Ice Making and Refrigerating Machinery, Ammonia compression system), Liverpool, England.

Kroeschell Bros. Ice Machine Co. (Ice Making and Refrigerating Machinery, Carbonic anhydride system), Chicago, Ill.

MacDonald, C. A. (Ice Making and Refrigerating Machinery, Ammonia compression system), Chicago, Ill., and Sydney, N. S. W., Australia.

Nason Manufacturing Co. (Ammonia and Steam Fittings), New York City, N. Y.

Newburgh Ice Machine and Engine Co. (Ice Making and Refrigerating Machinery, Ammonia compression system), Newburgh, N. Y.

Pennsylvania Iron Works Co. (Ice Making and Refrigerating Machinery, Ammonia compression system), Philadelphia, Pa.

Philadelphia Pipe Bending Works (Wrought Iron Coils and Bends), Philadelphia, Pa.

Remington Machine Co. (Ice Making and Refrigerating Machinery, Ammonia compression system), Wilmington, Del.

Ruemmeli Manufacturing Co. (Ice Making and Refrigerating Machinery, Gradirworks, Ice Cans, Fittings, etc.), St. Louis, Mo.

Siddely & Co. (Ice Making and Refrigerating Machinery, Ammonia absorption system), Liverpool, England.

Sterne & Co. (Ice Making and Refrigerating Machinery, Ammonia compression system), London, England.

Stevenson Co., Limited (Cold Storage Doors), Chester, Pa.

Tight Joint Co. (Ammonia Fittings), New York City, N. Y.

Triumph Ice Machine Co. (Ice Making and Refrigerating Machinery, Ammonia compression system), Cincinnati, Ohio.

Vilter Manufacturing Co. (Ice Making and Refrigerating Machinery, Ammonia compression system and Corliss Engines), Milwaukee, Wis.

Vogt, Henry, Machine Co. (Ice Making and Refrigerating Machinery, Ammonia absorption system), Louisville, Ky.

Vulcan Iron Works (Ice Making and Refrigerating Machinery, Ammonia compression system), San Francisco, Cal.

Wheeler Condenser and Engineering Co. (Water Cooling Towers), New York City, N. Y.

Wheeler Condenser and Engineering Co. (Auxiliary Devices for Increasing Steam Engine Economy), New York City, N. Y.

Whitlock Coil Pipe Co. (Coils and Bends, Feed-water Heaters), Elmwood, Conn.

Wolf Co., Fred W. (Ice Making and Refrigerating Machinery, Ammonia compression system), Chicago, Ill.

Wolf Co., Fred W. (Ammonia Fittings and Ice and Refrigerating Machinery Supplies), Chicago, Ill.

Wood, Wm. T., & Co. (Ice Tools), Arlington, Mass.

York Manufacturing Co. (Ice Making and Refrigerating Machinery, Ammonia compression system, York, Pa.

TOPICAL INDEX.

D

E

N

O

P

R.

S

T

Z

www.ingramcontent.com/pod-product-compliance
Lightning Source LLC
Chambersburg PA
CBHW051245300426
44114CB00011B/902

* 9 7 8 3 3 3 7 1 7 6 7 3 0 *